World Sustainability Series

Series Editor

Walter Leal Filho, European School of Sustainability Science and Research, Research and Transfer Centre "Sustainable Development and Climate Change Management", Hamburg University of Applied Sciences, Hamburg, Germany

Due to its scope and nature, sustainable development is a matter which is very interdisciplinary, and draws from knowledge and inputs from the social sciences and environmental sciences on the one hand, but also from physical sciences and arts on the other. As such, there is a perceived need to foster integrative approaches, whereby the combination of inputs from various fields may contribute to a better understanding of what sustainability is, and means to people. But despite the need for and the relevance of integrative approaches towards sustainable development, there is a paucity of literature which address matters related to sustainability in an integrated way.

Notes on the quality assurance and peer review of this publication

Prior to publication, the works published in this book are initially assessed and reviewed by an in-house editor. If suitable for publication, manuscripts are sent for further review, which includes a combined effort by the editorial board and appointed subject experts, who provide independent peer-review. The feedback obtained in this way was communicated to authors, and with manuscripts checked upon return before finally accepted. The peer-reviewed nature of the books in the "World Sustainability Series" means that contributions to them have, over many years, been officially accepted for tenure and promotion purposes.

Pranav Kumar Prabhakar · Walter Leal Filho
Editors

Preserving Health, Preserving Earth

The Path to Sustainable Healthcare

Springer

Editors
Pranav Kumar Prabhakar
Department of Research Impact
and Outcome
Lovely Professional University
Phagwara, Punjab, India

Walter Leal Filho
Faculty of Life Sciences
HAW Hamburg
Hamburg, Germany

ISSN 2199-7373 ISSN 2199-7381 (electronic)
World Sustainability Series
ISBN 978-3-031-60544-4 ISBN 978-3-031-60545-1 (eBook)
https://doi.org/10.1007/978-3-031-60545-1

This Springer imprint is published by the registered company Springer Nature Switzerland AG
The registered company address is: Gewerbestrasse 11, 6330 Cham, Switzerland

Preface

In an era defined by unprecedented environmental challenges and a growing awareness of the interconnectedness of human health and planetary well-being, the need for sustainable healthcare solutions has never been more urgent. The edited volume you hold in your hands, *Preserving Health, Preserving Earth: The Path to Sustainable Healthcare*, emerges as a beacon of hope and a call to action in navigating the complex intersection of health preservation and environmental stewardship.

This book represents a collaborative effort by leading experts and scholars from diverse fields, each offering unique insights and perspectives on the pressing issues facing modern healthcare systems. Through a multidisciplinary lens, the chapters delve into the intricate relationships between human health, environmental degradation, and the imperative for sustainable healthcare practices. As we confront the dual challenges of environmental degradation and the increasing burden of disease, it becomes abundantly clear that the health of our planet is intimately intertwined with the well-being of humanity. From the impacts of climate change on public health to the pollution of air, water, and soil, the chapters in this volume elucidate the myriad ways in which environmental factors shape our health outcomes. Yet, amidst these challenges, there exists a beacon of hope—a path forward toward sustainable healthcare practices that prioritize the health of both individuals and the planet. Through the exploration of renewable energy solutions, the promotion of green infrastructure in healthcare facilities, and the advocacy for policy reforms, this book charts a course toward a more resilient and equitable healthcare future.

We extend our deepest gratitude to the esteemed contributors who have lent their expertise and insights to this volume. Their dedication to advancing the dialogue on sustainable healthcare is reflected in the richness and depth of the chapters presented herein. To the readers—whether scholars, practitioners, policymakers, or concerned citizens—we invite you to engage with the ideas and proposals put forth in these pages.

Ultimately, *Preserving Health, Preserving Earth* serves as a testament to the collective resolve to safeguard both human health and the planet we call home. May this book inspire dialogue, spark innovation, and catalyze meaningful action toward a future where healthcare is not only accessible and equitable but also sustainable for generations to come.

With an unwavering commitment to the preservation of health and the Earth.

Phagwara, India — Pranav Kumar Prabhakar
Hamburg, Germany — Walter Leal Filho

Contents

Confronting the Health Impacts of Climate Change: A Comprehensive Exploration

Aniruddha Sen and Ayan Chatterjee

Abstract This review article extensively explores the intricate correlation between climate change and its profound effects on human health and well-being. On a global scale, there is an alarming trend of increase in ambient temperatures, erratic weather patterns, and the escalation of greenhouse gas emissions. In the Indian context, a nation highly susceptible to climate shifts, these effects are particularly pronounced due to its varied geography and dense population. The study examined the ramifications of alterations in both indoor and outdoor environments. Changes in climate lead to modified ventilation patterns, compromising indoor air quality. Concurrently, outdoor settings experience elevated levels of air pollution and heightened concentrations of allergenic pollen. These changes have significant health implications, exacerbating conditions such as asthma and contributing to the surge of cardiovascular ailments. Furthermore, the article delves into the challenges faced by individuals compelled to migrate for work, often landing in regions disproportionately affected by climate change. These migrating populations confront amplified health risks due to inadequate living conditions, restricted healthcare access, and exposure to extreme weather occurrences. The review also underscores the intricate interplay between climate change and its consequences on various bodily systems. Heatwaves strain thermoregulation mechanisms, resulting in heat-related illnesses. Vector-borne diseases expand their reach as shifting temperatures create favorable breeding environments for disease-carrying insects. Waterborne diseases surge due to altered precipitation patterns impacting water quality and availability. In conclusion, this comprehensive review underscores the urgent necessity for action to mitigate the health repercussions of climate change. It advocates for multi-faceted interventions addressing both indoor and outdoor settings, safeguarding susceptible populations, and fostering global collaboration to mitigate the escalating health effects of a changing climate.

A. Sen
All India Institute of Medical Sciences, Kunraghat, Gorakhpur 273008, Uttar Pradesh, India

A. Chatterjee (✉)
Department of Allied Health, School of Health and Medical Sciences, Adamas University, Barasat, Kolkata 700126, West Bengal, India
e-mail: ayan4189@yahoo.com

P. K. Prabhakar and W. Leal Filho (eds.), *Preserving Health, Preserving Earth*, World Sustainability Series, https://doi.org/10.1007/978-3-031-60545-1_1

Keywords Thermal modelling · Public health · Mitigation · Adaptation · Mobility

1 Introduction

The robust evidence of global climate change underscores an undeniable and alarming trend of warming temperatures over the past century. Each successive decade since the mid-1800s has demonstrated an unequivocal increase in global temperatures, with the decade from 2011 to 2020 standing out as the warmest ever recorded, surpassing pre-industrial levels by approximately 1.09 °C. This accelerating warming trend is directly linked to the escalating emissions of greenhouse gases, primarily stemming from anthropogenic activities such as the combustion of fossil fuels, deforestation, and intensive agricultural practices (IPCC 2007). The concentration of carbon dioxide in the atmosphere, a key driver of the greenhouse effect, has experienced an exponential surge. Presently exceeding 415 parts per million (ppm), this level is unprecedented in at least 800,000 years (Climate change 2007: the physical science basis 2007). The multifaceted impacts of these climatic shifts extend across ecological stability, infrastructure resilience, global economies, and most critically, human health (Friedlingstein et al. 2022). The Lancet Countdown 2021 report, a comprehensive analysis of climate change and health, provides critical insights supported by a thorough examination of 44 indicators across five key domains (USGCRP 2018).

These domains encompass direct impacts and vulnerabilities related to climate change, adaptation planning, mitigation actions and their public health co-benefits, economic considerations, and political and public engagement. The health implications of climate change are severe, with projections indicating that by 2030–2050, heat exposure, childhood undernutrition, malaria, and other climate-sensitive diseases could lead to over 250,000 additional deaths annually (Xiang et al. 2015). Mental health is also profoundly affected, as individuals grapple with the trauma associated with livelihood losses, forced migration, and community disruptions (Knowlton et al. 2014). Food security is under imminent threat, with estimates suggesting potential average yield losses of 25% for maize, 14% for sorghum, and 10% for millets in India by 2100 (Clayton 2021). The economic toll is substantial, with India projected to lose 5.8% of its GDP by 2100 due to the adverse impacts of climate change. Urgent and substantial actions are imperative to address these challenges. Deep cuts in greenhouse gas emissions, coupled with effective adaptation measures, are crucial not only to avert massive economic losses but also to safeguard the health and well-being of populations. The call for action is reinforced by the urgency to mitigate the profound impacts of climate change on health and well-being. Climate action plans must adopt a holistic approach, considering the health impacts across various economic sectors and vulnerable communities, to ensure a resilient and sustainable future (Venugopal et al. 2016).

2 Climate Change in India

India's diverse geography, encompassing the Himalayan glaciers, arid deserts, fertile plains, coastal regions and island chains, paired with a massive population exceeding 1.38 billion in 2021, renders it highly vulnerable. Multiple climatic changes are transpiring simultaneously across regions. Surface air temperatures have risen around 0.7 °C nationally from 1901 to 2018 (Xiang et al. 2014). The warming spikes further in cities due to the heat island effect—studies in Gujarat recorded over 6 °C higher night temperatures in urban areas. The warming is exacerbated by drastic changes in land use patterns. Forest lands and wetlands are being rapidly converted to agricultural land or urban built-up areas (Mohanty and Mohanty 2017). This change in land surface types significantly impacts energy flux between the land and atmosphere affecting weather patterns. Deforestation in the fragile Himalayan Mountain ecosystems is also triggering rainfall variability and soil erosion. Glacial retreat continues unabated due to warming—by 2050, one-third of the 4,040 Himalayan glaciers could disappear affecting perennial rivers. The monsoon rainfall is disrupted as well—though total rainfall shows no significant trend yet, there is an increase in extreme rainfall events and simultaneous decline in the number of rainy days, severely impacting India's largely rain-fed agriculture. Just 1 °C rise in summer temperatures is projected to lower crop yields of major cereals by 3–7% (Lundgren et al. 2013).

Intensification of tropical cyclones has led to the 2005 Mumbai floods and 2019 Cyclone Fani with wind speeds over 200 km/h (Lucas et al. 2014). Climate change also elevates air pollution levels which already contribute heavily to India's disease burden. In Delhi, higher temperatures over 1998–2018 enhanced particulate matter concentrations combining with local emissions (Cheng et al. 2022). The public health implications are massive—from surged heat-related mortality, changing patterns of vector-borne and water-borne diseases, along with mental health issues associated with livelihood disruptions and migration.

3 Indoor and Outdoor Environmental Changes

Indoor settings are being profoundly impacted by climate change in India. Traditional housing designs incorporated climate-responsive architectural elements like air pockets, open courtyards, induced ventilation and shading (Singh et al. 2011). However, modern buildings have moved towards sealed envelopes reliant on-air conditioning. These buildings have significantly altered indoor ventilation patterns, affecting air quality (Adam-Poupart et al. 2013). Studies reveal over 87% of urban homes exhibit poor ventilation, with concentrations of pollutants like CO_2, particulate matter, volatile organic compounds, and biological contaminants exceeding safe limits (Liang et al. 2019). This heavily impacts health given over 70% of time is spent indoors (Balakrishnan et al. 2011).

Use of biomass cooking fuels also degrades indoor air quality—almost 30% rural and 5% urban households rely on polluting fuels like firewood, dung cakes etc. exacerbating respiratory conditions (Government of India 2018). Outdoor urban air quality is deteriorating rapidly from local emissions coupled with regional climate factors. Over 90% residents breathe air exceeding national safety limits (Dash et al. 2017). Climate change serves to elevate pollution levels further through complex mechanisms. Particulate matter levels spike at higher temperatures as heat intensifies photochemical reactions and increases ground-level ozone. Pollen and mold proliferation also escalates in warmer weather with early onset of flowering cycles, longer pollen seasons, and higher production. Reports indicate allergic rhinitis and asthma hospitalization from pollen allergies are surging. Without urgent mitigating policies for cleaner energy, these synergistic health risks shall continue rising (Adam-Poupart et al. 2013).

4 Health Implications

Climate change serves to exacerbate several health conditions especially for marginalized communities in India. With elevated temperatures, humidity and air pollution levels, cases of heat cramps, heat exhaustion and life-threatening heat strokes are surging (Klepeis et al. 2001). Mortality spikes have been recorded during heatwaves—the 2015 heatwave caused over 2,300 deaths in India (Dhiman et al. 2010). Thermal discomfort is linked to lower worker productivity, performance deficits and occupational injuries (Patel et al. 2011). Respiratory conditions like asthma worsen due to combined exposure to allergens, ozone and particulate matter from weather changes and rising pollution (Bain et al. 2014). Studies indicate over 75% rise in wheezing, breathlessness, and asthma attacks during summer months (Caminade et al. 2019). Asthma hospitalization rates peak following dust storms or extreme precipitation which elevates pollen and mold in the air (Revi 2008).

Cardiovascular mortality and morbidity also escalate during heatwaves as well as following extreme cold events in winter (Borhade 2011). Air pollutants induce systemic inflammation, constrict blood vessels, and disrupt heart rhythms—daily emergency hospitalizations for acute heart attacks surge by around 44% per 10 $\mu g/m^3$ increase in same day PM2.5 levels. India has over 100 million seasonal migrants engaged in informal work. Agricultural distress from climate factors is spurring short-term mobility to urban areas (Ahmadalipour and Moradkhani 2018). However, cities concentrate health risks from heatwaves, disasters, and infections. Inadequate housing, nutrition and exclusion from health services amplify risks for migrants (Liu et al. 2003).

Their hazardous working conditions like in brick kilns, mines or sewers add to the burden. Women and children are especially vulnerable, facing threats of violence, adverse pregnancies, and lack of childcare. Climate change worsens these intersections of poverty, informality, mobility, and environmental risk (Pearson et al. 2010).

5 Complex Interplay with Bodily Systems

Rising temperatures due to climate change directly threaten thermoregulation mechanisms in the human body. Heatwaves are becoming more frequent and intense in urban regions causing thermal stress (Klepeis et al. 2001). Prolonged heat exposure overwhelms the body's cooling system potentially leading to multi-organ dysfunction and death among vulnerable groups like infants, elders, and outdoor workers (Kaur and Pandey 2021). Renal function is disrupted due to blood flow redistribution and dehydration while the gut barrier integrity is compromised permitting leak of endotoxins into blood circulation (Tiwari et al. 2014). The changing climate also enables expansion of vectors like mosquitoes, ticks and flies which transmit infectious diseases (Zhang et al. 2016). Warming allows accelerated vector reproduction rates due to shortened life cycles and expanded seasonal duration (Azhar et al. 2014). Heavy rainfall further offers more stagnant water breeding grounds amplifying populations. Cases of malaria, dengue, chikungunya, Japanese encephalitis, and other vector-borne diseases are rising (Kjellstrom et al. 2016). Similarly, waterborne infections spread due to extreme rainfall events which contaminate drinking water sources. Over half the Indian population faces high to extreme risk of diarrheal diseases due to unsafe water supply which is being further jeopardized by climate variability (Sheffield and Landrigan 2011). Cholera outbreaks in coastal cities are aggravated by warming oceans, higher humidity and floods from sea level rise and storms. Addressing health resilience to climate change necessitates tackling vulnerability to such disease transmission mechanisms (Agarwal et al. 2018).

6 Urgent Necessity for Action

Addressing the escalating public health impacts of climate change in India necessitates urgent multi-faceted interventions. Adaptation policies should encompass strengthening health systems, disease surveillance, resilient infrastructure, disaster preparedness and targeted safeguards for vulnerable groups. Heat action plans implemented in Ahmedabad and other cities incorporating early warning systems, public awareness and infrastructure modifications have demonstrated health benefits. However, intensified mitigation efforts simultaneously remain vital to curb emissions and avert catastrophic impacts projected this century (Wu et al. 2017). The National Action Plan on Climate Change launched in 2008 emphasized efficacy, equity, and environmental sustainability in climate actions across sectors. Visionary initiatives like India's renewable energy commitments, expanding forests through community participation and sustainable agriculture models display home-grown leadership (Roy et al. 2001).

However, realizing their full potential warrants transfer of green technologies and financing from industrialized states responsible for most emissions historically.

Article 4 of the UNFCCC and concept of 'common but differentiated responsibilities' compels such global collaboration, both on mitigation and support for climate change adaptation. Vulnerable low-income countries bear negligible culpability for climate change, but shoulder amplified health, economic and social consequences from business-as-usual. Urgent climate justice demands that polluting nations curb emissions while assisting developing regions build requisite resilience (Rumana et al. 2014).

7 Conclusion

Our comprehensive exploration of the intricate correlation between climate change and its profound effects on human health and well-being emphasizes the urgent necessity for multifaceted interventions. The evidence presented throughout this review underscores the severity of the health implications, with projections indicating alarming increases in heat exposure, childhood undernutrition, and climate-sensitive diseases. Mental health is also profoundly affected, with individuals grappling with the trauma associated with livelihood losses, forced migration, and community disruptions. The specific focus on India, a nation highly vulnerable to climate shifts due to its varied geography and dense population, reveals the multifaceted challenges faced by diverse regions. From the Himalayan glaciers to arid deserts, fertile plains, coastal areas, and island chains, each region experiences unique climate changes with far-reaching health consequences. Surface air temperatures have risen nationally, and the warming spikes further in urban areas due to the heat island effect. Glacial retreat disrupted monsoon patterns, and intensification of tropical cyclones pose significant threats to agriculture, water resources, and public health. Our exploration extends to the intricate interplay between climate change and various bodily systems. Rising temperatures directly threaten thermoregulation mechanisms, leading to heat-related illnesses and multi-organ dysfunction, particularly among vulnerable groups. Vector-borne diseases, including malaria, dengue, and chikungunya, are on the rise due to expanded breeding environments for disease-carrying insects. Waterborne infections escalate with extreme rainfall events, impacting over half of the Indian population facing high to extreme risks of diarrheal diseases. The review delves into the significant changes occurring in both indoor and outdoor environments. Modern buildings with sealed envelopes alter indoor ventilation patterns, resulting in poor air quality, especially in urban areas. Outdoor urban air quality is deteriorating rapidly due to local emissions and climate factors, exacerbating pollution levels, and contributing to respiratory conditions.

The complex interplay of climate change with health extends to the challenges faced by migrating populations, particularly those engaged in informal work, facing hazardous conditions, inadequate housing, and limited healthcare access. Recognizing the escalating public health impacts of climate change, urgent action is imperative. Adaptation policies should encompass strengthening health systems, disease surveillance, resilient infrastructure, disaster preparedness, and targeted safeguards for vulnerable groups. Heat action plans, incorporating early warning systems, public awareness, and infrastructure modifications, have demonstrated health benefits and should be expanded. Simultaneously, intensified mitigation efforts are vital to curb emissions and avert catastrophic impacts projected for this century.

India's efforts, such as the National Action Plan on Climate Change and visionary initiatives in renewable energy, afforestation, and sustainable agriculture, display home-grown leadership. However, achieving their full potential requires global collaboration, including the transfer of green technologies and financing from historically polluting nations. Climate justice demands that developed nations curb emissions while assisting developing regions in building requisite resilience. In essence, this comprehensive review advocates for immediate and sustained action to mitigate the escalating health effects of climate change. The call to address this global crisis extends beyond national boundaries, requiring collaborative efforts to safeguard the health and well-being of current and future generations.

References

Adam-Poupart A, Labreche F, Smargiassi A (2013) Climate change and occupational health and safety in a temperate climate: potential impacts and research priorities in Quebec, Canada. Ind Health 51:68–78

Agarwal AK, Singh AP, Gupta T, Agarwal RA, Sharma N, Rajput P et al (2018) Mutagenicity and cytotoxicity of particulate matter emitted from biodiesel-fueled engines. Environ Sci Technol 52:14496–14507

Ahmadalipour A, Moradkhani H (2018) Escalating heat-stress mortality risk due to global warming in the Middle East and North Africa (MENA). Environ Int 117:215–225

Azhar GS, Mavalankar D, Nori-Sarma A, Rajiva A, Dutta P, Jaiswal A et al (2014) Heat-related mortality in India: excess all-cause mortality associated with the 2010 Ahmedabad heat wave. PLoS ONE 9(3):e91831

Bain R, Cronk R, Hossain R, Bonjour S, Onda K, Wright J et al (2014) Global assessment of exposure to faecal contamination through drinking water based on a systematic review. Trop Med Int Health 19(8):917–927

Balakrishnan K, Ramaswamy P, Sambandam S, Thangavel G, Ghosh S, Johnson P et al (2011) Air pollution from household solid fuel combustion in India: an overview of exposure and health related information to inform health research priorities. Glob Health Action 4(10):3402

Borhade A (2011) Health of internal labour migrants in India: some reflections on the current situation and way forward. Asia Eur J 8:457–460

Caminade C, McIntyre KM, Jones AE (2019) Impact of recent and future climate change on vector-borne diseases. Ann N Y Acad Sci 1436(1):157–173

Cheng CY, Tseng YL, Huang KC, Chiu IM, Pan HY, Cheng FJ (2022) Association between ambient air pollution and emergency room visits for pediatric respiratory diseases: the impact of COVID-19 pandemic. Toxics 10(5):247

Clayton S (2021) Climate change and mental health. Curr Environ Health Rep 8(1):1–6

(2007) Climate change 2007: the physical science basis. Agenda 6(07):333

Dash SK, Dey S, Salunke P, Dalal M, Saraswat V, Chowdhury S, Choudhary RK (2017) Comparative study of heat indices in India based on observed and model simulated data. Curr World Environ 12:504–520

Dhiman RC, Pahwa S, Dhillon GPS, Dash AP (2010) Climate change and threat of vector-borne diseases in India: are we prepared? Parasitol Res 106(4):763–773

Friedlingstein P, O'sullivan M, Jones MW, Andrew RM, Gregor L, Hauck J et al (2022) Global carbon budget 2022. Earth Syst Sci Data Discuss 2022:1–159
Government of India (2018) India's sixth national report to the convention on biological diversity ministry of environment and forests
IPCC (2007) Fourth assessment report. Inter-Governmental Panel on Climate Change, Cambridge: Cambridge University Press
Kaur R, Pandey P (2021) Air pollution, climate change, and human health in Indian cities: a brief review. Front Sustain Cities 3:705131
Kjellstrom T, Briggs D, Freyberg C, Lemke B, Otto M, Hyatt O (2016) Heat, human performance, and occupational health: a key issue for the assessment of global climate change impacts. Annu Rev Public Health 37:97–112
Klepeis NE, Nelson WC, Ott WR, Robinson JP, Tsang AM, Switzer P et al (2001) The National Human Activity Pattern Survey (NHAPS): a resource for assessing exposure to environmental pollutants. J Expo Anal Environ Epidemiol 11(3):231–252
Knowlton K, Kulkarni SP, Azhar GS, Mavalankar D, Jaiswal A, Connolly M et al (2014) Development and implementation of South Asia's first heat-health action plan in Ahmedabad (Gujarat, India). Int J Environ Res Public Health 11(4):3473–3492
Liang M, Liang H, Rao Z, Hong X (2019) Characterization of polycyclic aromatic hydrocarbons in urban-rural integration area soil, North China: spatial distribution, sources and potential human health risk assessment. Chemosphere 1(234):875–884
Liu S, Krewski D, Shi Y, Chen Y, Burnett RT (2003) Association between gaseous ambient air pollutants and adverse pregnancy outcomes in Vancouver, Canada. Environ Health Perspect 111(14):1773–1778
Lucas RAI, Epstein Y, Kjellstrom T (2014) Excessive occupational heat exposure: a significant ergonomic challenge and health risk for current and future workers. Extrem Physiol Med 14:1–8
Lundgren K, Kuklane K, Gao C, Holmer I (2013) Effects of heat stress on working populations when facing climate change. Ind Health 51(1–13):2013
Mohanty G, Mohanty SK (2017) Physical strain of women workers in rice cultivation system. Int Res J Adv Eng Sci 2:51–56
Patel JR, Vora KP, Tripathi S, Zeng H, Tumpey TM, Katz JM et al (2011) Infection of lung epithelial cells with pandemic 2009 A(H1N1) influenza viruses reveals isolate-specific differences in infectivity and host cellular responses. Viral Immunol 24(2):89–99
Pearson JF, Bachireddy C, Shyamprasad S, Goldfine AB, Brownstein JS (2010) Association between fine particulate matter and diabetes prevalence in the U.S. Diabetes Care 3(10):2196–2201
Revi A (2008) Climate change risk: an adaptation and mitigation agenda for Indian cities. Environ Urban 20(1):207–229
Roy S, Ray MR, Basu C, Lahiri P, Lahiri T (2001) Abundance of siderophages in sputum: indicator of an adverse lung reaction to air pollution. Acta Cytol 45:958–964
Rumana HS, Sharma RC, Beniwal V, Sharma AK (2014) A retrospective approach to assess human health risks associated with growing air pollution in urbanized area of Thar Desert, western Rajasthan, India. J Environ Health Sci Eng 12:1–9
Sheffield PE, Landrigan PJ (2011) Global climate change and children's health: threats and strategies for prevention. Environ Health Perspect 119(3):291–298
Singh MK, Mahapatra S, Atreya SK (2011) Solar passive features in vernacular architecture of North-East India. Sol Energy 85(9):2011–2022
Tiwari S, Bisht DS, Srivastava AK, Pipal AS, Taneja A, Srivastava MK et al (2014) Variability in atmospheric particulates and meteorological effects on their mass concentrations over Delhi, India. Atmos Res 145:45–56
USGCRP (2018) Fourth national climate assessment. U.S. Global Change Research Program, Washington, DC
Venugopal V, Chinnadurai JS, Lucas RAI, Kjellstrom T (2016) Occupational heat stress profiles in selected workplace in India. Int J Environ Res Public Health 89:1–13

Wu H, Wang T, Riemer N, Chen P, Li M, Li S (2017) Urban heat island impacted by fine particles in Nanjing, China. Sci Rep 7:11422

Xiang JP, Pisaniello BD, Hansen A (2014) Health impacts of workplace heat exposure: an epidemiological review. Ind Health 52:91–101

Xiang J, Hansen A, Pisaniello D, Bi P (2015) Perceptions of workplace heat exposure and controls among occupational hygienists and relevant specialists in Australia. PLoS ONE 10(8):e0135040

Zhang S, Wang M, Ghan SJ, Ding A, Wang H, Zhang K et al (2016) On the characteristics of aerosol indirect effect based on dynamic regimes in global climate models. Atmos Chem Phys 16:2765–2783

Sustainable Practices in the Pharmaceutical Industry: Development and Adoption

Koyel Kar, Sailee Chowdhury, Priyanka Chakraborty, and Arpan Saha

Abstract Recently, organizations, customers, and governments have all begun to pay more attention to the pharmaceutical industry's sustainability. Concerns about integrating sustainability principles into creating new delivery systems, new products with lower environmental risks, waste recycling, reducing water use, eco-friendly manufacturing processes, and recyclable packaging have heightened interest in this subject. Pharmaceutical firms are experiencing a higher level of management complexity as a result of the need to maintain environmental, economic, and social sustainability as well as control costs. This has resulted in integrating economic reasoning into the management of sustainability challenges. The manufacturing environment has altered since the Government of India's Ministry of Environment and Forests first mandated non-financial reporting for corporate organizations in 1993. With most of the activity documented in the last decade, the regulatory framework for examining the environmental effects of chemical-based industrial facilities has been reinforced. The numerous manufacturing processes in pharmaceutical manufacturing facilities require a wide range of chemical syntheses. These procedures take a lot of energy, which results in substantial greenhouse gas emissions and waste production. This chapter explains how pharmaceutical companies are pursuing sustainability while taking into account societal economic and social well-being, environmental concerns, and other factors. The main goal is to provide people with a better knowledge of the sustainability trend that is currently affecting the pharmaceutical manufacturing industry and where it is going. There has been a discussion of the advantages of several strategies for achieving sustainability in pharmaceutical organizations. The readers will better grasp how pharmaceutical settings might adopt various practices to achieve sustainability, what different performance measures are accessible, and how reporting sustainability improves an organization's transparency and efficiency. Thus, with the following objectives in mind, the business and management studies

K. Kar (✉) · S. Chowdhury · P. Chakraborty
BCDA College of Pharmacy & Technology, 78/1 Jessore Road (S), Hridaypur, Barasat, Kolkata 700127, India, West Bengal
e-mail: koyel20@gmail.com

A. Saha
Bharat Pharmaceutical Technology, Amtali 799130, Tripura, India

P. K. Prabhakar and W. Leal Filho (eds.), *Preserving Health, Preserving Earth*, World Sustainability Series, https://doi.org/10.1007/978-3-031-60545-1_2

that have already been done on sustainability in the pharmaceutical industry have been reviewed.

Keywords Sustainability · Pharmaceutical industry · Sustainable development · Green chemistry · Environmental effects

1 Introduction

The adoption of sustainable practices is on the rise globally across several industrial sectors, according to an assessment of a wider range of literature. It was noted that as socio-ecological problems gain more attention, environmental initiatives are being included in a wider range of corporate strategies for manufacturing companies (Salwa et al. 2008). The aforementioned research concluded that the four main approaches to sustainable manufacturing were material, resource, eco-efficiency, and waste minimization. The relationship between complicated environmental problems and manufacturing processes has been acknowledged since the first UN conference on the environment was held in Stockholm in 1972. The Brundtland Commission defined sustainability as "meeting the needs of the present without compromising the ability of future generations to meet their own needs"; this gave rise to the idea. There are two goals this chapter seeks to accomplish. First and foremost, it seeks to improve knowledge of sustainability's development, associated practices, and incorporation into the business models often used in the pharmaceutical industry (Mélanie et al. 2013). The analysis from multiple research studies concluded that businesses are under pressure from stakeholders to demonstrate their accountability, transparency, and effectiveness of governance through corporate sustainability disclosures. It also aims to provide a brief overview of the industry's trends and contributions in light of the current Indian scenario. Consequently, integrating sustainability at the corporate level is a strategic choice that must be made to generate a sufficient level of commitment throughout the entire organization (Sabrina et al. 2016). In the manufacturing sector, if sustainability is the objective, the best way to achieve it is through process innovations or the adoption and integration of existing practices most efficiently. Sustainability in the industrial sector emphasizes the same fundamentals of ecologically friendly goods and services, but how this is accomplished varies depending on the industry. Sustainability may imply different things to the industrial sector than it does to the service sector. According to Paul and Stuart (1995), the environmental challenges of today are a result of the unsustainable industrial activity of the past (Paul and Stuart 1995).

Statistically proving that trade, information exchange, and technological access are the new avenues via which economic globalization is creating prospects for prosperity and improved quality of life. Technology and knowledge advancements are fostering economic growth, but they also hold the potential to mitigate dangers and challenges to the long-term viability of our economies, social structures, and environments. The purpose of this chapter is to provide an understanding of the severity and urgency of the dangers and challenges to the pharmaceutical industry's collective sustainability. This paper aims to communicate openly and clearly about other major sustainability drivers to a range of stakeholders, regardless of the size and location of the pharmaceutical organization. These drivers include the increasingly available choices and opportunities for investment decisions and research and development (R&D), which are oriented towards green manufacturing. The purpose of this chapter is to provide insight into and raise awareness of the sustainability-related practices and trends in the Indian pharmaceutical business. The findings that have been reported can assist the companies in making necessary adjustments and in producing a fair and impartial portrayal of practices connected to sustainability.

Pharmaceutical processes are quite complex when compared to other industries. For this reason, the pharmaceutical industries in India and around the world must incorporate sustainability into their overall corporate strategy and report on it to the public for public awareness. This is also among the factors that have made this industry a focus of sustainability efforts over the past 20 years (Esteban 2008). Below is the structure of this chapter. The content of the survey pertains to the specifics of the methodology employed in its compilation.

In this regard, the explanation of the environmental effects of pharmaceutical manufacturing processes adds further clarity. It encapsulates the movement of Indian pharmaceutical producers to embrace sustainability-related practices. An explanation of how the Indian industry has affected the environment opens the book. The main conclusions, key observations, and gaps in the literature are discussed in light of the chapter (Fig. 1).

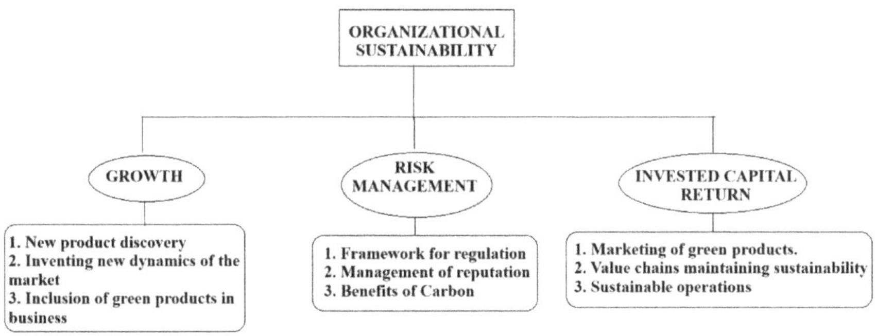

Fig. 1 Balance of sustainability with growth

2 Research Methodology Adopted to Review Research Articles

2.1 Design of Research

Many other documented research forms, such as reports, white papers, and working papers, were examined because the literature on sustainability and related practices in the pharmaceutical sector is not specifically limited to scholarly publications. Keywords from the disciplines of corporate social responsibility, green chemistry, life cycle assessment, sustainable manufacturing, and sustainability in the pharmaceutical business were looked up to gauge the trends. Major findings and potential recommendations have been provided following a review of multiple reports, articles, and online data sources as well as individual papers relating to sustainability in the pharmaceutical business and related areas overall.

In terms of environmental awareness, the years 2010 and 2014 were significant due to their careful consideration of environmental issues and the establishment of new rules and regulations both internationally and in India. It's interesting to note that the majority of scientific research papers, which include various case studies of the pharmaceutical industry, were published in reputable journals such as Business Strategy and The Environment, Green Chemistry, Journal of Cleaner Production, and many more. These journals present cases of process improvement in pharmaceutical manufacturing, including the use of eco-design concepts in early product development, the adoption of less hazardous materials, and alternative manufacturing processes. Together, this reduced waste production, and made sure that energy use and greenhouse gas emissions were reduced, hence reducing the environmental impact and offering financial benefits with improved stakeholder communication.

This chapter is exploratory and is based on a survey of the literature to investigate the practices used by the Indian and global pharmaceutical industries and the enhancements in performance that follow. Using the methods outlined in this chapter, a methodical review of various types of literature was conducted for this goal. Through the study of numerous journal articles, reports, and whitepapers, data in the form of statistical numbers was gathered. Tables were then used to provide a summary.

- Understanding the environmental, economic, and social components of sustainability and its reporting required studying the Global Reporting Initiative guidelines, which are standard.
- Research on the adoption of sustainability-related practices and their reporting has been identified.
- To investigate the knowledge around the adoption of sustainability and its reporting, a thorough assessment of the literature was conducted.
- The pharmaceutical business was selected for additional investigation based on a review of the literature to address qualitative concerns about sustainability relative to practices and their reporting.

- Indian pharmaceutical companies were selected and examined in light of the literature regarding industry practices in the pharmaceutical sector to address sustainability on a global scale.
- Reports were made on observations, important discoveries, gaps, and conclusions about qualitative sustainability challenges to attain sustainability in Indian pharmaceutical companies.

2.2 Trends of Sustainability Growth

Peer-reviewed scientific journal publications have shown that different reporting techniques have occasionally been proposed by various international agencies and governments. The aforementioned initiatives aim to advance, embrace, and include sustainable practices and thought processes. Additionally, they seek to facilitate relevant stakeholder communication across industrial sector organizations globally (Low et al. 2016). The majority of the material on this topic is therefore available as reports and whitepapers, in addition to research articles that are published in scholarly journals. This is a significant discovery. The voluntary guidelines that have been developed are an attempt to address sustainability.

3 Types of Sustainability Analysis

3.1 Social

When it comes to the social component of sustainability, the articles that came out of the review are fewer in number—six in all. This comparative scarcity emphasizes the critical need for further research on this element of sustainability. The analysis of this topic focuses on several concerns, including the function of employees, segments of the population that have access to pharmaceutical compounds, parts linked to supply chains, packaging of the products, and conduct of customers (Caligiuri et al. 2013; Quak et al. 2019; Nematollahi et al. 2018; Lorenzini et al. 2018; Chaar and Lee 2012).

Regarding the contribution of workers and their involvement in social issues, Caligiuri et al. investigate how many employees participate in corporate volunteer initiatives facilitated by NGOs (Caligiuri et al. 2013). In the context of pharmaceuticals, the article examines how these initiatives boost staff involvement, which promotes both the development of CRS policies and employee identification with corporate values. The study report by Quak et al., recommends the use of metrics to assess the capacity of pharmaceutical firms to ensure the public's access to medications. In particular, the authors emphasize prospects for employing the AtM Index, a tool thought to be able to sensitize pharmaceutical corporations towards the issue of ensuring better access to pharmaceuticals, in their study on drug access in emerging markets (Caligiuri et al. 2013).

To support better access performance, the research emphasizes the necessity to examine sharing arrangements across several stakeholders related to pharmaceuticals. Lorenzini et al. reviewed the significance of pharmaceutical packaging innovation and the role it plays in giving patients more opportunities to get appropriate and effective treatments (Lorenzini et al. 2018). Both the study by Chaar and Lee, which examines consumer behavior about various drug prescription systems, and the study by Sahu and Kohli, which examines the relationship between pharmaceutical companies and hospitals, make clear the necessity of meeting patients' needs by giving them centrality (Chaar and Lee 2012; Sahu and Kohli 2019).

3.2 *Environmental Sustainability*

The recurrent issue, the environmental aspect of pharmaceutical sustainability is very prominent. Cleaner production, eco-friendly supply chains, eco-friendly human resource management (HRM), and eco-friendly materials are the four main areas of study for the ecological component identified by the thematic study of the full papers. There are research articles addressing ecological challenges that specifically reference production in the pharmaceutical industry as part of the research stream looking at cleaner production, discussing the topic of the ecological impact of products throughout their life cycles and how this is disclosed by businesses (Veleva and Cue 2017). They point out that in the pharmaceutical industry, business transparency does not provide the provision of a holistic picture of this impact. For instance, the paper by Belkhir and Elmeligi, which emphasizes the establishment of a research deficit about this particular sector, emphasizes the necessity to pay attention to how pharmaceutical manufacturing emissions affect the environment (Kaenzig et al. 2011).

Based on an empirical study of several of the top pharmaceutical companies, the authors note that even though their analysis reveals that the pharmaceutical industry's carbon emissions are even higher than those of the automotive industry, which has received much more significant attention, this industry is still largely under the radar and has received little attention. This finding highlights the necessity for policies that give a thorough representation of the sector's carbon footprint. This necessity is also acknowledged in other studies that include both broad articles that analyze more specialized themes as well as papers that strive to understand how to assess efficiency in ecological terms in medicines (Belkhir and Elmeligi 2019). For instance, studies on the significance of ISO certifications and system standards can be found among the latter (Veleva et al. 2003). Li and Hamblin, for instance, underline how ISO14001 certifications are among the most crucial factors in figuring out how pharmaceutical companies use cleaner production practices in their analysis of the Chinese market (Marinkovic et al. 2016). Similar studies have been done in the pharmaceutical industry to attempt and understand the factors that influence entrepreneurs' attitudes toward carrying out green activities. The supply chain is the topic of the second area of research that we look at that relates to the environmental aspect (Li and Hamblin 2016). According to General Studies, one of the most important factors to think about is the necessity of addressing the environmental issue (Blum-Kusterer and Hussain 2001).

In their study, Singh et al. claim that there is increasing interest in concerns surrounding the management of pharmaceutical product waste and methods to address some of the challenges connected to, for example, the management of expired pharmaceuticals or the risk of pharmaceutical product-related water contamination (Kumar et al. 2019). Undeniably, choosing suppliers and other supply chain participants based on their contribution to ecological effect is a particularly crucial issue in this research (Singh et al. 2016).

For instance, Gardas et al. offer an empirical inquiry to comprehend the selection criteria for third-party logistics service providers in their study of the Indian market. The authors emphasize the role that logistics, in general, and these actors, in particular, play in reducing CO_2 emissions (Moradi and Jolai 2018). They draw attention to a particular research hole in the pharmaceutical industry, offer a framework of 14 criteria, and stress the importance of including other sustainability facets. The significance of a green perspective in HRM that emphasizes green HRM practices is one of the topics on which contemporary writing on pharmaceuticals has concentrated. To establish a zero-waste culture in the company, Veleva et al. emphasize the importance of human resources and the function of employee involvement (Gardas et al. 2019). The adoption of green HRM practices that can encourage pro-environmental behavior by employees is another pertinent subject that Saeed et al. discuss (Veleva et al. 2017). The writers discuss these practices in particular and contrast the pharmaceutical industry with other industries in terms of how they are implemented. According to the findings, it is more probable that sustainable behaviors will be adopted from an environmental standpoint when green HRM practices are integrated into a broader understanding of sustainability in the workplace. In other words, they appear to be associated with a broad strategy for sustainability. Similarly, Masri and Jaaron investigate green HRM practices and their effects on employees' environmental performance in the particular context of Palestinian pharmaceutical firms (Saeed et al. 2019).

Comparison between the pharmaceutical industry with other industries emphasizes the critical importance of personnel selection phases (compared to those of, for example, training) among various HRM practices, emphasizing the need to close a research gap regarding the construction of measurements of the carbon footprint of workers. Instead, Andersson et al. emphasize the importance of managers in motivating employees to behave in a particular way. The development of green materials that the pharmaceutical industry can use for production is the subject of a final line of inquiry (Masri and Jaaron 2017). These more comprehensive studies identify opportunities for the pharmaceutical industry by examining materials and prospective applications. The use of sustainable products like algae, which can be used to make therapeutic agents, or the use of waste from the production of marine resources, such as fish rather than other food waste, has been suggested in some articles in this field of study (Andersson et al. 2005; Sudhakar et al. 2019; Dave and Routray 2018).

Manda et al. suggest using cutting-edge membranes to remove pharmaceutical waste from water. There is only a small number of research in this area that looks at how environmental and social sustainability interact, as well as how environmental and economic sustainability interact (Mirabella et al. 2014). Azim and Azam examine social and environmental disclosure about the particular instance of Bangladeshi pharmaceutical firms. The authors draw attention to the backlog in the pharmaceutical industry in this area (Manda et al. 2014). The Weraikat et al. research suggests a study to confirm strategies to persuade customers to return unneeded medications.

Advantageous projections are also explored for the environment because drug dispersion is reduced (in line with a vision of environmental sustainability), but it is also advantageous from a social responsibility standpoint because it allows for the possibility of distributing these unused drugs to populations that lack access to them (Azim and Azam 2013). The analyzed studies show an optimization model for the integrated production and distribution planning of a supply chain, with the two (economic and environmental) objectives of cost minimization and greenhouse gas emissions minimization, as well as themes of reverse logistics, which refers to all operations related to the reuse of products and materials, in the supply chain (Weraikat et al. 2016).

3.3 Economical Sustainability

As studied by Aquino et al. the incorporation of an economic logic into management has occurred in many healthcare systems as a result of the need to balance two objectives, namely the need to reduce costs and the need to assure sustainability. On the one hand, this need is frequently seen as a way to minimize costs, leading to a reduction in services rather than increased efficiency, decreased waste, etc. On the other side, with the implementation of protocols and standards in service delivery, the enhancement of efficiency has been translated as the standardization of procedures (Jabbarzadeh et al. 2019).

The study by Aquino et al. explicitly investigates the opportunities provided by technology to reconcile standardization and personalization of healthcare within this context. More particularly, the report emphasizes the need and possibility of pharmaceutical businesses using 3D printing to create customized drugs. Key findings point to a potentially game-changing role for 3D printing in pharmaceutical manufacture and/or compounding as a tool for smarter healthcare (Aquino et al. 2018).

The term 'economic sustainability' is also brought up about patent activity. In contrast to their product patents, Azad et al. concentrate on significant pharmaceutical businesses in Bangladesh and their experience in process patent operations. The authors observe that using automation and the purchase of patents alone to achieve sustainability over the long term seems inappropriate. Thus, technical efficiency is assessed, and it is demonstrated that scale inefficiency is the primary cause of inefficiency in major pharmaceutical enterprises (Aquino et al. 2018).

It has been shown that this kind of inefficiency outperforms pure technical inefficiency. Growth is a proximate indicator of economic viability. The study demonstrates that exports, R&D spending, and prior-year earnings have a beneficial influence on the expansion of SMEs. However, for their long-term growth and survival plan, SMEs in the pharmaceutical business need to focus more on internationalization and value creation through R&D investment.

The financial performance of pharmaceutical firms is also taken into consideration, particularly those that use environmentally friendly manufacturing methods. The managerial issue raised has to do with weighing the costs and advantages of sustainable supply chain management and green manufacturing. In addition to tracking changes in companies' resource and financial performance, Menzel et al. examine annual and sustainability reports made public by businesses. This study concludes that there is no meaningful connection between green manufacturing and financial success; hence, the expected finding that a company's greener manufacturing leads to greater corporate performance is not supported by the results (Azad et al. 2018).

3.4 Comprehensive Approaches

Many papers offer a comprehensive viewpoint by considering all the sustainability factors that affect business sustainability at once (Menzel et al. 2010). Corporate sustainability, which has its roots in sustainability on a global scale, is translated to the business level and is defined as "meeting the needs of a firm's direct and indirect stakeholders (such as shareholders, employees, clients, pressure groups, communities, etc.), without compromising its ability to meet the needs of future stakeholders as well" (Steger et al. 2007). Leonard and Schneider (2004) take a comprehensive approach to sustainability in the pharmaceutical sector, not just considering people and the environment but also profitability. For sustainability activities in all their dimensions—environmental, social, and economic—to be monitored, valued, and controlled like conventional business drivers, they advocate aligning integrated sustainability programs with business goals. In their 2017 article, Chaturvedi et al. address how pharmaceutical corporations balance societal economic and social welfare with environmental sustainability. They suggest a paradigm for organizational sustainability, which may be used interchangeably with the term "corporate sustainability," in which there is a balance between economic growth and financial profits and societal and environmental costs. Vihari et al. (2019) expand organizational sustainability by introducing the concept of organizational learning. Business model innovation—more specifically, innovation in the value proposition, value generation, and value capture system—is also addressed as a significant factor in corporate (Dyllick and Hockerts 2002). Strategic partnerships with non-profit organizations can also improve corporate sustainability and ensure the value of sustainable initiatives (Hansen et al. 2010). Several articles on sustainable supply networks consider the economic, social, and environmental aspects of this diverse area. Pharmaceutical firms need to concentrate on sustainability issues as institutional pressure to produce more sustainably grows. Additionally, they should extend their responsibility for their activities from production to the entire supply chain with the additional goal of preventing financial losses, environmental and social risks, or scandals (Vihari et al. 2019). The notion that pharmaceutical enterprises must adhere to sustainable supply chain activities, such as sustainable design and development, strategic sourcing, sustainable product returns, and recycling, appears to have widespread support (Walker and Jones 2012). Pharmaceutical businesses must decide how committed they are to sustainable supply chain practices, according to Janatyan et al. (2018). They also suggest using a multi-objective model to construct a pharmaceutical distribution network by the primary sustainability dimensions. The possibility for pharmaceutical businesses to involve their suppliers in their sustainability activities is discussed by Villena and Gioia (2018). Like this, Villena (2019) investigates how sustainability criteria move throughout international supply chains. According to the study, a company's efforts to promote sustainability throughout its supply network are hampered by a lack of collaboration between procurement and internal and external stakeholders. As a result, the buyer's procurement unit must connect with the supplier's procurement unit to flow the buyer's sustainability requirements. Green HRM is linked to green supply chain management and has a beneficial impact on the triple bottom line of sustainability performance, in addition to strategic procurement and supplier engagement (Zahiri et al. 2017). Articles on corporate social responsibility (CSR) frequently incorporate the three sustainability dimensions. The first query relates to how pharmaceutical firms address sustainability concerns in their CSR reports. According to Min et al. (2017), most pharmaceutical companies presently engage in CSR by using the triple-bottom-line strategy of managing their businesses and having a good overall impact. The authors conclude that CSR, regardless of the size of the organization, provides value to corporate financial performance and should be seen as a long-term investment. They contend that pharmaceutical businesses should adopt a CSR strategy because it invests in stakeholder management and fosters partnerships that boost reputation and revenue. According to Schneider et al. (2010), there has been an increase in the breadth and depth of relevant operations across the pharmaceutical industry as they analyze the development of reported sustainability activity. Demir and Min (2019) analyze the consistency and disparities in pharmaceutical companies' CSR reporting. They find that while disclosures on established issues, like the environment and labor relations, exhibit some degree of standardization, those paying particular attention to delicate ones, like human rights and supply chain, are far from being standardized. Performance evaluation of intangibles related to CSR is taken into consideration while talking about it (Zaid et al. 2018). The eight pillars of sustainable operations—people, quality, safety, productivity, availability, environment, community, continuous improvement, and engineering excellence—are evaluated as part of Mistry's (2018) proposal for a performance measuring system.

4 Effect on the Environment Caused by the Pharmaceutical Industry

Human medications are on UNESCO's list of emerging pollutants. According to the "2030 Agenda for Sustainable Development" Goal Targets (Parisi and Hockerts 2011), their detection and removal represent the key stage. Drug levels in the environment are below what is considered therapeutic. Pharmaceuticals are found in amounts lower than 100 ng/L in water surfaces that receive treated wastewater (UNESCO (United Nations Educational, Scientific and Cultural Organization) 2020). Due to these low quantities, it is challenging to determine how harmful these substances are to both human health and the ecology. According to aus der Beek et al. (2016), most medications have not been thoroughly investigated for their potential long-term toxicity, environmental presence, or other impacts. Beta-blockers, antibiotics, anticancer medications, and endocrine disruptors, among other drug classes, have been shown to have devastating effects on the ecosystem, including increased mortality and impairment of the physiological and reproductive processes of aquatic species (Vumazonke et al. 2020; Nie et al. 2013). Furthermore, because it is hard to separate humans from nature, these consequences have a disastrous impact on human health. Due to the enormous number of medications and the difficulties in assessing the dangers associated with multi-compound exposure at low doses and over extended periods, the problem's scope is still largely unclear (Kümmerer 2019). 10% of currently available medications pose a potential environmental risk, according to the German Environment Agency (UBA). Despite the lack of developed detection methods for all pharmaceuticals that enter the ecosystem, some of them are overwhelmingly present and have been shown to have harmful impacts on ecosystems (Küster et al. 2010). Hormones, antibiotics, antidepressants, analgesics, anti-inflammatory drugs, and anticancer medications are some of these categories (Monteiro and Boxall 2010). Estrogens in the environment are a significant source of pollution (Bilal and Igbal 2019). Only the usage of birth control pills is responsible for the about 30,000 kg/yr of naturally occurring steroidal estrogens and an extra 700 kg/yr of manufactured estrogens that are excreted by the world's population. However, the amount of these hormones in the environment is mostly a result of the livestock industry, which uses a range of growth-regulating steroids extensively to boost the rate of meat production. Estrogens are unquestionably necessary for healthy human physiology, but if they build up in the environment and reach the human food chain, they can have major negative effects (Adeel et al. 2017). These hormones can disturb the physiology of both humans and animals and have an impact on regular reproduction (Trevino et al. 2015). Estrogens are also associated with a higher risk of breast cancer in women and prostate cancer in males. Given the increased usage of antibiotics during the COVID-19 pandemic, which resulted in the exhaustion of the final line of antibiotics, antimicrobial resistance is a concern for global public health (Lai et al. 2021). It has been noted that the use of antibiotics in human medicine, veterinary medicine, and agriculture is connected to the contamination of various environmental components, which led to an increase in antibiotic resistance and the occurrence of ecotoxicological effects (Zainaba et al. 2020). A growing environmental concern to public health is posed by patients' improper antibiotic disposal practices, which include flushing them into sewage systems (Anwar et al. 2020). Additionally, long-term antibiotic pollution exposure can have a negative influence on human health, particularly in people who already have chronic illnesses like obesity, diabetes, or asthma (Ianiro et al. 2016).

During the COVID-19 pandemic, antidepressant contamination considerably increased globally (Rabeea et al. 2021). Antidepressants have been found in both urban and rural water systems up to this point. Diverse antidepressants cause cytotoxicity, genotoxicity, altered stress responses, changes in weight and length, as well as liver and kidney damage, in a variety of aquatic animal species due to bioaccumulation in their tissues (Castillo-Zacarías et al. 2021). Antidepressant pollution (sertraline, fluoxetine) has affected human neurodevelopment and many mental problems since the human and animal habitats largely overlap (Abbey-Lee et al. 2018). Non-steroidal anti-inflammatory drugs, such as acetaminophen, acetylsalicylic acid, ibuprofen, diclofenac, and naproxen, are commonly found in soil, wastewater, surface water, drinking water, and groundwater in high concentrations that significantly contribute to environmental pollution (Tyumina et al. 2020). These medications have long-lasting ecotoxic effects because of how difficult it is for them to be biologically altered by the environment due to their stable chemical structure. By causing oxidative stress and interfering with the action of detoxification enzymes, they are known to predominantly induce diseases in the organs of invertebrates and vertebrates (Hodkovicova et al. 2022). These medications can also impair oocyte maturation via an unidentified mechanism, leading to hepatotoxicity and cardiovascular problems (Lister and Van Der Kraak 2009). These drugs can also impact oocyte maturation, resulting in hepatotoxicity and anomalies in the cardiovascular system, claim Lister and Van Der Kraak (2009) and Xia et al. (2017).

Pharmaceutical substances with the potential to be highly persistent and harmful in the environment include beta-blockers (Küster et al. 2010). While there is a dearth of information regarding their environmental adsorption, it is known that these medications have a somewhat high water solubility and have been found in surface waters at quantities of g/L. These substances are highly hydrolysis-resistant, bioavailable, and transportable in the environment. As a result, their buildup in the environment may have unintended effects on various living things. Metoprolol and propranolol are considered harmful substances to aquatic life by European Union Directive 93/67EEC. This shows the outcomes of tests done on *S. vacuolates*, a type of green alga (Maszkowska et al. 2014).

When anticancer medications are released into the environment, they disrupt fertility and significantly alter the genetic makeup of living things, which has an impact on the ecosystem. Anticancer medications limit the growth and division of cells. Anticancer medications are recommended in smaller doses, but even at ng/L concentrations, they have devastating effects, including harm to aquatic species through mutagenic, carcinogenic, and teratogenic effects. Cytostatic medications are frequently found in hospitals and pharmaceutical manufacturing wastewater because of their use and improper disposal. Cisplatin is one of the most dangerous medications, with anticancer drug detection in hospital wastewater occurring 58% of the time (Dan et al. 2021). Even at ng/L quantities, the presence of cisplatin in water-bodies can have harmful effects on aquatic life (Roque-Diaz et al. 2021). The surface water Watch List (WL) contains details on new pollutants that pose a serious risk to aquatic life and people when ingested through contaminated drinking water. All the pharmaceutical categories described above are included in this list (Gomez et al. 2022).

5 Environmental Impact and Sustainable Action

The pharmaceutical industry has demonstrated its significant presence in India, with over 50,000 registered units in the bulk and chemical sectors. Its usefulness is seen in the Indian market with a CAGR of 14.3%. The economy is predicted to reach USD 55 billion by 2025. 80% of the nation's medicine demand is met by its range of goods, which come in solid, liquid, and injectable forms. India has had the highest number of operating units in Asia approved by the US FDA. This facilitates the production of a wide range of bulk drugs by Indian pharmaceutical businesses and increases the export of these advanced bulk products across international borders. According to the Organisation of Pharmaceutical Producers of India, the pharmaceutical industry in India is the world's fourteenth largest by value and the third largest by volume, contributing around 2% of the total pharmaceutical production worldwide.

Even with the pharmaceutical industry's strong expansion, India is still in its infancy when it comes to adopting practices connected to sustainability and being transparent in its reporting for the benefit of stakeholders. Stakeholder demands force companies to demonstrate their responsibility, transparency, and efficacy in governance through corporate sustainability disclosures (Kolk and Perego 2010). By highlighting the significance of not just measuring and monitoring performance on ESG but also placing transparency at the core of the strategic operation, India's Standard and Poor's environmental, social, or corporate governance index of the National Stock Exchange took the lead in this regard. To promote investments in financially and environmentally sustainable enterprises, the Bombay Stock Exchange created the novel tool known as GREENEX, which evaluates a company's energy efficiency performance using publicly available energy and financial data. However, reporting and disclosure of these practices and their effects on a company's efficiency and values are still in their infancy in India.

Note that since its founding in India in 2003, the UN Global Compact Network has not required any reporting or disclosures. Nevertheless, the adoption of sustainability has received a lot of attention on a global scale throughout the past ten years. The pharmaceutical industry in India has also been the subject of fewer study reports in this area thus far. Reluctantly adopting environmental protection practices, Indian firms began reporting on sustainability only in 2001 (Sahay 2004). Till 2006, no pharmaceutical company had disclosed information about the environment or energy out of

50 served firms. The 100 most valued corporations in India's sustainability activities were examined. Significant variation in reporting across several sustainability-related indicators is revealed by their investigation. Only two of the nine pharmaceutical businesses were discovered to have disclosed environmental conservation activities and engaged in voluntary sustainability reporting. Comparing the Indian pharmaceutical sector to its international counterparts and other Indian sectors, it is evident that the former is less receptive to the adoption of sustainability and related practices.

Despite significant progress, environmental reporting remains non-standardized and non-comparable (Aparajita 2010). Oil & gas, together with the mining and metals industry, were among the primary initiators, yet there is a dearth of sustainability reporting, even though some are striving for environmental excellence. Dr. Reddy's Lab and Jubilant Organosys are the only two Indian pharmaceutical companies that have undertaken the initiative to integrate environmental, health, and safety concerns with their manufacturing and production processes through sustainability reporting. Thus, applying green chemistry can impact and enhance the pharmaceutical industry's environmental performance. Recently, studies have focused on adopting green chemistry, evaluating life cycles, taking environmental concerns into account, and scaling up manufacturing processes using flow technology to better assess the true sustainability of the pharmaceutical industry (Kralisch et al. 2013) (Table 1).

Table 1 Analysis of the Indian pharmaceutical industry by SWOT

Strength	Weakness	Challenges	Opportunities
Rapidly expanding developing market	Raising R&D output but declining	More stringent regulations	Advanced economy
Robust pool of highly qualified workers	Intensely competitive setting	Demands for attempts to reduce costs	Substantial FDI in the industry
The large production base of facilities licensed by the US FDA	Minimal expenditure in R&D	Key performance parameter benchmarking	An enhanced mechanism for managing information
Substantial home market	Initiatives for improvement are always optional	Enhanced correspondence with interested parties	Managing corporations
Positive research and development atmosphere	Apprehension about capital gains	Implementing new technology with less of an impact on the environment	Novel technology accessible for enhanced operational performance
Government programs to form commercial alliances	Lack of willingness to share intellectual data	Fulfilling global regulatory requirements	Streamlining processes to simplify work

6 Research Forfeits and Difficulties with Pharmaceutical Industry Sustainability

Currently, top management does not place much emphasis on decisions about sustainability in the organization, out of all the decisions involved in operating a business. To handle the sustainability component of decision-making in a dynamic corporate environment, it is imperative to engage all stakeholders and gain knowledge of their perspectives (Adisa et al. 2016). A pharmaceutical manufacturing company's operational strategy needs to incorporate sustainability-oriented initiatives comprehensively. Health and safety concerns, the environmental effects of greenhouse gas emissions during the manufacturing of APIs, and the growing consumer preference for environmentally friendly products are just a few of the well-established and up-and-coming factors that are major drivers of the issue of sustainability in the pharmaceutical sector. The growing concern over sustainability and related practices in API manufacturing is that these activities could have deadly consequences due to their environmentally intensive nature (Sanchez 2015).

First, aside from a small number, it was discovered that the majority of Indian pharmaceutical companies do not actively and comprehensively manage the sustainability issue. Furthermore, the reporting of their approach to resolving the same is opaque. Pharmaceutical companies have implemented several measures to integrate socially and ecologically responsible production practices on a worldwide scale; these have been sufficiently discussed in the sections that follow. While several large Indian corporations have succeeded in making their production processes sustainable, the industry has generally evolved in a slow but varied manner. It's interesting to note that international regulators and drug-importing businesses have been closely monitoring the Indian pharmaceutical sector's operations, which has contributed to the sector's apparent change towards enhancing its environmental performance. Using these procedures and disclosing them to stakeholders in a formal manner fosters dependable and solid corporate relationships (Goel 2010). For pharmaceutical businesses, information disclosure through enhanced formal reporting is unavoidable given the growing need for greater responsibility and openness from the companies. Accordingly, sustainability reporting serves as a crucial barometer for gauging a manufacturing company's shift to "green manufacturing". The literature study reveals that sustainability reporting and performance evaluation at different levels have developed independently, although with some noteworthy parallels. The transformation happened as a result of the necessity to support top management in making decisions and realizing long-term value creation. By observing sustainability alongside growth and profitability, the Indian pharmaceutical industry can learn from this evolution. Making decisions that are transparent and equitable is also aided by this (Amran and Keat 2014).

7 Opportunities for Future Research

The social impact of the Pharmaceutical Industry is self-evident: drugs and vaccines (but also other pharmaceutical products, such as medical devices) play a fundamental role in disease prevention, diagnosis, therapy, and fundamentally in the continuous improvement of people's quality of life. Despite this, we found few studies dealing with the social impact of the Pharmaceutical Industry. Hence, the social sustainability of the pharmaceutical industry could be studied from a two-way perspective: One is about market access to innovative drugs that contribute to gains in terms of well-being and another is about the accessibility of drugs to all sectors of the population among the different markets at a Global level (Milanesi et al. 2020).

A further emergent gap concerns the relationship between mature and emerging markets. The emphasis on emerging markets mainly concerns the production of drugs by local companies or subsidiaries of multinational companies, covering only one dimension of the production of Pharmaceuticals Globally. Furthermore, the analysis of emerging markets does not take into consideration the social dimension of sustainability, which is linked to market access to new drugs and access to medicines by all sectors of the population. There seems to be a disconnect between the interest in emerging markets from a production/environmental impact perspective and the objectives of ensuring wider access to care. The potential barriers to access to drugs, including cultural or regulatory ones, are a key topic to be explored (Bade et al. 2023).

Another under-investigated topic is the relationship between innovation and sustainability. On the one hand, the empirical evidence shows a change in the technological trajectory towards research and development of expensive molecules with disruptive effects on some diseases. These expensive molecules are in systems that have the objective of offering access to drugs by a universalistic logic. On the other hand, the literature seems to ignore this aspect for the most part. Therefore, a research gap has emerged in the study of the relationship between innovation and sustainability (Moermond et al. 2022).

Nowadays, one of the most useful concerns is waste management. As is well known, the issue of waste management has attracted attention in many industries. In the pharmaceutical sector, the study of waste management concerns different levels. At the level of the user, the recovery and disposal of expired or expired drugs and their packaging are the focus. At the production level, the management of production waste and attention to water use are of interest. The best manufacturing practices for the Pharmaceutical Industry almost exclusively focus on the correct procedure to produce medicines. However, medical devices, such as plasters, condoms, and diagnostic tools, deserve special mention, as they fall into several product categories, from electronics to the plastic industry, due to the integration of rubber. In addition to environmental pollution itself, the issue of waste disposal also involves far more serious implications, such as the proliferation in the environment of drug-resistant bacteria, which are one of the main threats to public health. Therefore, it is not only a matter relevant to the environment but also to social sustainability and the economic impact of waste management (Chaturvedi et al. 2017).

It has been noticed that little attention has been paid to the issue of economic sustainability and, above all, to the relationship between Pharmaceutical Companies and the economic sustainability of the health system. This area undoubtedly represents an opportunity for academic research. In recent years, the economic crisis has focused attention on the issue of health system sustainability, mainly concerning cost effectiveness, quality of care, and lately, patient involvement. The trade-off between effectiveness, efficiency, and sustainability reveals the complexity of health systems management. Within this context, different factors contribute to the chances of success by balancing the efficiency, effectiveness, and sustainability of health systems, and a leading role is certainly played by pharmaceutical companies. Other aspects must also be taken into consideration. In developed countries, average life expectancy has risen above the age of 80, and health science, new drugs, and technologies have improved the quality of life. Although unquestionably desirable, these aspects imply costs for the health system. Thus, the economic dimension of sustainability and the relationship between sustainable health systems (Pierzecka 2023).

The Major Findings of this study have been discussed in brief in this section. Among all the decisions for running an enterprise, the ones related to sustainability in the organization are presently not given much importance by the top management. Accordingly, sustainability-oriented approaches must be holistically integrated as part of the operational strategy of a pharmaceutical manufacturing enterprise. There are several established and emerging causes that act as key drivers to the issue of sustainability in the pharmaceutical sector like health and safety concerns, environmental impacts of GHG emission during API manufacturing, and increasing consumer preference for environment-friendly products among others. Sustainability and related practices associated with API manufacturing are emerging concerns requiring immediate attention because of the fatal consequences anticipated from such environmentally-intensive production activities.

Few significant findings emerged from the review undertaken to prepare this article. Firstly, it was learned that most of the Indian pharmaceutical companies are not managing the sustainability issue actively and holistically, barring a few. Secondly, there is no transparency in reporting how they are addressing the same. At the global level, various steps have been taken by pharmaceutical firms to incorporate environmentally and socially responsible production and they have been mentioned in sufficient detail in the subsequent sections. Some big Indian companies have reportedly achieved sustainability in their production process but the evolution of the sector on an overall basis has been rather slow yet diverse. Interestingly, the Indian Pharmaceutical sector's visible shift towards improving its environmental performance has also been driven by the fact that foreign regulators and drug-importing companies closely monitor its operation. With an increased demand for more transparency and accountability on the part of the companies, disclosure of information through improved formal reporting is inevitable for pharmaceutical companies.

Thus, sustainability reporting is an important indicator of the transition of a manufacturing company towards 'green manufacturing'. A review of the literature shows that performance measurement at various levels and sustainability reporting have evolved separately but with relevant similarities. The evolution occurred as a consequence of the need to help top management in decision-making and in achieving long-term value creation. This also supports making fair and transparent decisions (Moermond et al. 2022).

8 Conclusion

Given the mention of the major findings in brief, the review concluded with the following remarks: Sustainability-related thinking at the top management level and across the organization with its incorporation in the vision, mission, and routine practices have more positive outcomes on environmental, economic, and social welfare. Most organizations may follow externally specified codes of conduct or have developed internal governance policies but without focusing on transparency for stakeholders. External motivational factors like improvement of corporate image, marketing, competitive image, global reporting initiatives, and standard industry-specific guidelines have positive impacts on sustainability-related practices in the pharmaceutical sector. Internal motivational factors such as the proactive thinking of top management and its involvement in implementing sustainability-related issues, improvement on environmental performance, innovation, cost saving, Research and Development facilities, etc. positively influence the performance objective. Thus transparency in external and internal factors and communication with stakeholders must be integrated and coordinated in line with companies' strategy. Deciding performance objectives via an environmental management system (EMS) is more appropriate for selecting product design and greening of eco-efficient manufacturing processes. Application of green chemistry principles, and choosing the correct metrics have a greater impact on the selection and designing of environmentally benign products and processes. The lack of transparency in data availability reduces the credibility of studies with the use of various metrics and results in low-value addition among stakeholders too. Thus, it is necessary to continuously improve the consistency and transparency of the information and assumptions with the use of different sustainability-related practices to address environmental, economic, and social aspects. A deep understanding of the process and analysis of the stages associated with hot spot generation contributes to on-site reduction of waste, energy consumption, and related emissions. Key benefits such as waste reduction, mitigation of carbon emission, and legislative compliance affect significantly sustainability achievement in pharmaceutical companies and give more flexibility and competitive advantage. Actions for performance improvement such as commitment, punishment, incentives, and establishment of standard industry-specific guidelines have to be implemented at the global level. This, in return, will push the local legislation to act accordingly and collectively from sustainability thinking to action.

References

Abbey-Lee RN, Uhrig EJ, Garnham L, Lundgren K, Child S, Løvlie H (2018) Experimental manipulation of monoamine levels alters personality in crickets. Sci Rep 8:16211. https://doi.org/10.1038/s41598-018-34519-z

Adeel M, Song X, Wang Y, Francis D, Yang Y (2017) Environmental impact of estrogens on human, animal and plant life: a critical review. Environ Int 99:107–119. https://doi.org/10.1016/j.envint.2016.12.010

Adisa A, John B, Beatrice C, Samson K, Aziz E (2016) The global warming potential of the production and consumption of Kenyan tea. J Clean Prod 112(Part 5):4031–4040. https://doi.org/10.1016/j.jclepro.2015.07.029

Amran A, Keat OS (2014) Sustainability reporting: meeting stakeholder demands. Strat Dir 30(7):38–41

Andersson L, Shivarajan S, Blau G (2005) Enacting ecological sustainability in the MNC: a test of an adapted value-belief-norm framework. J Bus Ethics 59:295–305

Anwar M, Iqbal Q, Saleem F (2020) Improper disposal of unused antibiotics: an often-overlooked driver of antimicrobial resistance. Expert Rev Anti Infect Ther 18:697–699. https://doi.org/10.1080/14787210.2020.1754797

Aparajita G (2010) Information, direct access to farmers, and rural market performance in central India. Am Econ J Appl Econ 2(3):22–45

Aquino RP, Barile S, Grasso A, Saviano M (2018) Envisioning smart and sustainable healthcare: 3D Printing technologies for personalized medication. Futures 103:35–50

Aus der Beek T, Weber FA, Bergmann A, Hickmann S, Ebert I, Hein A, Küster A (2016 Apr) Pharmaceuticals in the environment—global occurrences and perspectives. Environ Toxicol Chem 35(4):823–835

Azad MAK, Munisamy S, Teng KK, Talib MBA, Saona P (2018) Productivity changes of the pharmaceutical industry in Bangladesh: does process patent matter? Global Bus Rev 19:1013–1025

Azim MI, Azam S (2013) Corporate sustainability reporting by pharmaceutical companies: is it what it seems to be? Corp Ownersh Control 11:754–765

Bade C, Olsacher A, Boehme P, Truebel H, Bürger L, Fehring L (2023) Study on sustainability in the pharmaceutical industry—an assessment of sustainability maturity and effects of sustainability measure implementation on supply chain security. Corp Soc Responsib Environ Manag 01–19

Belkhir L, Elmeligi A (2019) Carbon footprint of the global pharmaceutical industry and relative impact of its major players. J Clean Prod 214:185–194

Bilal M, Igbal HMN (2019) Persistence and impact of steroidal estrogens on the environment and their laccase-assisted removal. Sci Total Environ 690:447–459. https://doi.org/10.1016/j.scitotenv.2019.07.025

Blum-Kusterer M, Hussain SS (2001) Innovation and corporate sustainability: an investigation into the process of change in the pharmaceuticals industry. Bus Strat Environ 10:300–316

Caligiuri P, Mencin A, Jiang K (2013) Win-win-win: the influence of company-sponsored volunteerism programs on employees, NGOs, and business units. Pers Psychol 66:825–860

Castillo-Zacarías C, Barocio ME, Hidalgo-Vázquez E, Sosa-Hernández JE, Parra-Arroyo L, López-Pacheco IY (2021) Antidepressant drugs as emerging contaminants: occurrence in urban and non-urban waters and analytical methods for their detection. Sci Total Environ 757:143722. https://doi.org/10.1016/j.scitotenv.2020.143722

Chaar BB, Lee J (2012) Role of socioeconomic status on consumers attitudes towards DTCA of prescription medicines in Australia. J Bus Ethics 105:447–460

Chaturvedi U, Sharma M, Dangayach GS, Sarkar P (2017) Evolution and adoption of sustainable practices in the pharmaceutical industry: an overview with an Indian perspective. J Clean Prod 184:01–34

Dan L, Hongxing C, Hongsong L, Schlenkc D, Mue J, Lacorte S (2021) Anticancer drugs in the aquatic ecosystem: environmental occurrence, ecotoxicological effect, and risk assessment. Environ Int 153:106543. https://doi.org/10.1016/j.envint.2021.106543

Dave D, Routray W (2018) Current scenario of Canadian fishery and corresponding underutilized species and fishery byproducts: a potential source of omega-3 fatty acids. J Clean Prod 180:617–641

Demir M, Min M (2019, Jun) Consistencies and discrepancies in corporate social responsibility reporting in the pharmaceutical industry. Sustain Account Manag Policy J 10(2):333–364

Dyllick T, Hockerts K (2002) Beyond the business case for corporate sustainability. Bus Strat Environ 11:130–141. https://doi.org/10.1002/bse.323

Esteban D (2008) Strengthening corporate social responsibility in the pharmaceutical industry. J Med Mark 8:77–79. https://doi.org/10.1057/palgrave.jmm.5050126

Gardas BB, Raut D, Narkhede RBE (2019) Analyzing the 3PL service provider's evaluation criteria through a sustainable approach. Int J Prod Perform Manag 68:958–980

Goel P (2010) Triple bottom line reporting: an analytical approach for corporate sustainability. J Financ, Account, Manag 1(1):27–42

Gomez CL, Marinov D, Sanseverino I, Navarro CA, Niegowska CM, Rodriguez P (2022) Selection of substances for the 4th Watch list under the water framework directive. Publications Office of the European Union, Luxembourg

Hansen J, Ruedy R, Sato M, Lo K (2010, Dec) Global surface temperature change. Rev Geophys 48(4)

Hodkovicova N, Hollerova A, Blahova J, Mikula P, Crhanova M, Karasova D (2022) Non-steroidal anti-inflammatory drugs caused an outbreak of inflammation and oxidative stress with changes in the gut microbiota in rainbow trout (Oncorhynchus mykiss). Sci Total Environ 849:157921. https://doi.org/10.1016/j.scitotenv.2022.157921

Ianiro G, Tilg H, Gasbarrini A (2016) Antibiotics as deep modulators of gut microbiota: between good and evil. Gut 65:1906–1915. https://doi.org/10.1136/gutjnl-2016-312297

Jabbarzadeh A, Haughton M, Pourmehdi F (2019) A robust optimization model for efficient and green supply chain planning with postponement strategy. Int J Prod Econ 214:266–283

Janatyan N, Zandieh M, Alem-Tabriz A, Rabieh M (2018) Designing sustainable distribution network in pharmaceutical supply chain: a case study. Int J Supply Oper Manag 5: 122e133

Kaenzig J, Friot M, Margni M, Jolliet O (2011) Using life cycle approaches to enhance the value of corporate environmental disclosures. Bus Strat Environ 20:38–54

Kolk A, Perego P (2010) Determinations of the adoption of sustainability assurance statements: an international investigation. Bus Strat Environ 19:182–198

Kralisch D, Denise O, Kressirer S, Staffel C, Sell I, Krtschil U, Loeb P (2013) Bridging sustainability and intensified flow processing within process design for sustainable future factories. Green Process Synth 2(5):465–478. https://doi.org/10.1515/gps-2013-0058

Kumar A, Zavadskas EK, Mangla SK, Agrawal V, Sharma K, Gupta D (2019) When risks need attention: adoption of green supply chain initiatives in the pharmaceutical industry. Int J Prod Res 57:3554–3576

Kümmerer K (2019) From a problem to a business opportunity design of pharmaceuticals for environmental biodegradability. Sustain Chem Pharm 12:100136. https://doi.org/10.1016/j.scp.2019.100136

Küster A, Alder AC, Escher BI, Duis K, Fenner K, Garric J (2010) Environmental risk assessment of human pharmaceuticals in the European Union: a case study with the β-blocker atenolol. Integr Environ Assess Manag 6:1–523. https://doi.org/10.1897/IEAM_2009-050.1

Lai CC, Chen SY, Ko WC, Hsueh PR (2021) Increased antimicrobial resistance during the COVID-19 pandemic. Int J Antimicrob Agents 57:106324. https://doi.org/10.1016/j.ijantimicag.2021.106324

Leonard T, Schneider J (2004, May). Integrated sustainability in the pharmaceutical industry. Corp Environ Strategy: Int J Sustain Bus 11(5):119–128

Li X, Hamblin D (2016) Factors impacting on cleaner production: case studies of Chinese pharmaceutical manufacturers in Tianjin. China J Clean Prod 131:121–132
Lister AL, Van Der Kraak GJ (2009) Regulation of prostaglandin synthesis in ovaries of sexually mature zebrafish (Danio rerio). Mol Reprod Dev 76:1064–1075. https://doi.org/10.1002/mrd.21072
Lorenzini G, Mostaghel R, Hellstr€om D (2018) Drivers of pharmaceutical packaging innovation: a customer-supplier relationship case study. J Bus Res 88:363–370
Low YS, Halim I, Adhitya A, Chew W, Sharratt P (2016) A systematic framework for the design of an environmentally sustainable pharmaceutical supply chain network. J Pharm Innov 11:250–263. https://doi.org/10.1007/s12247-016-9255-8
Manda BMK, Worrell E, Patel MK (2014) Innovative membrane filtration system for micropollutant removal from drinking water—prospective environmental LCA and its integration in business decisions. J Clean Prod 72:153–166
Marinkovic V, Bekcic S, Pejovic G, Sibalija T, Majstorovic V, Tasic L (2016) An approach to TQM evaluation in pharma business. TQM J 28:745–759
Masri HA, Jaaron AAM (2017) Assessing green human resources management practices in Palestinian manufacturing context: an empirical study. J Clean Prod 143:474–489
Maszkowska J, Stolte S, Kumirska J, Łukaszewicz P, Mioduszewska K, Puckowski A (2014) Beta-blockers in the environment: part II. Ecotoxicity study. Sci Total Environ 493:1122–1126. https://doi.org/10.1016/j.scitotenv.2014.06.039
Mélanie D, Michael RO, Peter DB (2013) Sustainable manufacturing tactics and cross-functional factory modeling. J Clean Prod 42:31–41. https://doi.org/10.1016/j.jclepro.2012.11.008
Menzel V, Smagin J, David F (2010) Can companies profit from greener manufacturing? Meas Bus Excell 14:22–31
Milanesi M, Runfola A, Guercini S (2020) Study on pharmaceutical industry riding the wave of sustainability: review and opportunities for future research. J Clean Prod 261:01–12
Min S, Lee B, Yoon S (2017, Sept) Deep learning in bioinformatics. Brief Bioinform 18(5):851–869
Mirabella N, Castellani V, Sala S (2014) Current options for the valorization of good manufacturing waste: a review. J Clean Prod 65:28–41
Moermond CTA, Puhlmann N, Brown AR, Owen SF, Ryan J, Snape J, Venhuis BJ, Kummerer K (2022) GREENER pharmaceuticals for more sustainable healthcare. Environ Sci Technol Lett 09:699–705
Monteiro SC, Boxall ABA (2010) Occurrence and fate of human pharmaceuticals in the environment. Rev Environ Contam Toxicol 202:53–154. https://doi.org/10.1007/978-1-4419-1157-5_2
Moradi M, Jolai F (2018) Purchasing planning and order allocation in the pharmaceutical sustainable supply chain using the theoretical graph (GT-MP-DM) (case study: supplying the clotting factor for patients with hemophilia). Int J Supply Oper Manag 5:361–378
Nematollahi M, Hosseini-Motlagh S-M, Ignatius J, Goh M, Saghafi NM (2018) Coordinating a socially responsible pharmaceutical supply chain under periodic review replenishment policies. J Clean Prod 172:2876–2891
Nie XP, Liu BY, Yu HJ, Liu WQ, Yang YF (2013) Toxic effects of erythromycin, ciprofloxacin, and sulfamethoxazole exposure to the antioxidant system in Pseudokirchneriella subcapitata. Environ Pollut 172:23–32. https://doi.org/10.1016/j.envpol.2012.08.013
Parisi C, Hockerts KN (2011) Causal maps and the performance measurement of CSR related intangibles: a case study. Int J Learn Intellect Cap 8:290–309. https://doi.org/10.1504/IJLIC.2011.041075
Patel RV, Mistry BM, Shinde SK, Syed R, Singh V, Shin HS (2018, Jul) Therapeutic potential of quercetin as a cardiovascular agent. Europ J Med Chem 155:889–904
Paul S, Stuart H (1995) Creating sustainable corporations. Bus Strat Environ 4(3):154–165
Pierzecka EC (2023) Pharmaceutical companies in the light of the idea of sustainable development—an analysis of selected aspects of sustainable management. Sustainability 15:8889

Quak S, Heilbron J, Meijer J (2019) Ranking, coordination, and global governance: the case of the Access to Medicine Index. Bus Polit 21:172–204

Rabeea SA, Merchant HA, Khan MU, Kow CS, Hasan SS (2021) Surging trends in prescriptions and costs of antidepressants in England amid COVID-19. DARU J Pharm Sci 29:217–221. https://doi.org/10.1007/s40199-021-00390-z

Roque-Diaz Y, Sanadar M, Han D, López-Mesas M, Valiente M, Tolazzi M, Melchior A, Veclani D (2021 Oct) The dark side of platinum based cytostatic drugs: from detection to removal. Processes 9(11):1873

Sabrina E, Romana R, Rupert JB (2016) Exploring the integration of corporate sustainability into strategic management: a literature review. J Clean Prod 112:2833–2850. https://doi.org/10.1016/j.jclepro.2015.08.031

Saeed BB, Afsar B, Hafeez S, Khan I, Tahir M, Afridi MA (2019) Promoting employee-environmental behavior through green human resource management practices. Corp Soc Responsib Environ Manag 26:424–438

Sahay A (2004) Environmental reporting by Indian corporations. Corp Soc Responsib Environ Manag 11(1):12–22

Sahu K, Kohli S (2019) Performance improvement tool for the medicines manufacturing pharmaceutical companies under sustainable practices. Int J E-Entrep Innov 09:35–48

Salwa HA, Rashid SE, Philip L (2008) A comparison of four sustainable manufacturing strategies. Int J Sustain Eng 1:214–229. https://doi.org/10.1080/19397030802513836

Sanchez MA (2015) Integrating sustainability issues into project management. J Clean Prod 96:319–330

Schneider A, Friedl MA, Potere D (2010, Aug) Mapping global urban areas using MODIS 500-m data: new methods and datasets based on 'urban ecoregions'. Remote Sens Environ 114(8):1733–1746

Singh RK, Kumar R, Kumar P (2016) Strategic issues in pharmaceutical supply chains: a review. Int J Pharm Healthc Mark 10:234–257

Steger U, Ionescu-Somers A, Salzmann O (2007) The economic foundations of corporate sustainability. Corp Govern 7:162–177. https://doi.org/10.1108/14720700710739804

Sudhakar MP, Kumar BR, Mathimani T, Arunkumar K (2019) A review of bioenergy and bioactive compounds from microalgae and macroalgae-sustainable energy perspective. J Clean Prod 228:1320–1333

Trevino LS, Wang Q, Walker CL (2015) Hypothesis: activation of rapid signaling by environmental estrogens and epigenetic reprogramming in breast cancer. Reprod Toxicol 54:136–140. https://doi.org/10.1016/j.reprotox.2014.12.014

Tyumina EA, Bazhutin GA, Cartagena Gómez AD, Ivshina IB (2020) Nonsteroidal anti-inflammatory drugs as emerging contaminants. Microbiology 89:148–163. https://doi.org/10.1134/S0026261720020125

UNESCO (United Nations Educational, Scientific and Cultural Organization) (2020) Emerging pollutants in water and wastewater. https://en.unesco.org/emergingpollutantsinwaterandwastewater

Veleva V, Cue BW (2017) Benchmarking green chemistry adoption by "big pharma" and generics manufacturers. Benchmark 24:1414–1436

Veleva V, Hart M, Greiner T, Crumbley C (2003) Indicators for measuring environmental sustainability: a case study of the pharmaceutical industry. Benchmark 10:107–119

Veleva V, Bodkin G, Todorova S (2017) The need for better measurement and employee engagement to advance a circular economy: lessons from Biogen's "zero waste" journey. J Clean Prod 154:517–529

Vihari NS, Rao MK, Doliya P (2019) Organizational learning as an innovative determinant of organizational sustainability: an evidence-based approach. Int J Innov Manag 23. https://doi.org/10.1142/S1363919619500191

Villena VH (2019, May) The missing link? The strategic role of procurement in building sustainable supply networks. Prod Oper Manag 28(5):1149–1172

Villena VH, Gioia DA (2018, Nov) On the riskiness of lower-tier suppliers: managing sustainability in supply networks. J Oper Manag 1(64):65–87
Vumazonke S, Khamanga SM, Ngqwala NP (2020) Detection of pharmaceutical residues in surface waters of the eastern Cape province. Int J Environ Res Public Health 17:4067. https://doi.org/10.3390/ijerph17114067
Walker H, Jones N (2012) Sustainable supply chain management across the UK private sector. Supply Chain Manag 17:15–28. https://doi.org/10.1108/13598541211212177
Weraikat D, Zanjani MK, Lehoux N (2016) Two-echelon pharmaceutical reverses supply chain coordination with customer incentives. Int J Prod Econ 176:41–52
Zahiri B, Zhuang J, Mohammadi M (2017) Toward an integrated sustainable-resilient supply chain: a pharmaceutical case study. Transp Res Part E Logist Transp Rev 103:109–142. https://doi.org/10.1016/j.tre.2017.04.009
Zaid AA, Jaaron AAM, Talib BA (2018) The impact of green human resource management and green supply chain management practices on sustainable performance: an empirical study. J Clean Prod 204, 965–979. https://doi.org/10.1016/j.jclepro.2018.09.062
Zainaba SM, Junaidb M, Xu N, Malik RN (2020) Antibiotics and antibiotic-resistant genes (ARGs) in groundwater: a global review on dissemination, sources, interactions, environmental and human health risks. Water Res 187:116455. https://doi.org/10.1016/j.watres.2020.116455
Zhu XX, Tuia D, Mou L, Xia GS, Zhang L, Xu F, Fraundorfer F (2017, Dec) Deep learning in remote sensing: a comprehensive review and list of resources. IEEE Geosci Remote Sens Mag 5(4):8-36

Climate Change and Health

Prabhakar Singh, Mohammad Murtaza Mehdi, Saqib Hassan, Kunal Biswas, and V. Ramesh Kumar

Abstract The global phenomenon of climate change has emerged as one of the most critical challenges of the twenty-first century, exerting far-reaching effects on various aspects of human life. The environmental shifts and public health outcomes has garnered increasing attention from researchers, policymakers, and healthcare professionals. Over the past decade, extreme weather events have increased by 43%, leading to disruptions in healthcare infrastructure, food systems and diseases transmission. Climate change manifests through alterations in temperature, precipitation patterns, and extreme weather events amplifying various health risks. The geographical range of vector-borne diseases like malaria and dengue fever expands as warming temperatures extend the breeding seasons of disease-carrying insects. Waterborne diseases become more common as rising temperatures alter water quality and facilitate the proliferation of pathogens. Vulnerable populations and socioeconomically disadvantaged communities bears the brunt of these health impacts. Climate change and mental health has also drawn more attention. The psychological toll of witnessing and surviving climate-related disasters, contributes to elevated stress, anxiety, and depression levels. Comprehensive and multidisciplinary measures are urgently required to reduce the negative health effects of a changing environment. This book chapter explores the complex relationship between climate change and health, delving into the multifaceted ways, and highlights the comprehensive strategies to mitigate the adverse impacts of climate change and safeguard human health on a global scale.

Keywords Climate change · Diseases · Health · Pollution

P. Singh (✉) · S. Hassan · V. R. Kumar
Department of Biotechnology, Sathyabama Institute of Science and Technology, Chennai 600119, Tamil Nadu, India
e-mail: prabhakar.singh.bte@sathyabama.ac.in

M. M. Mehdi
Department of Biochemistry, School of Bioengineering and Biosciences, Lovely Professional University, Phagwara, Punjab, India

K. Biswas
Centre for Nanoscience and Nanotechnology, International Research Centre, Sathyabama Institute of Science and Technology, Chennai 600119, Tamil Nadu, India

P. K. Prabhakar and W. Leal Filho (eds.), *Preserving Health, Preserving Earth*, World Sustainability Series, https://doi.org/10.1007/978-3-031-60545-1_3

1 Introduction

The term "climate change" describes the gradual modification of Earth's mean temperature and weather patterns, which is mostly caused by human activity and natural disasters. Increased frequency and severity of heat waves, changed precipitation patterns, rising sea levels, melting of glaciers and ice sheets, changes in ecosystems and habitats, and variations in the frequency and severity of severe weather conditions like hurricanes, droughts, and excessive precipitation are all consequences of climate change. Significant effects of these changes will be felt by the environment, economics, public health, and general well-being. Global temperatures are gradually rising as a result of human activities like deforestation, industrial process and burning fossil fuels that release greenhouse gases into the atmosphere. These activities also increase the accumulation of greenhouse gases such as carbon dioxide (CO_2), methane (CH_4), and nitrous oxide (N_2O), which has an impact on society and the environment via global warming.

Climate change has been identified as the most significant global health concern of the 21st century due to its adverse impact on the physical environment, ecosystems, and human interactions with them (Perera et al. 2020; Romanello et al. 2021; Vicedo-Cabrera et al. 2021). The health consequences range from infectious disorders brought on by poor hygiene and an overabundance of microorganisms to early deaths from natural disasters (Lemery et al. 2021; Romanello et al. 2021). According to several research, by 2030, the adverse impact of climate change will likely be a contributing factor in 700,000 annual deaths, on top of the 400,000 deaths it currently causes annually (McKinnon 2012). Governments should invest and modulate their policy in climate change associated with public health research, finance linked to climate-resilient health systems, and collaborate to implement regulations that reduce climate change, encourage public health, and track progress over the next 15 years. These recommendation are based on according to the ten recommendations made by the multidisciplinary Commission on Climate Change and Health of The Lancet in 2015 (Watts et al. 2018).

Residents across varying altitudes and in countries with differing income levels face the peril of climate change irrespective of their age or financial standing. While the adverse health effects of climate change have been extensively recorded, the existing evidence has mainly emphasized health consequences linked to specific individual weather conditions (Murray et al. 2020; Zhao et al. 2021). Previous reports have indicated that heatwaves may increase the mortality rates for both cardiovascular and respiratory ailments (Zhao et al. 2019). Nevertheless, there is a scarcity of studies that have delineated the connections among various climatic factors or their combined effects with non-environmental elements on human health. Elucidating these pathways is pivotal in comprehending the repercussions of climate change on human health and devising strategies for mitigation and adaptation.

The altering climate patterns due to global warming have resulted in a heightened occurrence, strength, and duration of extreme weather events and natural disasters worldwide. These events, influenced by climate change, detrimentally impact diverse health outcomes through intricate pathways. Examples include suboptimal ambient temperatures, rising sea levels, and various other conditions. They illustrate the interconnected pathways among different climate-altering occurrences such as suboptimal temperatures, wildfires, and floods, affecting a wide spectrum of health outcomes. These events contribute to health issues spanning non-communicable diseases, infections, injuries, mental disorders, food insecurity, and more, affecting human health through a multitude of channels.

2 Effects of Climate Change on Human Health

2.1 Heat-Related Illnesses and Mortality

In comparison to 1986–2005 baseline, record temperatures in 2020 led to a new high of 3.1 billion more person-days of heatwave exposure among adults over 65 and 626 million more person-days affecting children under the age of one. In terms of 2021, those who were 65 years of age or younger, as well as those who were disadvantaged socially, were most impacted by the June record-breaking temperatures exceeding 40 °C in the Pacific Northwest regions of the United States and Canada. These temperatures would have been nearly unthinkable in the absence of human-caused climate change. According to statistics, 485 incidents of heat-related illnesses and injuries at work occurred in Australia during the 11-year period between 1997 and 2007. In a similar vein, Washington, DC saw 480 cases of heatstroke in 11 years, from 1995 to 2005. Between 1992 and 2006, heat stress claimed the lives of sixty-eight farmers in the United States. 389 people died in Japan from heat shock between 1989 and 2012 (Heidari et al. 2020).

One of the most serious and severe illnesses that can affect a person's health is heat stroke, which happens when the body temperature hits 103 °F (43.9 °C). This condition concurrently affects the function of many organs in the body and can be fatal in a matter of hours (Mohammadyan and Sepehr 2010). One of the climatic and environmental conditions-related difficulties that can have a major impact on the problems and complications that influence human health is heat stress. Heat stress, in general, refers to a group of internal and environmental thermal variables that lead to illness in humans. Wind temperature, velocity of air, radiant heat, humidity, and body temperature are among the exterior elements determining the severity of heat stress; internal factors include metabolism, environmental compatibility, and internal body temperature. The 2021 report is released in conjunction with the 26th Conference of the Parties (COP26) of the UN Framework Convention on Climate Change, where nations are under pressure to fulfill the goal of the Paris Agreement, which is to keep the average global temperature rise to 1.5 °C, and to raise the necessary funds so that every nation can effectively combat climate change (Heidari et al. 2020).

2.2 Increased Prevalence of Vector-Borne Diseases

Due to the inherent sensitivity of arthropod vector biological research and vector-borne infection transmission to climate, effects on diseases transmitted by vectors have been a topic of interest since the medical consequences associated with world-wide climate change were first investigated. A variety of studies and viewpoints notion that there would be a major impact of climate change on the risks of vector-borne diseases (Ogden and Lindsay 2016). The main mosquito-borne illnesses that have been the subject of this debate are dengue and malaria, which are spread from person to person by the bites of mosquitoes and do not involve hosts that are animals (i.e., that they are not zoonoses). However, there has recently been a push for a more comprehensive assessment of how climate change is affecting the variety of vector-borne illnesses (Ogden et al. 2014).

The contribution of the vector is usually oversimplified in ecological theories of vector-borne illness transmission, which may be useful in deriving general principles of transmission of diseases (Dobson 2004). However, in order to evaluate present and future disease risks and to provide operationally relevant early forecasting tools, projections of models of diseases transmitted by vectors (particularly those predicting the implications of climate change) must be precise in terms of space and time. Comprehending the biology of the vectors is essential for these models.

The primary causes of infectious disease development or re-emergence include changes in human populations, the environment, land use, agricultural methods, habitat, and climate (Reiter 2001). On the other hand, given their inherent susceptibility to weather and climate, arthropod-transmitted infectious diseases might be the most susceptible to the effects of climate change among all infectious diseases. Rising temperatures, especially at higher latitudes; altered precipitation, making certain regions more vulnerable to droughts; increased climatic variability; and extreme weather events, such as strong storms, intense heat waves, torrential downpours, etc., are all predicted changes to our climate (Pachauri et al. 2014).

There are various ways in which these climate changes could contribute to the emergence and resurgence of vector-borne diseases: (i) the migration of vectors and vector-borne diseases poleward as temperate zones' climates warm and become more favorable for these species. If temperatures rise too high for these species, a poleward contraction of their most equatorial limits may also occur in tandem with this process (Brownstein et al. 2005); (ii) increased probability and frequency of tropical and subtropical vector-borne diseases spreading and endemic establishment into presently temperate regions due to a combination of (a) higher temperatures in the receiving location, which increase the survival of vectors and vector-borne pathogens; (b) rising vector and vector-borne pathogen abundance in tropical and subtropical source locations; (c) rising import rates of vector-borne infections, especially those from tropical and subtropical regions, as a result of increased human migration brought on by climate change (McMichael et al. 2012); (iv) the appearance and establishment of novel genotypes of vector-borne infections as a result of modifications in animal host and vector dynamics brought on by climate change;

(iii) the resurgence of endemic vector-borne diseases linked to rising temperatures and weather and climatic variability (Ogden et al. 2007). Vector mortality rates are directly impacted by temperature. Vectors have optimal temperatures for survival since in the lab, the death rates of ticks (Eisen et al. 2014), adult and larval dipteran vectors (Bayoh and Lindsay 2004; Christiansen-Jucht et al. 2014) rise at low and high temperatures. The operating temperature of the water in which pupal and larval dipteran vectors reside affects their mortality rates, and this temperature is influenced by changes in ambient temperature throughout the day (Paaijmans et al. 2010). However, how well populations of vectors survive at the boundaries of geographical areas where conditions for temperature are within optimal parameters will depend on their ability to find shelter from temperature extremes and the ability of their habitat and/or environment to provide such refuges (Eisen et al. 2014).

For ixodid ticks, who spend the majority of their lives in the soil's surface layers or litter layer, where they slowly transition from one phase of life to the next or just wait out bad weather, the effects of refuges are especially significant in their geographic ranges. According to Ogden et al. (2004), the surface layer of soil in temperate forests regulates daily fluctuations in air temperature by being warmer than the surrounding air during the winter. Every phase of the tick survive in this safe environment during bad weather; those that do not will probably perish. However, only adolescent dipteran vectors are able to seek shelter (Ogden et al. 2004). Although urban areas can serve as havens for adult mosquitoes, allowing populations to survive outside of their typical geographic ranges (Lima et al. 2016), the direct impacts of temperature on vector survival might not have as much of an influence on tick populations as they would on dipteran populations. All stages of all vectors, with the exception of some mosquito species' eggs, are eradicated by desiccation (Aboagye-Antwi and Tripet 2010; Benedict et al. 2010; Sutherst and Bourne 2006). Additionally, vectors are only able to survive in environments that: (i) support sources of water where adult dipteran vectors and ticks can rehydrate; and (ii) offer immature stages of dipteran vectors aquatic environments. Adult female mosquitoes not only need blood meals to supply the nutrients needed for egg production, but they can also replenish their fluids by consuming nectar or drinking standing water (Benedict et al. 2010). Ticks, on the other hand, do not require standing water because they can hydrate from humid environments by producing hygroscopic saliva to the litter layer surroundings and then re-ingesting it after it gets saturated (Bowman and Sauer 2004). The survival of juvenile dipterans can be impacted by heavy rains, and tick mortality rates are expected to rise with deep and ongoing flooding. The effects of climate change will probably affect dipteran-borne and tick-borne diseases differently. While tick populations react to spatiotemporal means in climate instead of climate variability, dipteran populations can adapt quickly to short-term alterations to weather and climate. Due to these distinctions, changes in the risks of tick-borne diseases will primarily be long-term in nature with minimal interannual volatility, but changes in the risks of dipteran-borne diseases may exhibit both short- and long-term epidemic characteristics. These various features highlight the need for more thorough, vector- and specific to a disease data and models for forecasting changing vector-borne illness hazards for animal and public health programs, and they should be taken into account when planning ahead for altering vector-borne disease threats with climate change.

In Canada, Lyme disease was made nationally reportable in 2009; however, prior to that year, data were only accessible from individual provinces, although most provinces had been reporting cases of Lyme disease since prior to 2002. Around 2004, the tick vector Ixodes scapularis began to migrate north from the USA into eastern and central Canada, potentially linked to climate change. This marked the beginning of the outbreak of Lyme disease in Canada (Ogden et al. 2014). Since then, as the vector's geographic range has grown, the amount of cases has risen consistently annually, with some minor interannual variation (such as the gap between 2013 and 2012, which may have been caused by a particularly cold winter in 2012 that reduced tick survival). An unprecedented number of Culex tarsal is mosquitoes emerged in the Prairie Provinces in 2007 as a result of increased spring rainfall, which offered a large number of places for larval development and the right temperatures at the same time. High temperatures during the summer of 2012 was linked to an outbreak in eastern Canada caused by Culex pipiens/restuans. Through direct impacts on the biological functioning of the vectors and the spread cycles, or indirectly through socioeconomic mechanisms impacting the susceptibility of humans and animals to such diseases and our ability to prevent and control them, it is likely that changes in the climate will enhances the effect of vector-borne illnesses on human and animal health worldwide (Ogden and Lindsay 2016).

2.3 Air Quality Degradation and Respiratory Issues

Various acute and chronic illnesses such as chronic obstructive pulmonary disease (COPD), lung cancer and cardiovascular disease has been associated with air pollution. Approximately 4.2 million fatalities has been reported in 2016 were attributed due to ambient air pollution. Ambient air pollution has been responsible for roughly 16% of lung cancer fatalities, 25% of deaths from COPD, roughly 17% of deaths from ischemic heart problem and stroke, and roughly 26% of deaths due to respiratory infections globally. Air pollution is responsible for roughly 43% of COPD deaths, 29% of lung cancer fatalities, 25% of deaths from ischemic heart disease, and 24% of deaths from stroke worldwide (Watanabe et al. 2021).

Children and elderly adults with chronic illnesses are among the most vulnerable demographic groups. The prevalence of pollution-related respiratory illnesses is well-established in industrialized regions, and the duration of exposure to air pollutants and pollution is positively connected with the severity of these illnesses. The fields of pulmonology and allergology have seen a great deal of research with the goal of enhancing the diagnosis, management, and avoidance of respiratory illnesses brought on by toxins. The primary line of protection against the development and spread of illnesses brought on by air pollution is the respiratory system (Fig. 1).

The quantity of pollutants ingested and their degree of cellular deposition determine the severity and extent of the harm. Short-term consequences of air pollution

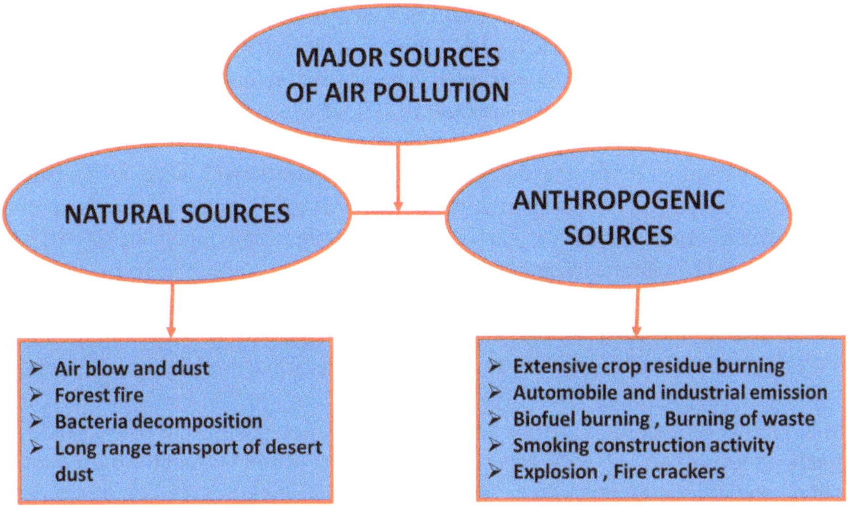

Fig. 1 Major source of air pollution

include respiratory discomfort, lung function abnormalities, and respiratory-related deaths. On the other side, long-term consequences include harm to lung development, lung cancer, COPD, and the potential emergence of allergies and asthma. Apart from lung ailments that are directly brought on by environmental contaminants, there has also been a noticeable rise in certain infectious diseases that are linked to the effects of air pollution on the climate, flora, and wildlife (Watanabe et al. 2021). Presently, air pollution and climate modulation have become major global health concerns. There are risks to human health associated with the growing metropolitan areas experiencing extreme climate events (heavy rains, high temperatures, floods, and droughts). It has been investigated how aerosols and PM concentrations affect human health and the potentially harmful effects of inhaled particles that enter the respiratory system.

2.4 Food and Water Insecurity Leading to Malnutrition

Education, nutrition, and survival have all improved in the last few decades. However, the Sustainable Development Goals, which aim to end poverty and malnutrition by 2030, are still far from being met (Organization 2021). Overeating, food insecurity, poverty, and restricted access to health care are the main factors contributing to malnutrition in all age groups. Globalization and industrialization of the food system cause a double burden of malnutrition, with overweight and obesity and negative impacts on the climate coexisting with undernutrition (i.e., wasting, stunting, and deficiencies in micronutrients). The primary underlying causes of child malnutrition in addition to the social, economic, and political issues influencing food security and nutrition are being exacerbated by climate change and the COVID-19 epidemic. Globally, climate change has a detrimental impact on factors that determine health, such as availability to water and agriculture. These consequences exacerbate problems between countries related to social and economic inequality by affecting groups that may be marginally responsible for the issue (Munzert and Jankin 2019). Due to inadequate management and distribution, the global food system has not been able to significantly enhance the second sustainability target on hunger; the COVID-19 epidemic has brought the system's fragility to light.

It is anticipated that the consequences of climate change would intensify health risks to people, the environment, and the earth. Because of its effects on the agro-ecosystem, climate change may have an impact on food production and security of nutrition (Myers et al. 2017). Nutrient deficiencies, chronic undernutrition, and susceptibility among the most food-insecure population groups are made worse by effects on the amount, quality, accessibility, and cost of foods from the agricultural, fishery, and animal sectors (Awiti 2022; Munzert and Jankin 2019; Swinburn et al. 2019). Reduced fish catches, livestock output, crop yields, and forest ecosystem services are some of the ways that global warming could upset the environment. According to Frieler et al. (2015), its impact has already had a negative impact on global yields of staple crops like maize and wheat, which are crucial for food security, and it are going to continue to do so even at moderate warming levels (Frieler et al. 2015). It has been demonstrated that higher CO_2 levels reduce the amount of protein in crops like barley, wheat, rice, and potatoes. Moreover, pests that pose a hazard to crop production may benefit from climate change, appearing sooner and in places where they were previously unable to establish themselves (Uleberg et al. 2014). As an illustration, consider the 2019–2020 outbreak of desert locusts in multiple East African nations, which started in the Arabian Peninsula with the intense tropical storm Mekunu (Salih et al. 2020). Additionally, it is anticipated that climate change would alter microbial populations and species interactions, leading to an increase in foodborne and waterborne illnesses (Cissé 2019; Leggieri et al. 2021; Levy et al. 2018). According to WHO predictions, diarrheal illness will account for an additional 48,000 deaths in children under 15 by 2030 and an additional 330,000 fatalities by 2050 as a result of climate change.

There are also notable impacts on animal health, biodiversity, and cattle productivity. According to recent studies, dairy cows in southern European countries that experience heat stress may lose up to 5.5 kg of milk per cow per day. In sub-Saharan African countries, severe drought has resulted in 20–60% reductions in animal populations over the past 20 years (Change) Similar to how acidity, salinity, and oxygen concentrations in the aquatic environment change due to warming, fisheries are also impacted. These modifications might cause fish species' maximum body weights to drop, which would reduce their ability for capture (Perry et al. 2005; Pörtner 2008). Furthermore, it is probable that warming causes the iron and long-chain omega fatty acid concentrations in seafood to decrease. Given their crucial roles in the ecosystem—such as providing a clean and consistent water supply, preventing landslides and land degradation, providing or improving the habitat of aquatic and terrestrial life, and producing a variety of goods for sale or domestic use—the threat posed by climate change to forests and trees is another extremely concerning factor. According to studies, there has been a significant increase in tree mortality as a result of warming and changes in precipitation (Allen et al. 2010). The loss of tree species will also limit the ecosystem's potential for certain activities (Gamfeldt et al. 2013; Hooper et al. 2012).

2.5 *Mental Health Challenges Due to Climate-Related Stressors*

Human physical and mental disorders may arise directly or indirectly from the effects of climate change on vulnerable societies and biological subjects. Studies in the field of psychiatry on mental illnesses associated with climate change are conspicuously lacking, nevertheless. It has been demonstrated that the effects of climate change on mental health occur at different times. There are significant differences in the phenomenological of the impacts of climate change; certain mental diseases are more particular to unusual climate conditions, while others are more prevalent. A shortage of access to knowledge, resources, and protection, as well as direct exposure and increased vulnerability due to geographic conditions, are further effects of climate change on many population groups. It would be important to note that in certain studies, the relationship between climatic occurrences and mental health issues has been explained by introducing concepts that have just lately been developed: ecoanxiety, ecoguilt, ecological psychology, ecological grief, solastalgia, biospheric worry, etc.

There's a growing corpus of research on the relationship between mental health and climate change (Berry et al. 2010; Bourque and Cunsolo Willox 2014). An extraordinary event may have an impact on mental health before, during, or even after it happens (Association 2017). The effects of climate change on mental health can range from mild feelings of stress and discomfort to severe problems, including depression, post-traumatic stress disorder, anxiety, and suicidal thoughts (Blanc et al. 2019; Organization 2017, 2021; Ursano et al. 2017). Additional repercussions might include how people and communities are affected in their day-to-day experiences, perceptions, and ability to cope with the effects of the changing climate and its ramifications. Many persons who are subjected to weather-related or climate-related natural disasters suffer from stress and major mental health implications. Climate change may result in an increase in the frequency of some natural disasters. Posttraumatic stress disorder (PTSD), depression, general anxiety, higher levels of substance abuse or misuse, and suicidal thoughts are well-known reactions to catastrophic incidents that involve life disruption, such as the loss of life, social support, resources, social networks, and or extensive relocation (Kim 2016).

Studies have indicated a strong correlation between peritraumatic experiences and acute stress during and shortly following a disaster, which is predicted to cause PTSD to develop (Gruebner et al. 2017). Subsequently, survivors experience additional effects like diminished everyday activities and a loss of their "sense of place." These ailments may have an effect and increase the likelihood of mental health issues. People become anxious, nervous, unhappy, and helpless when they hear news about climate change. Different kinds of psychopathological reactions to these occurrences are brought about by the tangible influence of those alterations in life. In a nutshell, acute effects are any extreme occurrences (e.g. hurricanes, wildfires, floods, etc.) that instantly put defenseless and defenseless people at risk of mental harm. Sub-acute repercussions include strong feelings felt by those who indirectly observe the impacts of climate change, worry over the possibility that humans along with other species will survive, and, ultimately, a feeling of being ceased, confused, and passive. Massive social and community effects that escalate competition for scarce resources into violence, migration and displacement, post-disaster management, and ongoing environmental stress are examples of long-term effects (Acharibasam and Anuga 2018; Hsiang et al. 2013).

3 Vulnerable Populations

The IPCC has established that environmental degradation caused by human activity and global environmental changes has negatively impacted people's quality of life, with varying consequences on different social groups. The effects of climate change have disproportionately impacted the wellbeing of the weaker and marginalized communities and groups, such as women, the elderly, the poor, disabled persons, and Black and Indigenous people (Ngcamu 2023). Ronoh et al. (2015) contend that a sizable population of disabled children worldwide—roughly seven million—are prone to natural catastrophes and are negatively impacted by them (Ronoh et al. 2015). Malak et al. (2020), who claimed that vulnerable populations suffer negative effects from climatic variability and that children and the elderly are more likely to be affected than young adults, corroborated the earlier findings (Malak et al. 2020). Furthermore, women's lack of political, economic, and physical autonomy has gotten worse during and after climate-related disasters, which is linked to already-existing socioeconomic vulnerabilities (Benjamin and Thomas 2020).

3.1 Women

According to Brown et al. (2012), women are more sensitive to the effects of climate change and variability, as they are mostly dependent on agriculture in rural areas (Brown et al. 2012). For rural women, agriculture is the only source of security in their lives, families, and communities in the face of extreme poverty. Food insecurity has worsened in nations like Bangladesh due to a combination of wealth inequality and vulnerability to the catastrophic effects of climate change (Tanjeela and Rutherford

2018). The majority of Bangladesh's climate casualties are impoverished women. Gender inequality makes these women more vulnerable than men in both rural and urban settings, and food insecurity has influenced the resilience and vulnerability of communities that are susceptible to climate-related disasters (Shamsuddoha 2012; Tanjeela and Rutherford 2018).

Food hierarchy may raise the danger of hunger and water shortages during and after catastrophes for women and children, especially if they are pregnant or nursing (Cutter 2017). Similar to this, financial reliance and economic disparity make it harder for children, women, the elderly, and those with disabilities to recover after disasters. Typically, they lack the capacity to autonomously decide whether to leave or remain in disaster-affected places (Ayeb-Karlsson et al. 2020). According to Ayeb-Karlsson et al. (2020), women are typically restricted to their homes and their mobility is at risk during disasters, unlike their male (masculine) colleagues who are able and flexible in their movements (Ayeb-Karlsson et al. 2020). The negative effects of climate change also have a significant impact on children and newborns because they put their education, safety, and family security at danger. They also increase the risk of diseases like diarrhea, heat stress, and dehydration. According to a 2018 IPCC report, different populations are affected differently by climate change, and in certain cases, it causes large-scale migration. The World Bank issued a warning, citing observations that women became more vulnerable to physical and sexual harassment during and after catastrophes, as well as an increase in domestic abuse and its severe consequences (Ngcamu 2023).

The United Nations Environment Programme (UNEP) and the United Nations Department of Political and Peacebuilding Affairs (UNDPPA), 2020, Gender, Climate and Security Report demonstrated how governments treat vulnerable populations in a gender-blind manner, making them less able to withstand the effects of climate change. Women's vulnerability has increased as a result, and they have been exposed to new risks and insecurities without being given the opportunity to learn coping mechanisms. Since they are the ones who provide for the necessities of their homes and communities, women are disempowered when it comes to resource access, according to the report (Zaidi and Fordham 2021). This is true even if they are the ones who suffer the most from the effects of climatic changes (Datey et al. 2023).

3.2 Elderly

The study by Malak et al. (2020) that examined older people's vulnerability to a cyclone in Bangladesh's coastal area revealed that it was almost excessive. The elderly are more susceptible to a cyclone than young adults due to a number of characteristics that the authors discovered, including fewer resources, decreased physical stamina, and declining mental ability. The elderly rely on young adults for care, have little training, limited access to health services, and low social awareness. According to a number of academics, older persons are impacted by climate-related disasters in the Global South at every stage of the disaster and are less equipped to mitigate its effects. Furthermore, the elderly suffer terrible losses that are important to their way of life, such as the loss of friendships and culture (traditional practices, indigenous knowledge they possess) as a result of death and relocation (Harvey 2019; Malak et al. 2020).

3.3 Persons with Disabilities and Effect's on Climate Change

People with impairments, including those who are albinist, are another particularly susceptible population that has been identified by research as the most discriminated against group both during and after climate and environmental changes. Their absence of having access to education (Kosanic et al. 2022), as consistently shown in the research, family insecurities, and illnesses (Goldhagen et al. 2020) are the observed causes of their severe sensitivity to climate change. People who have albinism are primarily affected by the long-term effects of climate change because of their increased vulnerability to factors such as rising temperatures, marginalization and discrimination, lack of access to resources (Zaidi and Fordham 2021), a lack of immediate assistance (Stein and Stein 2022a, b), and being disregarded by decision-makers and providers of services (Jodoin et al. 2020). As a result, children with impairments are particularly affected by the catastrophic effects of climate change since they have less access to schooling (Kosanic et al. 2022), which jeopardizes their health and the security of their families.

4 A Climate Change and Preventive Strategies

One of the trickiest problems the world is currently experiencing is climate change. It is a multifaceted issue that spans several fields, economics, including science, politics, society, and moral and ethical issues. It is a worldwide issue that has localized effects and will last for millennia. Since we have already committed to a certain degree of climate change, combating it requires a two-pronged strategy:

- Reducing greenhouse gas emissions and maintaining atmospheric concentrations of gases that trap heat ("mitigation").
- Making the necessary adjustments to prepare for the impending climate change.

4.1 Mitigation

In order to mitigate climate change, less heat-trapping greenhouse gas emissions must be released into the atmosphere. This can be done in two ways: either by reducing the sources of these emissions—such as the burning of fossil fuels for transportation, electricity, or heat—or by improving the "sinks"—such as soil, forests, and oceans—that absorb and store these emissions. Mitigation aims to prevent major human disturbances with Earth's climate, stabilize greenhouse gas concentrations long enough for ecosystems to naturally adjust to changing conditions, guarantee that food production is not jeopardized, and promote sustainable economic growth. The following are a few mitigation strategies that can be used to stop the rise in pollution emissions:

- Make energy-efficient practices
- A higher percentage of renewable energy
- The use of electricity in industrial processes
- Effective transportation methods: bicycles, shared automobiles, electric public transportation, etc.
- Emissions markets and the carbon tax

4.2 Adaptation

Living in a changing environment requires adjusting to the climate, either now or in the future. Reducing our exposure to the negative consequences of climate change—such as rising sea levels, more severe storms, and food insecurity—is the aim. This also includes optimizing the use of any advantageous opportunities brought about by climate change (e.g., longer growing seasons or higher yields in specific locations). Regarding adaptation strategies, a number of actions can be taken to lessen vulnerability to the effects of climate change:

- More secure infrastructures and locations for facilities
- Reforestation and restoration of landscape (natural landscape)
- Adaptable and varied farming to be ready for natural disasters
- Investigation and development of potential disasters, behavior of temperature, etc.
- Precautionary and preventive actions (health concerns, evacuation preparations, etc.)

5 Strategies for Strengthening a Climate-Resilient Health System

Climate change is particularly likely to affect the health sector. However, there aren't enough laws, initiatives, and programs in place to improve the health system's resilience and ability to adapt to climate change, particularly in developing nations. The detrimental consequences of climate change on people's health are made worse by insecure and non-resilient health systems. Thus, the goal of this study was to pinpoint tactics and fixes to increase the health system's resistance to climate change. The results empower managers and policymakers in the health sector to create all-encompassing plans for creating health systems that are climate resilient (Table 1).

The five-step CDC Building Resilience Against Climate Effects (BRACE) framework enables health authorities to create plans and initiatives to assist communities in being ready for the effects of climate change on their health.

- **Anticipate Climate Impacts and Assessing Vulnerabilities**: Determine the extent of the effects of climate change, the possible health consequences that may result from them, and the people and places that are most at risk (Mosadeghrad et al. 2023).
- **Project the Disease Burden**: Calculate the extra cost of health consequences related to climate change (Hsu et al. 2017).
- **Assess Public Health Interventions**: Determine which health actions are most appropriate for the observed health impacts that require the greatest consideration (Anderson et al. 2017).
- **Develop and Implement a Climate and Health Adaptation Plan**: Create a documented adaption plan that is updated on a regular basis. Distribute the plan and supervise its execution (Cameron).
- **Evaluate Impact and Improve Quality of Activities**: Analyze the procedure. Assess the worth of the knowledge gained and the work done (Marinucci et al. 2014) (Fig. 2).

5.1 Public Health into Climate Change Policy and Planning

The Intergovernmental Panel on Climate Change (IPCC) released its sobering report, Global Warming of 1.5 Degrees, in 2018. It highlights the significantly greater risks that 2.0 C of additional warming poses to human populations than the 1.5 C aspirational target set by the 2015 Paris Agreement. The report also emphasizes the need for "rapid and far-reaching transitions in land, infrastructure, urban, and industrial systems," including aggressive targets for negative emissions. According to IPCC simulations, Earth might see further warming of 1.5 C as early as 2030. For society to successfully address both the issues of decarbonizing the economy and planning for increasingly harsh climate conditions, the next ten years will be critical. For public health to accomplish both of these objectives in the coming ten years, it will be essential and require more leadership (Fox et al. 2019). Although there are global causes of climate change, the effects on health are intrinsically local since they occur where individuals live, work, learn, play, and travel. Certain local populations are significantly more sensitive than others due to factors such as geography, exposure, and susceptibility to health consequences (Berry et al. 2018). Due to their proximity to the effects of climate change and, in certain situations, the slower pace of national action on the issue, local sub-national governments have taken the lead in establishing climate change policies in several locations (Woodward and Samet 2018). Reducing greenhouse gas (GHG) emissions from energy, transportation, industry,

Table 1 Strategies for strengthening a climate-resilient health system (Mosadeghrad et al. 2023)

Governance and leadership	Creating a plan to adapt to climate change and promote national health
	Involving the government in climate change and health
	Adjusting health sector rules to mitigate the consequences of climate change to public health
	Developing and carrying out a thorough policy for health and climate change
	Raising healthcare executives' knowledge of the evolving risks
	Creating a crisis management approach based on moral principles
	Planning for health using climate vulnerability indices
	Putting pressure on governments to divest themselves of fossil fuels
	Collaborating PHC with organizations that address climate change
	Enabling important players from several sectors to operate locally
	Improving cross-disciplinary and global cooperation
	Devolution of authority towards decentralization of management
	Putting in place an iterative procedure to manage and keep an eye on the health concerns associated with climate change
	Making health, social, financial, and environmental performance reports mandatory
	Become a member of a group like the Climate and Health Alliance
	Developing a Climate Change Framework
Financing	Providing satisfactory capital
	Enhancing research funding
	Increasing the availability of long-term foreign funding
	Calculating the expenses of taking and not taking health-protective action
	Increasing local government funding for projects addressing climate change
	Growing dependence on "cost recovery" for services provided by the public sector
Medical products and technologies	Utilizing a skilled and long-lasting workflow
	Hiring and educating medical staff in epidemiology and biostatistics

(continued)

Table 1 (continued)

	Participating in community microbiological water testing with nursing personnel
	Enhancing health workers' abilities at various levels through training
	Enhancing training offerings to increase professionals' awareness
	Including instruction on climate change in university training programs
	Including a range of stakeholders
	Top doctor on addressing climate change in daily practice
	Creating health education materials for locals
	Recognizing susceptible professional associations
	Raising awareness of the hazards of infectious diseases at work
	Creating appropriate protection gear and clothes
	Educating people about health protection against climate change
	Ensuring healthcare workers have access to resources and funding to help them while they handle emergencies
Health information systems	Estimating the effects of climate change and evaluating the health system's capabilities and vulnerabilities
	Stepping up surveillance to focus on climate-sensitive illnesses and the causes of their risk
	Studies on the effects of climate change on health and enhance health care' ability to adapt
	Creating environmental risk early warning systems
	Maintaining worldwide databases on climate plans that are easily accessible, coordinated, and updated
	Examining long-term multidisciplinary socioeconomic, health, and climate variables
	Putting in place backup data systems
	Combining data from veterinary, ecological, environmental, and epidemiological sources
	Ensuring sufficient gathering of high-quality data
	Advancing the use of interpolation and proxy metrics in situations where data might not be available
	Creating spatial analysis technologies that are more capable of integrative analysis than the GIS software that is available now
	Improving cross-jurisdictional data and information sharing
	Enhancing the promptness of laboratory testing accessibility and outcomes

(continued)

Table 1 (continued)

	Including health into evaluations of climate change-related losses and damages
	Assembling and sharing top techniques from prosperous nations
	Enhancing channels of communication between meteorological services, the health sector, and other stakeholders
	Building capacity for climate change-specific project creation, management, and evaluation
	Creating and recommending plans for health adaption
Service delivery	Creating basic healthcare-based strategies to deal with climate change's short- and long-term implications
	Expanding the availability of mental health treatments
	Enhancing the capability and standard of healthcare facilities and services
	Normative guidelines for primary healthcare provision
	Giving vulnerable persons and regions priority
	Encouraging wellness initiatives and energy-efficient home practices
	Establishing health advisory platforms to enhance instruction and services
	Expanding the availability of medications to address elevated health hazards
	Creating a plan for municipal heatwave preparedness
	Getting to necessary equipment, such water pumps and power generators
	Implementing water and energy efficiency initiatives to modify employee behavior
	Offering incentives to lower building and transportation energy consumption
	Disposing of waste in a safe manner
	Educating patients on the effects of climate change and how they relate to human health
	Putting money into energy-efficient and renewable energy technologies
	Reducing needless maintenance and using fewer consumables
	Creating toolkits to lessen the damage that operating room procedures cause to the environment
	Moving medical institutions that are under risk actively
	Giving everyone access to a "essential services package"
	Enhancing programs for chronic illness self-management

(continued)

Table 1 (continued)

	Keeping an eye on community clients
	Altering one's diet to include less meat and more foods from food gardens
	Adding a thorough scenario study of the climate landscape to the current health programs
	Increasing funding for risk management of health emergencies
	Improving the ability of the health system to respond quickly to emergencies related to specific diseases
	Revision and enhancement of emergency risk communication tactics

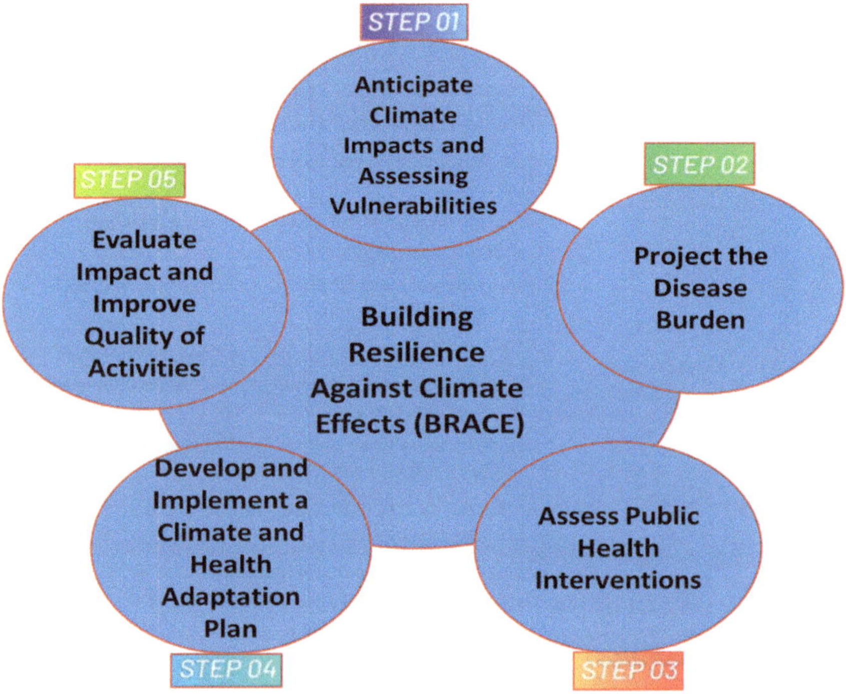

Fig. 2 Five-step CDC Building Resilience Against Climate Effects (BRACE) framework

and land use/agriculture has been the appropriate emphasis of public planning and policy formulation for climate change; public health has frequently been marginalized in these planning efforts (Araos et al. 2016; Aylett 2015; Georgeson et al. 2016). Public health, however, is an ideal leader of adaptation initiatives because the objective of many climate-adaptive planning is to safeguard human health at grassroots levels where impacts occur. Furthermore, the public health effects of climate change "are an essential component of the justification for climate action," as stated by Woodward and Samet, making them a possible catalyst for GHG reductions as well (Woodward and Samet 2018).

The topic of public health has recently started to receive more attention when it comes to climate change policy. Major reports like the Lancet Countdown: Tracking Progress on Health and Climate Change (Watts et al. 2017, 2018). The World Health Organization (WHO) identified as the Paris Agreement "a fundamental public health agreement" (Organization 2021). As a result, there is growing recognition that public health plays a crucial role in both lowering GHG emissions and preparing for a warmer, less stable climate (Watts et al. 2018; Woodward and Samet 2018). However, there are still a number of obstacles in the way of public health fulfilling its potential as a leader (Austin et al. 2016; Workman et al. 2018).

The conceptual and ethical foundation for a public health approach to climate change, based on the ten "essential public health services" that direct and inform public health practice, was laid out in a seminal paper published ten years ago by scientists from the CDC (Frumkin et al. 2008). The aforementioned analysis has revealed various ways in which current public health initiatives are already aiding in the formulation of policies related to climate change. It has also identified novel initiatives in the field of public health that have emerged in response to the difficulties of climate change, and it has identified the gaps and requirements for ensuring public health in the face of an increasingly warming and unpredictable future climate. The authors arrived to the conclusion that numerous public health interventions currently in place may effectively tackle the growing effects of climate change. However, they specifically recognized the need for improved coordination between government organizations and between governments along with non-government players. The three fundamental roles of public health, namely evaluation, policy creation, and assurance, were initially suggested by an Institute of Medicine (IOM) group as a framework for organizing public health initiatives (Health 1988).

5.1.1 Assessment

As the second as well as the third of the fundamental functions, public health assurance is set up and policy development is informed by the data collection, analysis, investigation, and identification of public health problems that are part of public health assessment (Health 1988). Assessment entails (i) monitoring health status in relation to climate-sensitive outcomes and (ii) investigating and diagnosing health effects associated with climate-related hazards.

Monitor Health

Three categories of data—(i) risk from the environment, (ii) Susceptibility of population and (iii) disease-related observation—were identified by Frumkin et al. as being necessary for monitoring the state of health in the context of a modulation in climate (Frumkin et al. 2008).

Investigate and Diagnose

Because of initiatives and programs being conducted in a variety of non-health sectors, health impact assessments (HIAs) have emerged as an additional tool for assessment in identifying climate hazards and health risks. The five steps of the HIA process are screening, scoping, evaluation, reporting, and monitoring (Organization 2021). It is a collaborative, flexible, multidisciplinary approach that incorporates the knowledge and experience of many public health and other professionals, officials, and community members (Brown et al. 2011). It can also be used to predict the effects of climate change on communities and to find ways that adaptation or mitigation efforts may have co-benefits for population health (Brown et al. 2014, 2011). Going forward, there is a chance to find trade-offs and synergies in available policy alternatives by advancing the development and application of the multi-sectoral method supported by HIA.

5.1.2 Policy Development

The second stage of the core functions cycle, known as policy formulation, involves identifying and preparing the actions that will be implemented and tracked throughout the assurance phase. It builds upon the diagnosis from assessment. In addition to government initiatives to create laws, rules, and processes, policy creation also includes voluntary actions taken by private parties and public outreach initiatives. Policy development encompasses the following when combined with the 10 fundamental public health amenities in the framework of climate change: (i) educating, informing, and encouraging populations toward climate resilience; (ii) forming partnerships to foresee and address health threats from climate change; and (iii) developing health policies that support adaptation and mitigation of the effects of climate change (Fox et al. 2019).

5.2 *Assurance*

The third phase of the essential functions cycle, known as assurance (Fig. 1), is an application phase that follows policy creation and involves putting policies into effect through programs, laws, workforce training, and other initiatives. The direct response loop is completed by assurance, which involves assessment, monitoring, and diagnosis carried out during the assessment phase. In the complex perspective of climate variation and human health, this latter link offers the iterative potential for dynamic adjustment—needed for integrated, adaptive management. Assurance includes the following when combined with the 10 important public health services from a climate-risk perspective: (i) upholding laws pertaining to diseases and impacts that are susceptible to climate change; (ii) connecting to and providing care in the context of climate hazards; (iii) guaranteeing a skilled workforce to respond to health threats related to climate change; and (iv) assessing climate health interventions (Fox et al. 2019).

5.3 Research

Research was needed ten years ago to identify population susceptibility, forecast health impacts, establish the relationship between climatic hazards and health outcomes, and define risk reduction methods, including their cost and effectiveness (Frumkin et al. 2008). Since then, a number of studies have made an effort to present a broad picture of the field's needs and research activities. In order to safeguard health from climate change, the World Health Organization (WHO) recognized five domain of research in 2009: (i) risk assessment; (ii) finding the best interventions; (iii) directing health-promoting adaptation and mitigation efforts in other sectors; (iv) enhancing decision support with suitable tools; and (v) calculating the costs of health protection from climate change (Organization 2009). In order to evaluate the indication base in these five WHO importance research areas, Hosking and Campbell-Lendrum conducted a scoping review. They discovered that a significant amount of research had been done on the relationship between climate hazards and health outcomes, but they also noted a dearth of studies on the effectiveness of interventions (Hosking and Campbell-Lendrum 2012). Sheehan et al. also noted the ongoing need for intervention evaluation and the creation of evidence-based best practices in the US environment (Sheehan et al. 2017). Trends in the quantity of papers on health and the climate from 1990 to 2014 were examined by Verner et al. (2016). The amount of literature on health and the climate has grown significantly over time, but it has lagged behind other industries like energy and transportation. The authors discovered that there was a dearth of research on the effects of malnutrition, non-communicable diseases, and mental health in particular, as well as the effects in low- and middle-income nations. They therefore suggest strengthening research capacity, especially in the global South (Verner et al. 2016). Public health has made strides in all of the key service areas related to its basic tasks, but it has, in many respects, been "playing catch-up" to other sectors that have more experience with climate change planning. Building on this success, the public health community essentially address the pressing requirements of the ensuing ten years, which include decarbonizing the economy and strengthening resilience against harsher climate conditions. The area of public health will be able to realize its strengths as a valuable and proactive partner in the design and execution of climate policy if the updated CFES model is implemented.

6 Conclusion

Global climate change is having a significant effect on people's health. Climate change directly as well as indirectly affects human health in multifaceted ways. As our planet experiences rising temperatures, extreme weather events, altered ecosystems, and changing disease patterns, the health impacts on individuals and communities have become increasingly evident. Heat-related illnesses, such as heatstroke and dehydration, are becoming more prevalent as temperatures soar. Additionally, the spread of infectious diseases like malaria, dengue fever, and Lyme disease is expanding due to changing geographical ranges of disease vectors. Poor air quality resulting from increased air pollution and wildfires contributes to respiratory problems, cardiovascular diseases, and other health issues. Furthermore, disruptions in food and water supply chains due to changing climates pose risks of malnutrition and waterborne diseases. Vulnerable populations are disproportionately affected by these health impacts. Addressing climate change and its health implications requires a multilayered approach involving collaboration between governments, policymakers, healthcare professionals, communities, and individuals. Education, advocacy, and public awareness campaigns play a pivotal role in fostering behavioral changes that promote environmental sustainability and resilience to climate-related health risks. Mitigation strategies to reduce greenhouse gas emissions, transitioning to renewable energy sources, implementing sustainable land and water management practices, enhancing healthcare infrastructure, and developing early warning systems for extreme weather events are crucial steps in safeguarding public health against the adverse effects of climate change. The relationship between climate change and human health is a critical and complex issue that demands immediate attention. By taking proactive measures to mitigate climate change and adapt to its consequences, we can protect human health, preserve ecosystems, and create a more sustainable future for generations to come.

References

Aboagye-Antwi F, Tripet F (2010) Effects of larval growth condition and water availability on desiccation resistance and its physiological basis in adult Anopheles gambiae sensu stricto. Malar J 9:1–11

Acharibasam JW, Anuga SW (2018) Psychological distance of climate change and mental health risks assessment of smallholder farmers in Northern Ghana: is habituation a threat to climate change? Clim Risk Manag 21:16–25

Allen CD, Macalady AK, Chenchouni H, Bachelet D, McDowell N, Vennetier M, Kitzberger T, Rigling A, Breshears DD, Hogg ET (2010) A global overview of drought and heat-induced tree mortality reveals emerging climate change risks for forests. For Ecol Manage 259(4):660–684

Anderson H, Brown C, Cameron L, Christenson M, Conlon K, Dorevitch S, Dumas J, Eidson M, Ferguson A, Grossman E (2017) Climate and health intervention assessment: evidence on public health interventions to prevent the negative health effects of climate change. Climate and health technical report series. BRACE Midwest and Southeast Community of Practice. Climate and Health Program, Centers for Disease Control and Prevention

Araos M, Austin SE, Berrang-Ford L, Ford JD (2016) Public health adaptation to climate change in large cities: a global baseline. Int J Health Serv 46(1):53–78
Association AP (2017) Mental health and our changing climate: impacts, implications, and guidance. APA, Washington DC
Austin SE, Biesbroek R, Berrang-Ford L, Ford JD, Parker S, Fleury MD (2016) Public health adaptation to climate change in OECD countries. Int J Environ Res Public Health 13(9):889
Awiti AO (2022) Climate change and gender in Africa: a review of impact and gender-responsive solutions. Front Clim 101
Ayeb-Karlsson S, Kniveton D, Cannon T (2020) Trapped in the prison of the mind: notions of climate-induced (im)mobility decision-making and wellbeing from an urban informal settlement in Bangladesh. Palgrave Commun 6(1):1–15
Aylett A (2015) Institutionalizing the urban governance of climate change adaptation: results of an international survey. Urban Clim 14:4–16
Bayoh MN, Lindsay SW (2004) Temperature-related duration of aquatic stages of the Afrotropical malaria vector mosquito Anopheles gambiae in the laboratory. Med Vet Entomol 18(2):174–179
Benedict M, Sandve S, Wilkins E, Roberts J (2010) Relationship of larval desiccation to Anopheles gambiae Giles and An. arabiensis Patton survival. J Vector Ecol 35(1):116–123
Benjamin L, Thomas A (2020) Gender and climate justice-implications for policy formation in the Caribbean region. Loyal Law Rev 66:329
Berry HL, Bowen K, Kjellstrom T (2010) Climate change and mental health: a causal pathways framework. Int J Public Health 55:123–132
Berry P, Enright PM, Shumake-Guillemot J, Villalobos Prats E, Campbell-Lendrum D (2018) Assessing health vulnerabilities and adaptation to climate change: a review of international progress. Int J Environ Res Public Health 15(12):2626
Blanc J, Spruill T, Butler M, Casimir G, Jean-Louis G (2019) 0885 is resilience a protective factor for sleep disturbances among earthquake survivors? Sleep 42:A356
Bourque F, Cunsolo Willox A (2014) Climate change: the next challenge for public mental health? Int Rev Psychiatry 26(4):415–422
Bowman AS, Sauer J (2004) Tick salivary glands: function, physiology and future. Parasitology 129(S1):S67–S81
Brown D, Chanakira RR, Chatiza K, Dhliwayo M, Dodman D, Masiiwa M, Muchadenyika D, Mugabe P, Zvigadza S (2012) Climate change impacts, vulnerability and adaptation in Zimbabwe. International Institute for Environment and Development
Brown H, Spickett J, Katscherian D (2014) A health impact assessment framework for assessing vulnerability and adaptation planning for climate change. Int J Environ Res Public Health 11(12):12896–12914
Brown HL, Proust K, Spickett J, Capon A (2011) The potential role of health impact assessment in tackling the complexity of climate change adaptation for health. Health Promot J Austr 22(4):48–53
Brownstein JS, Holford TR, Fish D (2005) Effect of climate change on Lyme disease risk in North America. EcoHealth 2:38–46
Cameron LL Climate and health intervention assessment
Christiansen-Jucht C, Parham PE, Saddler A, Koella JC, Basáñez M-G (2014) Temperature during larval development and adult maintenance influences the survival of Anopheles gambiae ss. Parasit Vectors 7:1–10
Cissé G (2019) Food-borne and water-borne diseases under climate change in low-and middle-income countries: further efforts needed for reducing environmental health exposure risks. Acta Trop 194:181–188
Cutter SL (2017) The forgotten casualties redux: women, children, and disaster risk. Glob Environ Chang 42:117–121
Datey A, Bali B, Bhatia N, Khamrang L, Kim SM (2023) A gendered lens for building climate resilience: narratives from women in informal work in Leh, Ladakh. Gend Work Organ 30(1):158–176

Dobson A (2004) Population dynamics of pathogens with multiple host species. Am Nat 164(S5):S64–S78

Eisen L, Monaghan AJ, Lozano-Fuentes S, Steinhoff DF, Hayden MH, Bieringer PE (2014) The impact of temperature on the bionomics of Aedes (Stegomyia) aegypti, with special reference to the cool geographic range margins. J Med Entomol 51(3):496–516

Fox M, Zuidema C, Bauman B, Burke T, Sheehan M (2019) Integrating public health into climate change policy and planning: state of practice update. Int J Environ Res Public Health 16(18):3232

Frieler K, Levermann A, Elliott J, Heinke J, Arneth A, Bierkens M, Ciais P, Clark D, Deryng D, Döll P (2015) A framework for the cross-sectoral integration of multi-model impact projections: land use decisions under climate impacts uncertainties. Earth Syst Dyn 6(2):447–460

Frumkin H, Hess J, Luber G, Malilay J, McGeehin M (2008) Climate change: the public health response. Am J Public Health 98(3):435–445

Gamfeldt L, Snäll T, Bagchi R, Jonsson M, Gustafsson L, Kjellander P, Ruiz-Jaen MC, Fröberg M, Stendahl J, Philipson CD (2013) Higher levels of multiple ecosystem services are found in forests with more tree species. Nat Commun 4(1):1340

Georgeson L, Maslin M, Poessinouw M, Howard S (2016) Adaptation responses to climate change differ between global megacities. Nat Clim Chang 6(6):584–588

Goldhagen JL, Shenoda S, Oberg C, Mercer R, Kadir A, Raman S, Waterston T, Spencer NJ (2020) Rights, justice, and equity: a global agenda for child health and wellbeing. Lancet Child Adolesc Health 4(1):80–90

Gruebner O, Lowe SR, Sykora M, Shankardass K, Subramanian S, Galea S (2017) A novel surveillance approach for disaster mental health. PLoS ONE 12(7):e0181233

Harvey E (2019) Climate change and the loss of cultural heritage

Health IoMCftSotFoP (1988) The future of public health, vol 88. National Academy Press

Heidari H, Mohammadbeigi A, Khazaei S, Soltanzadeh A, Asgarian A, Saghafipour A (2020) The effects of climatic and environmental factors on heat-related illnesses: a systematic review from 2000 to 2020. Urban Clim 34:100720

Hooper DU, Adair EC, Cardinale BJ, Byrnes JE, Hungate BA, Matulich KL, Gonzalez A, Duffy JE, Gamfeldt L, O'Connor MI (2012) A global synthesis reveals biodiversity loss as a major driver of ecosystem change. Nature 486(7401):105–108

Hosking J, Campbell-Lendrum D (2012) How well does climate change and human health research match the demands of policymakers? A scoping review. Environ Health Perspect 120(8):1076–1082

Hsiang SM, Burke M, Miguel E (2013) Quantifying the influence of climate on human conflict. Science 341(6151):1235367

Hsu W-H, Hwang S-A, Saha S, Schramm PJ, Van Zutphen AR, Conlon KC, Lin S (2017) Projecting climate-related disease burden: a case study on methods for projecting respiratory health impacts

Jodoin S, Ananthamoorthy N, Lofts K (2020) A disability rights approach to climate governance. Ecol Law Q 47:73

Kim EJ (2016) US Global Change Research Program: the impacts of climate change on human health in the United States: a scientific assessment. Taylor & Francis, Washington, DC, p 312. https://health2016.globalchange.gov/downloads

Kosanic A, Petzold J, Martín-López B, Razanajatovo M (2022) An inclusive future: disabled populations in the context of climate and environmental change. Curr Opin Environ Sustain 55:101159

Leggieri MC, Toscano P, Battilani P (2021) Predicted aflatoxin B1 increase in Europe due to climate change: actions and reactions at global level. Toxins 13(4):292

Lemery J, Knowlton K, Sorensen C, Linstadt H (2021) Commentary on Covid-19, climate change, and human health. In: Global climate change and human health: from science to practice

Levy K, Smith SM, Carlton EJ (2018) Climate change impacts on waterborne diseases: moving toward designing interventions. Curr Environ Health Rep 5:272–282

Lima A, Lovin DD, Hickner PV, Severson DW (2016) Evidence for an overwintering population of Aedes aegypti in Capitol Hill neighborhood, Washington, DC. Am J Trop Med Hyg 94(1):231

Malak MA, Sajib AM, Quader MA, Anjum H (2020) "We are feeling older than our age": vulnerability and adaptive strategies of aging people to cyclones in coastal Bangladesh. Int J Disaster Risk Reduct 48:101595

Marinucci GD, Luber G, Uejio CK, Saha S, Hess JJ (2014) Building resilience against climate effects—a novel framework to facilitate climate readiness in public health agencies. Int J Environ Res Public Health 11(6):6433–6458

McKinnon M (2012) Climate vulnerability monitor: a guide to the cold calculus of a hot planet. Estudios Gráfcos Europeos, SA, Spain, p 331

McMichael C, Barnett J, McMichael AJ (2012) An ill wind? Climate change, migration, and health. Environ Health Perspect 120(5):646–654

Mohammadyan M, Sepehr P (2010) Design of cool spot and assessment of its effect on WBGT index among furnace workers' position in Shimi Madani industry in Hamadan. J Maz Univ Med Sci 20(76):2–7

Mosadeghrad AM, Isfahani P, Eslambolchi L, Zahmatkesh M, Afshari M (2023) Strategies to strengthen a climate-resilient health system: a scoping review. Glob Health 19(1):62

Munzert S, Jankin S (2019) The 2019 report of The Lancet Countdown on health and climate change: ensuring that the health of a child born today is not defined by a changing climate. Lancet 394(10211):1836–1878

Murray CJ, Aravkin AY, Zheng P, Abbafati C, Abbas KM, Abbasi-Kangevari M, Abd-Allah F, Abdelalim A, Abdollahi M, Abdollahpour I (2020) Global burden of 87 risk factors in 204 countries and territories, 1990–2019: a systematic analysis for the Global Burden of Disease Study 2019. Lancet 396(10258):1223–1249

Myers SS, Smith MR, Guth S, Golden CD, Vaitla B, Mueller ND, Dangour AD, Huybers P (2017) Climate change and global food systems: potential impacts on food security and undernutrition. Annu Rev Public Health 38:259–277

Ngcamu BS (2023) Climate change effects on vulnerable populations in the Global South: a systematic review. Nat Hazards 118(2):977–991

Ogden NH, Lindsay LR (2016) Effects of climate and climate change on vectors and vector-borne diseases: ticks are different. Trends Parasitol 32(8):646–656

Ogden N, Lindsay L, Beauchamp G, Charron D, Maarouf A, O'callaghan C, Waltner-Toews D, Barker I (2004) Investigation of relationships between temperature and developmental rates of tick Ixodes scapularis (Acari: Ixodidae) in the laboratory and field. J Med Entomol 41(4):622–633

Ogden NH, Bigras-Poulin M, O'callaghan CJ, Barker IK, Kurtenbach K, Lindsay LR, Charron DF (2007) Vector seasonality, host infection dynamics and fitness of pathogens transmitted by the tick Ixodes scapularis. Parasitology 134(2):209–227

Ogden NH, Radojevic M, Wu X, Duvvuri VR, Leighton PA, Wu J (2014) Estimated effects of projected climate change on the basic reproductive number of the Lyme disease vector Ixodes scapularis. Environ Health Perspect 122(6):631–638

Organization WH (2009) Protecting health from climate change: global research priorities

Organization WH (2017) Policy options on mental health: a WHO-Gulbenkian Mental Health Platform collaboration

Organization WH (2021) United Nations Children's Fund (UNICEF) & World Bank (2021) Levels and trends in child malnutrition: UNICEF/WHO/The World Bank Group joint child malnutrition estimates: key findings of the 2021 edition. World Health Organization. https://apps.who.int/iris/handle/10665/341135

Paaijmans KP, Imbahale SS, Thomas MB, Takken W (2010) Relevant microclimate for determining the development rate of malaria mosquitoes and possible implications of climate change. Malar J 9:1–8

Pachauri RK, Allen MR, Barros VR, Broome J, Cramer W, Christ R, Church JA, Clarke L, Dahe Q, Dasgupta P (2014) Climate change 2014: synthesis report. Contribution of Working Groups I, II and III to the fifth assessment report of the Intergovernmental Panel on Climate Change. IPCC

Perera A, Nik VM, Chen D, Scartezzini J-L, Hong T (2020) Quantifying the impacts of climate change and extreme climate events on energy systems. Nat Energy 5(2):150–159
Perry AL, Low PJ, Ellis JR, Reynolds JD (2005) Climate change and distribution shifts in marine fishes. Science 308(5730):1912–1915
Pörtner H-O (2008) Ecosystem effects of ocean acidification in times of ocean warming: a physiologist's view. Mar Ecol Prog Ser 373:203–217
Reiter P (2001) Climate change and mosquito-borne disease. Environ Health Perspect 109(suppl 1):141–161
Romanello M, McGushin A, Di Napoli C, Drummond P, Hughes N, Jamart L, Kennard H, Lampard P, Rodriguez BS, Arnell N (2021) The 2021 report of the Lancet Countdown on health and climate change: code red for a healthy future. Lancet 398(10311):1619–1662
Ronoh S, Gaillard J, Marlowe J (2015) Children with disabilities and disaster preparedness: a case study of Christchurch. Kōtuitui: N Z J Soc Sci Online 10(2):91–102
Salih AA, Baraibar M, Mwangi KK, Artan G (2020) Climate change and locust outbreak in East Africa. Nat Clim Chang 10(7):584–585
Shamsuddoha M (2012) Displacement and migration from climate hot-spots in Bangladesh: causes and consequences. ActionAid Bangladesh
Sheehan MC, Fox MA, Kaye C, Resnick B (2017) Integrating health into local climate response: lessons from the US CDC Climate-Ready States and Cities Initiative. Environ Health Perspect 125(9):094501
Stein PJ, Stein MA (2022a) Climate change and the right to health of people with disabilities. Lancet Glob Health 10(1):e24–e25
Stein PJ, Stein MA (2022b) Disability, human rights, and climate justice. Hum Rights Q 44(1):81–110
Sutherst R, Bourne A (2006) The effect of desiccation and low temperature on the viability of eggs and emerging larvae of the tick, Rhipicephalus (Boophilus) microplus (Canestrini) (Ixodidae). Int J Parasitol 36(2):193–200
Swinburn BA, Kraak VI, Allender S, Atkins VJ, Baker PI, Bogard JR, Brinsden H, Calvillo A, De Schutter O, Devarajan R (2019) The global syndemic of obesity, undernutrition, and climate change: the Lancet Commission report. Lancet 393(10173):791–846
Tanjeela M, Rutherford S (2018) The influence of gender relations on women's involvement and experience in climate change adaptation programs in Bangladesh. SAGE Open 8(4):2158244018812620
Uleberg E, Hanssen-Bauer I, van Oort B, Dalmannsdottir S (2014) Impact of climate change on agriculture in Northern Norway and potential strategies for adaptation. Clim Change 122:27–39
Ursano R, Morganstein J, Cooper R (2017) Resource document on mental health and climate change. APA Resour Doc 1
Verner G, Schütte S, Knop J, Sankoh O, Sauerborn R (2016) Health in climate change research from 1990 to 2014: positive trend, but still underperforming. Glob Health Action 9(1):30723
Vicedo-Cabrera AM, Scovronick N, Sera F, Royé D, Schneider R, Tobias A, Astrom C, Guo Y, Honda Y, Hondula D (2021) The burden of heat-related mortality attributable to recent human-induced climate change. Nat Clim Chang 11(6):492–500
Watanabe M, Noma H, Kurai J, Kato K, Sano H (2021) Association with ambient air pollutants and school absence due to sickness in schoolchildren: a case-crossover study in a provincial town of Japan. Int J Environ Res Public Health 18(12):6631
Watts N, Adger WN, Ayeb-Karlsson S, Bai Y, Byass P, Campbell-Lendrum D, Colbourn T, Cox P, Davies M, Depledge M (2017) The Lancet Countdown: tracking progress on health and climate change. Lancet 389(10074):1151–1164
Watts N, Amann M, Ayeb-Karlsson S, Belesova K, Bouley T, Boykoff M, Byass P, Cai W, Campbell-Lendrum D, Chambers J (2018) The Lancet Countdown on health and climate change: from 25 years of inaction to a global transformation for public health. Lancet 391(10120):581–630
Woodward AJ, Samet JM (2018) Climate change, hurricanes, and health. In: vol 108. American Public Health Association, pp 33–35

Workman A, Blashki G, Bowen KJ, Karoly DJ, Wiseman J (2018) The political economy of health co-benefits: embedding health in the climate change agenda. Int J Environ Res Public Health 15(4):674

Zaidi RZ, Fordham M (2021) The missing half of the Sendai framework: gender and women in the implementation of global disaster risk reduction policy. Prog Disaster Sci 10:100170

Zhao Q, Li S, Coelho MS, Saldiva PH, Hu K, Huxley RR, Abramson MJ, Guo Y (2019) The association between heatwaves and risk of hospitalization in Brazil: a nationwide time series study between 2000 and 2015. PLoS Med 16(2):e1002753

Zhao Q, Guo Y, Ye T, Gasparrini A, Tong S, Overcenco A, Urban A, Schneider A, Entezari A, Vicedo-Cabrera AM (2021) Global, regional, and national burden of mortality associated with non-optimal ambient temperatures from 2000 to 2019: a three-stage modelling study. Lancet Planet Health 5(7):e415–e425

Healing with Care: A Roadmap to Sustainable Healthcare

Keshav Singh, Vivek Kumar Srivastava, and Ashish Shukla

Abstract "Sustainable Healthcare" explores the crucial junction between healthcare and sustainability and provides a thorough examination of tactics that advance both human well-being and ecological harmony. This chapter discusses the necessity of integrating medical services with environmentally responsible concepts as it navigates the changing landscape of healthcare practises. The chapter explores the complex interrelationships between patient care, resource management, technological integration, and environmental stewardship through a multidimensional perspective. The way towards a more sustainable healthcare system is illuminated by the innovative projects, case studies, and best practises that are highlighted. This chapter serves as an invaluable resource for healthcare professionals, lawmakers, and academics working to pave the way for a healthier future built upon a foundation of care and compassion by emphasising the holistic idea of healing embracing people, communities, and the earth.

Keywords Ecology · Practices · Sustainable · Environment · Healthcare

1 Introduction

In a time of unparalleled progress in medical science, achieving long-term, comprehensive health has become essential. This chapter explores the complex interrelationships between patient care, resource management, technological integration, and environmental stewardship, weaving an intricate tapestry that is sure to transform. This innovative journey breaks through traditional boundaries, providing a multifaceted view that acknowledges the intrinsic interdependence of these crucial components forming the framework of contemporary healthcare.

The fundamental insight that patient care is not a solo endeavor but rather a cooperative symphony, entwining with the prudent management of limited resources, the smooth integration of state-of-the-art technologies, and a diligent commitment to

K. Singh (✉) · V. K. Srivastava · A. Shukla
Raja Balwant Singh Engineering Technical Campus, Bichpuri, Agra, India
e-mail: Keshav037@gmail.com

P. K. Prabhakar and W. Leal Filho (eds.), *Preserving Health, Preserving Earth*, World Sustainability Series, https://doi.org/10.1007/978-3-031-60545-1_4

environmental stewardship, is at the core of this story. The complex interplay between these elements serves as the foundation for a sustainable healthcare paradigm, in which the quest for healing is balanced with conscientious resource management and environmental consciousness (Jameton and McGuire 2002).

As we follow this roadmap, it becomes clear that providing sustainable healthcare is a journey rather than a destination. It calls for a dynamic equilibrium that strikes a balance between the needs of the individual, the demands placed on resources, the integration of transformative technologies, and an unwavering dedication to environmental preservation. As mentioned by McKee, the need for sustainability in the face of a rapidly evolving global healthcare landscape is one that transcends the conventional boundaries of medical practice (McKee 2018). As a stark warning to confront the urgent issues endangering the resilience of healthcare systems across the globe, the necessity of sustainable healthcare becomes apparent.

2 Patient-Centric Paradigm: Beyond Treatment to Holistic Care

The idea of patient care experiences a profound metamorphosis in the dynamic world of contemporary healthcare, moving beyond conventional paradigms to embrace a more patient-centric, holistic approach. The way ideas about patient care have evolved is indicative of a transformative process that has redefined the relationship between a clinician and a patient and acknowledged the complexity of personal well-being (Zahid et al. 2021).

2.1 Evolving Notions of Patient Care

Reiterating the need of empathy and compassion is fundamental to the changing ideas about patient care. More and more healthcare professionals are realizing the importance of comprehending and appreciating the psychological and emotional components of a patient's experience. By developing empathy, professionals can improve patient outcomes and overall quality of care by fostering a more understanding and supportive atmosphere (Yu et al. 2022). In addition to treating physical illnesses, compassionate care also tends to people's emotional and mental health, which supports a patient-centric philosophy.

2.2 *Patient-Centered Approaches to Sustainable Well-Being*

The patient-centric paradigm aims to address the more expansive idea of well-being in addition to treating particular illnesses. Patients' individual needs and preferences are taken into consideration when making decisions about their care, and they are increasingly seen as collaborators in this process. This method acknowledges that social determinants of health, lifestyle choices, and the patient's unique objectives all play a role in achieving lasting well-being in addition to medical therapies (Nykänen and Seppälä 2012). Healthcare professionals can better customize interventions to the patient's values by using patient-centered approaches, which will lead to a more thorough and long-lasting improvement in well-being.

3 Balancing Act: Resource Management in Healthcare

Healthcare resource management is a fine balancing act, especially when dealing with the constraints of limited resources (Gowen et al. 2006). Innovative ways to allocation are necessary to enable optimal and equitable delivery of healthcare services while addressing ethical considerations, a hard endeavour of wellbeing.

3.1 *The Challenge of Finite Resources*

Globally, healthcare systems struggle with the issue of limited resources. Budgetary restrictions and the rising demand for healthcare services mean that careful allocation is required to guarantee efficacy and efficiency. Lack of resources affects several areas, such as staff, facilities, and medical equipment (Figueroa et al. 2019). As a result, healthcare practitioners must look for ways to optimize results while working within the constraints at hand.

3.2 *Innovations in Resource Allocation*

Innovative approaches to resource allocation are being used by healthcare organizations to meet the issues posed by limited resources. Predictive analytics is one example of a data-driven strategy that helps find patterns and distribute resources according to the requirements of particular populations. Technological innovations also allow healthcare professionals to reach a wider audience and maximize resource use, such as telemedicine and remote monitoring. When administrators, policymakers, and healthcare experts work together, creative models that improve resource efficiency are developed (World Health Organization 2002).

3.3 *Ethical Considerations in Resource Management*

Resource management in healthcare is inherently linked to ethical considerations, requiring careful decision-making to ensure fairness and justice. Ethical dilemmas arise when healthcare providers must prioritize certain treatments, interventions, or patient populations due to resource constraints (Penney 2020). Principles such as distributive justice guide decisions to allocate resources equitably, ensuring that all individuals have a fair opportunity to access necessary healthcare services. Transparency, accountability, and inclusivity become crucial in the ethical framework of resource management.

Healthcare resource management is a fine balancing act that calls for a diverse strategy. Adopting innovative resource allocation strategies while adhering to ethical guidelines to promote justice and fairness is necessary to address the problem of limited resources. Healthcare systems may maximize results for people and communities while navigating resource restrictions by implementing patient-cantered and sustainable approaches. Building robust and adaptable healthcare systems that can fairly and ethically address the different requirements of people depends on the continuous evolution of resource management techniques.

4 Technological Integration: Shaping the Future of Healthcare

A revolutionary era in patient care has been ushered in by the incorporation of technology, which has redefined patient care, encouraged state-of-the-art developments, and prompted critical consideration of ethical and practical ramifications (Zahid et al. 2021). Technology is having a significant and wide-ranging impact on healthcare as it develops.

4.1 *The Impact of Technology on Patient Care*

Patient care has undergone a technological revolution that has improved accessibility, effectiveness, and quality. Healthcare providers may share information more easily and efficiently thanks to electronic health records (EHRs), which promotes more individualized and coordinated care (Turley et al. 2011). Healthcare services are now more widely available thanks to telemedicine, which offers remote monitoring and consultations. Patients are given the ability to actively participate in their health management using wearable technology and mobile health applications, which supports early detection and preventive care. Healthcare professionals can make better judgments and achieve better results in diagnosis and treatment because to real-time data analytics.

4.2 Cutting-Edge Innovations in Healthcare

Modern inventions brought about by technology's quick development are changing the way healthcare is provided. Predictive analytics uses artificial intelligence (AI) and machine learning algorithms to help with disease diagnosis, treatment planning, and personalized medicine. Systems for robotic surgery improve the accuracy and productivity of surgical operations (Vishwakarma et al. 2023). Applications for virtual reality (VR) and augmented reality (AR) can be found in patient education, medical training, and even pain management. Medical gadgets are connected via the Internet of Things (IoT) to enable continuous monitoring and provide useful data for prompt actions.

4.3 Ethical and Practical Considerations in Technology Adoption

Even while technology is advancing healthcare in ways never seen before, there are moral and practical questions raised by its use. Concerning patient privacy, data security, and the possibility of bias in AI systems, ethical conundrums could occur (Brunton 2017). It's critical to strike a balance between protecting sensitive data and using technology to enhance patient care. In order to guarantee fair healthcare delivery, it is also necessary to address the digital divide and discrepancies in access to technology. To properly handle these issues, ethical frameworks and rules for technology adoption are crucial.

Healthcare's future is being shaped by technological integration, which is transforming patient care, advancing cutting-edge discoveries, and inspiring careful analysis of the moral and practical ramifications. Healthcare professionals, legislators, and the general public must work together to navigate the changing landscape as technology develops to make sure that technological integration upholds the core values of patient-centred, ethical, and compassionate care rather than diminishing them (Brunton 2017). Resolving the conflict between innovation and moral obligation is essential to achieving the full potential of technology to improve and revolutionize healthcare for all.

5 Interconnected Realities: Patient Care and Resource Optimization

The convergence of resource optimization and patient care in the ever-changing healthcare landscape has emerged as a key area of focus for building long-lasting and efficient healthcare systems. This complex relationship necessitates a careful balance, whereby improving patient outcomes is smoothly incorporated with effective resource management. This section looks at case studies that illustrate integrated healthcare systems, analyzes the opportunities and challenges that come with optimization, and explores the crucial elements of finding synergies between patient care and resource efficiency.

The connection between resource optimization and patient care is to find and take advantage of synergies that can increase the effectiveness of each component. This entails utilizing data analytics, streamlined procedures, and technology breakthroughs to improve patient experiences while also optimizing resource utilization efficiency. The use of electronic health records (EHRs), for example, makes information interchange easier, cuts down on duplication, and guarantees that resources are distributed exactly where and when they are needed (Tapuria et al. 2021). In order to reconcile the two seemingly incompatible priorities of individualized patient care and resource efficiency, creative solutions must be developed and implemented.

5.1 Case Studies in Integrated Healthcare Systems

Excellent examples of successfully merging resource optimization and patient care are integrated healthcare systems. The observable advantages of this kind of integration are illustrated by case studies from different healthcare organizations. The Kaiser Permanente model, for example, places a strong emphasis on early intervention and preventive care, which lowers hospital admission rates and improves patient outcomes (Albers Mohrman and Kanter 2012). The patient-centered approach at the Mayo Clinic emphasizes teamwork among medical specialists, which results in streamlined procedures and effective use of available resources. These case studies highlight how integrated systems can improve patient care while allocating resources as efficiently as possible.

5.2 Challenges and Opportunities in Optimization

The path to optimizing resources and harmonizing patient care is not without difficulties, despite the obvious benefits. Healthcare providers frequently encounter challenges like financial limitations, resistance to adopting new technologies, and interoperability problems. Another challenge is balancing cost-effective solutions with patient preferences. But these difficulties also present chances for development (Soekhai et al. 2019). An increasingly connected and efficient healthcare ecosystem can be achieved by making investments in interoperable systems, encouraging healthcare providers to adopt new technologies, and investigating alternative funding sources.

The interdependent realities of resource optimization and patient care demand a strategic, all-encompassing approach to healthcare management. Achieving sustainable healthcare systems that prioritize patient well-being while optimizing resource utilization requires finding synergies between these two facets. By examining case studies in integrated health systems, we can learn about effective models that strike a balance between these related priorities. Understanding optimization's opportunities and challenges enables the creation of creative solutions that will influence healthcare's future and establish a positive balance between patient care and resource efficiency (Ahmadi-Javid et al. 2017).

6 Building Bridges: Technology, Environment, and Patient-Centered Care

The convergence of environmental sustainability, patient-centred care, and technology is changing the way we approach and provide healthcare services in the ever-evolving healthcare landscape. Creating connections between these components is essential to creating an all-encompassing, cutting-edge healthcare ecosystem. This article examines how to integrate technology with patient-centred methods, how technology contributes to environmental sustainability, and how crucial it is to find a balance in order to provide genuinely comprehensive healthcare (Douglas and Douglas 2005).

6.1 Integrating Technology with Patient-Centric Approaches

The goal of integrating technology into healthcare is to improve patient care, not just digitize procedures. Healthcare practitioners are able to customize services to meet the needs of each patient by utilizing cutting-edge technologies like wearables, telehealth, and artificial intelligence (Andersen et al. 2019). For example, telehealth enables remote consultations, which eases access to healthcare while lessening the strain on physical facilities. Continuous monitoring is made possible by wearable technology, giving patients greater control over their medical care. Personalized treatment plans are made possible by artificial intelligence's assistance with data analysis. The integration of technology and patient-centred approaches results in a more customized and responsive healthcare delivery experience (Bohr and Memarzadeh 2020).

6.2 Technology's Role in Environmental Sustainability

The healthcare sector's environmental impact is growing along with it. When used wisely, technology can be a key component in reducing this effect. Because electronic health records (EHRs) eliminate the need for paper documentation, fewer resources are used. Reduced patient travel results in lower carbon emissions, which can be attributed to remote monitoring and telehealth services (Sherman et al. 2020). Furthermore, energy use in healthcare facilities is optimized by smart infrastructure management systems. Healthcare providers can maintain high-quality patient care while also supporting the larger goal of sustainability by adopting eco-friendly technologies.

6.3 Striking a Balance for Comprehensive Healthcare

Achieving comprehensive healthcare requires finding a careful balance between patient-centred care, environmental sustainability, and technology. This entails coordinating technology development with patient choices and environmental stewardship. For example, it is crucial to make sure that patients can easily access and comprehend the advanced diagnostic tools that are being used. Adoption of new technologies should be accompanied by environmental impact assessments at the same time to encourage environmentally friendly solutions (Pinho-Gomes et al. 2022). Healthcare professionals can make sure that patient well-being and technological advancements go hand in hand by carefully navigating this triad.

Creating connections between the environment, technology, and patient-centred care is essential to determining how healthcare will develop in the future. The whole healthcare experience is improved when technology and patient-centric methods are combined, becoming more efficient, individualized, and accessible. In addition, utilizing technology to solve issues related to environmental sustainability supports an all-encompassing and accountable healthcare system. Achieving comprehensive healthcare that prioritizes patient well-being and environmental stewardship while also embracing innovation requires striking the right balance between these interconnected elements (Sherman et al. 2020). The future of healthcare emerges from this complex interplay as a harmonious fusion of environmental awareness, technological advancement, and patient-centred care.

7 Policy Perspectives: Shaping Sustainable Healthcare Frameworks

Healthcare sustainability is now a must, not a choice, and policy frameworks are essential in guiding the sector in the direction of a more sustainable future (Ahmed et al. 2021). This section highlights the crucial role that government and policy play in bringing about revolutionary change, examines international perspectives on healthcare sustainability, and digs into the regulatory approaches to sustainable healthcare.

7.1 Regulatory Approaches to Sustainable Healthcare

The foundation of sustainable healthcare practices is effective regulation. Around the world, governments are realizing more and more how important it is to pass laws and enforce regulations that promote ecologically friendly practices in the healthcare industry (Porter and Linde 1995). Strict waste management policies, energy efficiency requirements for medical facilities, and financial incentives for implementing environmentally friendly technology are examples of regulatory strategies. Regulatory frameworks minimize the ecological footprint of the healthcare industry by providing a clear and favourable environment for healthcare providers to adopt sustainability measures.

7.2 International Perspectives on Healthcare Sustainability

Global sustainability is a challenge, and different countries have important insights into effective approaches. In order to lower carbon emissions, some nations are leading the way in integrating renewable energy sources into the infrastructure of healthcare (Vaishnavi and Suresh 2023). Others concentrate on environmentally conscious standards compliance in the supply chain through sustainable procurement practices. The international exchange of best practices facilitates the learning from successful initiatives and adaptation of policies by the global healthcare community to suit the specific needs and challenges of different regions.

7.3 The Role of Government and Policy in Driving Change

Governments have an important role in how the healthcare sector is steered toward sustainability. Governments can enact policies that prioritize environmentally friendly practices and thereby spark transformative change. Investments in environmentally responsible technologies by healthcare providers are tangibly motivated by financial incentives, tax breaks, and grants for sustainable healthcare initiatives (Mulimani 2017). Policy frameworks can also promote innovation that is advantageous to the environment and the healthcare sector by supporting research and development into green healthcare solutions. Therefore, the role of policy and government goes beyond simple regulation to include actively promoting and aiding in the transition to sustainable healthcare practices.

8 Challenges and Opportunities in the Journey Towards Sustainable Healthcare

In the context of the world's environmental problems, the pursuit of sustainable healthcare is both morally and strategically required. This adventure is not without its challenges and achievements, though. This article examines the obstacles in the way of sustainable healthcare, the new avenues for development, and the priceless lessons to be learned from both achievements and setbacks.

8.1 Identifying Roadblocks in Sustainable Healthcare

8.1.1 Financial Barriers

Investing in green technologies or infrastructure upgrades can be expensive initially, and healthcare organizations, particularly those with limited resources, may find it difficult to support sustainable initiatives (Bray et al. 2023).

8.1.2 Resistance to Change

When introducing new, sustainable practices in the healthcare industry, a conservative approach may cause opposition. Adoption of novel, environmentally friendly technologies may be impeded by established practices and conventional approaches (Schroeder et al. 2012).

8.1.3 Complex Regulatory Landscape

A major challenge is navigating a complicated web of rules and standards (Guzzo et al. 2020). Healthcare providers have a number of rules to follow and have to make sure sustainable practices meet national and international standards.

8.1.4 Limited Awareness and Education

Advancement may be hampered by a lack of knowledge and comprehension regarding how healthcare practices affect the environment. In order for patients, administrators, and healthcare professionals to understand the value of sustainable healthcare and how they can contribute, education may be necessary (Hussain et al. 2018).

8.2 *Emerging Opportunities for Improvement*

8.2.1 Technological Advancements

New opportunities for sustainable healthcare are made possible by the quick development of technology. Energy-efficient medical equipment, telemedicine, smart infrastructure management, and other innovations present chances to enhance healthcare delivery while lessening environmental impact (Mehrotra and Joshi 2021).

8.2.2 Collaborative Initiatives

There is a significant chance for shared responsibility when healthcare providers, governments, non-profits, and the private sector work together (Hearld et al. 2016). Collaborations can offer resources, knowledge, and a coordinated strategy to tackle sustainability issues.

8.2.3 Research and Development

Breakthroughs can be achieved by funding research and development aimed at green technologies and sustainable healthcare solutions (Schroeder et al. 2012). Ongoing R&D efforts can spur positive change in everything from energy-efficient facility design to environmentally friendly medical devices.

There are many obstacles in the way of achieving sustainable healthcare, but these obstacles also present chances for advancement and creativity. The trajectory of sustainable healthcare is shaped by identifying obstacles, taking advantage of new opportunities, and making collective learnings from both successes and failures. In order to successfully navigate this path, the healthcare sector must cultivate a culture of collaboration, make educational investments, and maintain flexibility in the face of changing opportunities and challenges. In the end, achieving sustainable healthcare is a dynamic process that calls for commitment, ingenuity, and a dedication to having a positive influence on the planet and the well-being of patients.

9 Towards a Sustainable Future: Actionable Strategies and Recommendations

Beyond theoretical aspirations, a strategic and practical approach is needed to move healthcare towards a sustainable future. This section highlights the significance of empowering stakeholders in the pursuit of sustainable well-being, examines the creation of a blueprint for sustainable healthcare, and provides doable implementation steps.

9.1 Crafting a Blueprint for Sustainable Healthcare

9.1.1 Assessment of Current Practices

Start by carrying out a comprehensive analysis of the supply chain procedures, energy usage, and waste management of current healthcare practices. The assessments will serves as the basis for developing a blueprint for sustainable healthcare (Grol et al. 2013).

9.1.2 Setting Clear Objectives and Targets

Establish quantifiable, targeted sustainability goals that are consistent with the organization's core principles and overarching objectives. Setting specific goals makes it possible to track progress and keep track of accomplishments.

9.1.3 Integration of Green Technologies

Accept cutting-edge approaches and green technologies to lessen the environmental effect of healthcare operations. Technology can make a big difference in achieving sustainability goals, from smart facility management systems to energy-efficient medical equipment (Franco et al. 2017).

9.1.4 Adoption of Sustainable Procurement Practices

Choose vendors and goods that meet environmentally friendly criteria in order to put sustainable procurement practices into effect. Give preference to suppliers who are devoted to lowering their carbon footprint and encouraging social responsibility.

9.2 *Practical Steps for Implementation*

9.2.1 Employee Training and Engagement

Provide healthcare workers with the information and abilities they need to advocate for sustainable practices (Luxford et al. 2011). Encourage employees to actively participate in sustainability initiatives by fostering a culture of environmental responsibility and awareness.

9.2.2 Waste Reduction and Recycling Programs

Develop extensive programs for recycling and waste reduction in healthcare facilities. The reduction of single-use plastics, recycling programs, and the appropriate disposal of medical waste all make a substantial contribution to sustainability efforts.

9.2.3 Energy Conservation Measures

Invest in conservation efforts, renewable energy sources, and energy-efficient infrastructure. Significant energy savings can be achieved by taking easy measures like improving appliance usage, HVAC controls, and lighting systems.

9.2.4 Community Outreach and Education

Engage the community in order to expand sustainability efforts outside of the healthcare facility. Organize educational initiatives to encourage people to adopt environmentally friendly behaviors and to increase public awareness of sustainable healthcare practices.

Developing a sustainable future for healthcare necessitates a multidimensional strategy that includes proactive stakeholder engagement, realistic implementation, and strategic planning. The healthcare sector can achieve environmental responsibility and promote well-being by creating a blueprint for sustainable healthcare, putting sensible measures into place, and empowering stakeholders. By adopting these practical approaches, healthcare organizations and practitioners improve not just the health of their patients but also the health of the environment, clearing the path for a more resilient and sustainable future.

10 Conclusion: A Call to Collective Action

The quest for sustainable healthcare is both necessary and continuous, necessitating the cooperation of all players in the healthcare system. Let's review the main points that highlight the importance of this mission and talk about the current initiatives being taken to achieve sustainable well-being.

10.1 Recapitulation of Key Themes

Important connections between patient care, resource optimization, technology, environmental responsibility, and policy frameworks have been made clear by the conversation around sustainable healthcare (Wang 2012). It highlights how these components must work in harmony to create a healthcare system that not only addresses patients' immediate needs but also protects the environment for coming generations.

The examination of obstacles and possibilities highlighted the complexity of this life-changing experience. Every challenge offers a chance for innovation and advancement, whether it comes in the form of financial constraints or opposition to change. The provision of sustainable healthcare necessitates strategic planning, teamwork, and an unwavering dedication to progress.

10.2 The Ongoing Journey Towards Sustainable Healthcare

The path to sustainable healthcare is an ongoing evolution rather than a final destination. It necessitates a continuous dedication to learning, flexibility, and incorporating cutting-edge methods and technologies (Fleiszer et al. 2015). Even though there has been progress, much work remains. As we move forward, it is critical to acknowledge accomplishments, draw lessons from mistakes, and adjust plans to take into account changing conditions in the healthcare and environmental sectors.

Initiatives to reduce waste, integrate green technologies, and use sustainable procurement methods are all positive moves. Moreover, the cross-border exchange of ideas and best practices highlights the necessity for collaborative efforts by demonstrating the global nature of the challenge.

10.3 Empowering Every Stakeholder in the Healthcare Ecosystem

The foundation of sustainable healthcare is empowerment. Each and every stakeholder from patients and the larger community to legislators and healthcare professionals plays a crucial part in determining the direction of the sector. Sustainable practices advance when healthcare professionals are equipped with knowledge, patients are made aware of the issue, and policymakers are presented with innovative ideas.

Acknowledging the interdependence of stakeholders cultivates a perception of collective accountability (McKimm et al. 2020). A collaborative culture is created through inclusive decision-making, open communication, and providing rewards and recognition. The spirit of cooperation is what propels a long-term healthcare system.

In summary, there is a strong need for group action. It calls on the public, professionals, lawmakers, and healthcare organizations to come together in support of sustainable healthcare. Every action we take as we overcome obstacles, grasp opportunities, and draw lessons from past mistakes helps to create a future that is more resilient, environmentally conscious, and patient-focused.

Sustainable healthcare is a shared responsibility that requires teamwork and goes beyond individual efforts. It is not an isolated goal. In order to move the healthcare sector toward a future where well-being is measured not only in patient outcomes but also in the health of the planet we call home, let this call to action reverberate in every decision, policy, and practice.

References

Ahmadi-Javid A, Jalali Z, Klassen KJ (2017) Outpatient appointment systems in healthcare: a review of optimization studies. Eur J Oper Res 258(1):3–34

Ahmed SU, Ali A, Memon AH, Hussain A (2021) Policy framework for implementation of sustainable procurement practices in healthcare industry. Environ Dev Sustain 23:3929–3945

Albers Mohrman S, Kanter MH (2012) Chapter 3 designing for health: learning from Kaiser Permanente. In Organizing for Sustainable Health Care (pp. 77–111). Emerald Group Publishing Limited

Andersen TO, Bansler JP, Kensing F, Moll J, Mønsted T, Nielsen KD, ... Svendsen JH (2019) Aligning concerns in telecare: three concepts to guide the design of patient-centred E-health. Comp Supp Coop Work (CSCW), 28:1039-1072

Bohr A, Memarzadeh K (2020) The rise of artificial intelligence in healthcare applications. In *Artificial Intelligence in healthcare* (pp. 25–60). Academic Press

Bray L, Meznikova K, Crampton P, Johnson T (2023) Sustainable healthcare education: a systematic review of the evidence and barriers to inclusion. Med Teach 45(2):157–166

Brunton M (2017) Risking the sustainability of the public health system: ethical conundrums and ideologically embedded reform. J Bus Ethics 142(4):719–734

Douglas CH, Douglas MR (2005) Patient-centred improvements in health-care built environments: perspectives and design indicators. Health Expect 8(3):264–276

Figueroa CA, Harrison R, Chauhan A, Meyer L (2019) Priorities and challenges for health leadership and workforce management globally: a rapid review. BMC Health Serv Res 19(1):1–11

Fleiszer AR, Semenic SE, Ritchie JA, Richer MC, Denis JL (2015) The sustainability of healthcare innovations: a concept analysis. J Adv Nurs 71(7):1484–1498

Franco A, Shaker M, Kalubi D, Hostettler S (2017) A review of sustainable energy access and technologies for healthcare facilities in the Global South. Sustain Energy Technol Assess 22:92–105

Gowen CR III, McFadden KL, Tallon WJ (2006) On the centrality of strategic human resource management for healthcare quality results and competitive advantage. J Manag Devel 25(8):806–826

Grol R, Wensing M, Eccles M, Davis D (Eds) (2013) Improving patient care: the implementation of change in health care. John Wiley & Sons

Guzzo D, Carvalho MM, Balkenende R, Mascarenhas J (2020) Circular business models in the medical device industry: paths towards sustainable healthcare. Resour Conserv Recycl 160:104904

Hearld LR, Bleser WK, Alexander JA, Wolf LJ (2016) A systematic review of the literature on the sustainability of community health collaboratives. Med Care Res Rev 73(2):127–181

Hussain M, Ajmal MM, Gunasekaran A, Khan M (2018) Exploration of social sustainability in healthcare supply chain. J Clean Prod 203:977–989

Jameton A, McGuire C (2002) Toward sustainable health-care services: principles, challenges, and a process. Int J Sustain High Educ 3(2):113–127

Luxford K, Safran DG, Delbanco T (2011) Promoting patient-centered care: a qualitative study of facilitators and barriers in healthcare organizations with a reputation for improving the patient experience. Int J Qual Health Care 23(5):510–515

McKee M (2018) Global sustainable healthcare. Medicine 46(7):383–387

McKimm J, Redvers N, El Omrani O, Parkes MW, Elf M, Woollard R (2020) Education for sustainable healthcare: leadership to get from here to there. Med Teach 42(10):1123–1127

Mehrotra R, Joshi K (2021) Healthcare sustainability through technological innovations. In *Advances in Environment Engineering and Management: Proceedings of the 1st National Conference on Sustainable Management of Environment and Natural Resource Through Innovation in Science and Technology* (pp 341–349). Springer International Publishing

Mulimani P (2017) Green dentistry: the art and science of sustainable practice. Br Dent J 222(12):954–961

Nykänen P, Seppälä A (2012) Collaborative approach for sustainable citizen-centered health care. Crit Iss Devel Sustain E-health Solut, 115–134

Penney D (2020) Ethical considerations for short-term global health projects. J Midwifery Womens Health 65(6):767–776

Pinho-Gomes AC, Yoo SH, Allen A, Maiden H, Shah K, Toolan M (2022) Incorporating environmental and sustainability considerations into health technology assessment and clinical and public health guidelines: a scoping review. Int J Technol Assess Health Care 38(1):e84

Porter M, Van der Linde C (1995) Green and competitive: ending the stalemate. Dynam Eco-Effic Econ Environ Regul Compet Adv 33:120–134

Schroeder K, Thompson T, Frith K, Pencheon D (2012) Sustainable healthcare. John Wiley & Sons

Sherman JD, Thiel C, MacNeill A, Eckelman MJ, Dubrow R, Hopf H, ... Bilec MM (2020) The green print: advancement of environmental sustainability in healthcare. Res Conserv Recycl 161:104882

Soekhai V, Whichello C, Levitan B, Veldwijk J, Pinto CA, Donkers B, ... de Bekker-Grob EW (2019) Methods for exploring and eliciting patient preferences in the medical product lifecycle: a literature review. Drug Disc Today, 24(7):1324-1331

Tapuria A, Porat T, Kalra D, Dsouza G, Xiaohui S, Curcin V (2021) Impact of patient access to their electronic health record: systematic review. Inform Health Soc Care 46(2):194–206

Turley M, Porter C, Garrido T, Gerwig K, Young S, Radler L, Shaber R (2011) Use of electronic health records can improve the health care industry's environmental footprint. Health Aff 30(5):938–946

Vaishnavi V, Suresh M (2023) Modelling the factors in implementation of environmental sustainability in healthcare organisations. Manag Environm Qual: an Intern J 34(1):137–158
Vishwakarma LP, Singh RK, Mishra R, Kumari A (2023) Application of artificial intelligence for resilient and sustainable healthcare system: systematic literature review and future research directions. Intern J Prod Res, 1–23.
Wang F (2012) Measurement, optimization, and impact of health care accessibility: a methodological review. Ann Assoc Am Geogr 102(5):1104–1112
World Health Organization. Noncommunicable Disease, & Mental Health Cluster (2002) Innovative care for chronic conditions: building blocks for action: global report. World Health Organization
Yu CC, Tan L, Le MK, Tang B, Liaw SY, Tierney T, ... Low JA (2022) The development of empathy in the healthcare setting: a qualitative approach. BMC Med Educ 22(1):1-13
Zahid A, Poulsen JK, Sharma R, Wingreen SC (2021) A systematic review of emerging information technologies for sustainable data-centric health-care. Int J Med Inform 149:104420

Community Engagement and Education for Eco-Conscious Health

Abhinay Thakur and Ashish Kumar

Abstract As the global focus on climate change and ecological preservation intensifies, healthcare practitioners are uniquely positioned to drive change by educating patients, families, and communities about the intersection of health and the environment. The chapter explores strategies for effective community engagement, emphasizing the importance of tailored approaches that consider cultural, socioeconomic, and demographic factors. It underscores the significance of clear communication, collaboration with community leaders, and the utilization of various platforms, from workshops and seminars to online resources and social media campaigns. The chapter discusses the potential benefits of integrating environmental awareness into healthcare curricula, empowering future generations of professionals to incorporate sustainability principles into their practices. Moreover, the chapter addresses challenges healthcare professionals might encounter in these endeavors, such as resistance to change and information overload. It offers guidance on navigating these obstacles and highlights the positive impact that grassroots efforts can have on fostering a collective sense of responsibility for the environment.

Keywords Community · Health · Environment · Engagement · Education · Sustainability

1 Introduction

In recent years, there has been a notable evolution in societal awareness regarding the intricate relationship between environmental sustainability and public health. Across the globe, various initiatives demonstrate this growing recognition and the concerted

A. Thakur
Division of Research and Development, Lovely Professional University, Phagwara, Punjab 144411, India

A. Kumar (✉)
Nalanda College of Engineering, Bihar Engineering University, Department of Science, Technology and Technical Education, Government of Bihar, Nalanda 803108, India
e-mail: drashishchemlpu@gmail.com

P. K. Prabhakar and W. Leal Filho (eds.), *Preserving Health, Preserving Earth*, World Sustainability Series, https://doi.org/10.1007/978-3-031-60545-1_5

efforts to address environmental challenges in healthcare settings and beyond (Trung Thanh et al. 2021; Dickinson 2005). For instance, Australia's Sustainable Hospitals Program focuses on reducing energy consumption, minimizing waste generation, and promoting environmentally friendly practices in healthcare facilities nationwide (Dickinson 2005; Boyd et al. 2023). Similarly, the Lancet Countdown on Health and Climate Change provides annual reports highlighting the health impacts of climate change, raising awareness among policymakers, healthcare professionals, and the public on a global scale (Ansari et al. 2024a). In Europe, countries like Germany, Sweden, and the Netherlands have established Green Hospitals Networks, fostering collaboration and knowledge sharing among hospitals committed to reducing their environmental footprint. In India, the National Clean Air Programme (NCAP) targets air pollution, a significant public health concern in many cities, through regulatory measures, technology adoption, and public awareness campaigns. Furthermore, the World Health Organization's Healthy Cities Program encourages urban planning and development initiatives worldwide to create healthier, more sustainable urban environments. This heightened recognition signifies a growing understanding of the profound impact that environmental factors exert on human health outcomes (Ansari et al. 2024b; Caban et al. 2019). As communities grapple with escalating environmental challenges such as pollution, climate change, and resource depletion, there has been a concurrent realization that these issues directly affect not only the health of ecosystems but also the well-being of individuals and communities. In response to this paradigm shift, healthcare practitioners have emerged as pivotal figures in the effort to address environmental concerns and promote sustainable practices. Leveraging their expertise in health promotion, disease prevention, and patient education, healthcare professionals are uniquely positioned to raise awareness about the health implications of environmental degradation. They play a central role in educating individuals and communities about how environmental factors, such as air and water quality, access to green spaces, and exposure to pollutants, can impact human health. Moreover, healthcare practitioners are increasingly recognized for their ability to advocate for eco-conscious behavior change within their communities (Moro 2022; Nandavar et al. 2023). By leveraging their trusted status and authoritative voice within the healthcare system, practitioners can effectively communicate the urgency of adopting sustainable practices to mitigate environmental risks and safeguard public health. Whether through patient consultations, community outreach programs, or policy advocacy efforts, healthcare professionals are instrumental in driving societal shifts towards more environmentally responsible behaviors and policies. With their extensive training and expertise in human health, they are well-equipped to effectively communicate the complex relationship between environmental factors and public health outcomes. This credibility enables healthcare practitioners to play a crucial role in educating individuals and communities about the significant impacts of environmental issues such as air and water pollution, climate change, and exposure to toxins.

By providing clear, evidence-based information, healthcare professionals empower individuals to understand the potential health risks associated with environmental degradation. They can explain how air pollution contributes to respiratory diseases, how contaminated water sources can lead to waterborne illnesses, and how climate change exacerbates health problems such as heat-related illnesses and the spread of infectious diseases (Chan et al. 2020). This education is essential for raising awareness and helping individuals recognize the importance of taking action to protect both their health and the environment. Furthermore, healthcare professionals empower individuals to make informed decisions and adopt behaviors that promote both personal well-being and environmental sustainability. By arming individuals with knowledge about the health impacts of their actions, healthcare practitioners enable them to make conscious choices that minimize their exposure to environmental hazards and reduce their ecological footprint (Zhou and Yuen 2024; Biehl et al. 2019). This may involve simple lifestyle changes such as reducing energy consumption, using eco-friendly products or advocating for cleaner air and water policies in their communities. Due to their direct interactions with patients and their involvement in community health initiatives, healthcare professionals are well-positioned to recognize patterns of environmental injustice and the disproportionate burden of environmental hazards borne by vulnerable populations.

By engaging with patients and communities, healthcare professionals can uncover the complex interplay between social determinants of health and environmental factors that contribute to health inequities. They may observe, for example, how low-income neighborhoods are disproportionately affected by industrial pollution or lack access to clean water and green spaces. These observations enable healthcare practitioners to understand the root causes of environmental health disparities and tailor interventions to address the specific needs of affected communities. Furthermore, healthcare professionals play a crucial role in advocating for solutions that promote environmental justice and equitable access to clean air, water, and green spaces (Biehl et al. 2019; Wu and Ma 2024; An et al. 2024). Through their advocacy efforts, they can raise awareness about environmental injustices, mobilize community support, and collaborate with policymakers to implement policies and initiatives that address systemic inequalities. This may involve advocating for stricter environmental regulations, investing in infrastructure improvements, or supporting community-driven initiatives to mitigate environmental hazards. Ultimately, the goal of healthcare professionals' efforts in addressing environmental health disparities is to improve overall community well-being and reduce disparities in health outcomes. By advocating for environmental justice and equitable access to resources, healthcare practitioners contribute to creating healthier, more resilient communities where all individuals have the opportunity to thrive (Peiró-Signes et al. 2022; Ahmad et al. 2024). Through their commitment to promoting health equity and environmental sustainability, healthcare professionals play a critical role in advancing social justice and building a more just and equitable society.

2 Understanding Eco-Conscious Health

2.1 Definition and Components of Eco-Conscious Health

Eco-conscious health refers to a holistic approach to health that recognizes the intricate interplay between environmental factors and human well-being. It emphasizes the importance of adopting sustainable practices and behaviors that not only promote personal health but also contribute to the health of the planet. Environmental awareness is a foundational aspect of eco-conscious health, encompassing individuals' mindfulness of their impact on the environment and their efforts to minimize their ecological footprint (Virginie et al. 2022; Parreño-Rodriguez et al. 2023). This awareness extends to various aspects of daily life, including resource consumption, waste generation, and the broader environmental consequences of everyday choices. Individuals with eco-conscious health strive to make informed decisions that prioritize environmental sustainability while also considering their well-being and that of future generations. Health promotion is another key pillar of eco-conscious health, which seeks to foster physical, mental, and emotional well-being while simultaneously advancing environmental sustainability. This holistic approach to health recognizes the interconnectedness between human health and the health of the planet (Moro 2022; González-Hidalgo et al. 2022). Activities that promote personal health, such as outdoor exercise, nutritious and sustainable diet choices, and stress reduction techniques, also contribute to environmental conservation and resilience. For instance, Bhutan's Gross National Happiness Index places environmental sustainability at the forefront of its development model, emphasizing the interconnectedness between human well-being and environmental health. Similarly, the Blue Zones Project draws inspiration from communities worldwide where longevity and vitality are common, promoting sustainable lifestyle practices such as plant-based diets and active living to enhance both individual health and environmental resilience. Sweden's Vision Zero Initiative and the Transition Towns Movement further showcase the integration of sustainability into various sectors, from road safety to community resilience, fostering environments that prioritize human and environmental health (Rahman et al. 2024). Additionally, organizations like the Wellbeing Economy Alliance advocate for economic systems that prioritize well-being over GDP growth, recognizing the essential link between a thriving society and a sustainable planet.

Central to eco-conscious health is the adoption of sustainability practices that align with principles of environmental stewardship and resource conservation. This involves a deliberate effort to reduce energy consumption, conserve water, and minimize waste generation through recycling and composting. Additionally, supporting the use of environmentally friendly products and services further contributes to the preservation of natural resources and the mitigation of environmental degradation. By integrating these practices into their daily lives, individuals not only reduce their ecological footprint but also serve as role models for sustainable living within their communities. Moreover, eco-conscious health emphasizes the importance of community engagement and collective action in addressing environmental challenges (Tien et al. 2024; Shimul et al. 2024). Recognizing that environmental issues are inherently social, individuals with eco-conscious health actively participate in community initiatives, advocate for sustainable policies, and collaborate with others to promote environmental stewardship. This may involve volunteering for local environmental organizations, participating in community clean-up events, or lobbying policymakers for stronger environmental regulations. By fostering a sense of shared responsibility and mobilizing collective efforts, communities can work together to create positive change and build a more sustainable future for all. Through collaboration and cooperation, individuals can amplify their impact and address environmental issues on a broader scale. Ultimately, eco-conscious health encourages a holistic approach to well-being that not only considers individual health but also recognizes the interconnectedness between human health and the health of the planet (Fouillouze et al. 2023; Brink and Wamsler 2019).

2.2 Relationship Between Environmental Factors and Public Health Outcomes

Environmental factors exert a profound influence on public health outcomes across various domains, encompassing air quality, water quality, climate change, and the built environment. In terms of air quality, the detrimental effects of poor air quality are evident, stemming from sources such as industrial pollution, vehicle emissions, and indoor air contaminants. Exposure to pollutants in the air can contribute to a range of health issues, including respiratory problems, cardiovascular diseases, and exacerbation of pre-existing conditions (Lozano 2012; Lam and Law 2019). This poses a significant public health concern, particularly in urban areas with high levels of pollution. In terms of air quality, cities like Delhi, India, and Beijing, China, have faced significant challenges due to high levels of air pollution resulting from industrial emissions, vehicular traffic, and agricultural burning. This pollution has been linked to respiratory illnesses, cardiovascular diseases, and premature mortality among residents. Similarly, in regions like Sub-Saharan Africa, access to clean water remains a pressing issue, with millions of people at risk of waterborne diseases due to contaminated water sources (Akenji 2014; Liu et al. 2023). The lack of proper sanitation and infrastructure exacerbates these challenges, leading to outbreaks of diseases such as cholera and dysentery. In Latin America, deforestation and urbanization have contributed to habitat loss and the spread of vector-borne diseases such as malaria and dengue fever. Similarly, water quality plays a crucial role in shaping public health outcomes, as contaminated water sources can pose serious health risks. Waterborne diseases such as cholera, dysentery, and gastrointestinal infections are prevalent in communities where access to clean water is limited or compromised. This disproportionately affects underserved populations, highlighting the importance of ensuring access to safe and clean drinking water for all.

Climate change represents another critical environmental factor that impacts public health on a global scale. The consequences of climate change are far-reaching and multifaceted, ranging from extreme weather events and heat-related illnesses to food and water shortages and the spread of infectious diseases (Zhou and Yuen 2024). Vulnerable populations, including children, the elderly, and those living in low-income communities, are particularly susceptible to the health impacts of climate change, underscoring the urgent need for mitigation and adaptation strategies. Furthermore, the built environment plays a significant role in shaping public health outcomes by influencing factors such as physical activity levels, access to healthy foods, transportation options, and exposure to environmental hazards. Urban planning and infrastructure design can either support or hinder health-promoting behaviors, with well-designed communities fostering active lifestyles and equitable access to resources, while poorly planned environments may exacerbate health inequities and increase vulnerability to environmental risks.

2.3 Importance of Sustainable Practices for Individual and Community Well-Being

Adopting sustainable practices is imperative for promoting both individual and community well-being due to several compelling reasons. Firstly, many sustainable practices offer direct health benefits that contribute to overall well-being. For instance, adopting a plant-based diet rich in fruits, vegetables, and whole grains has been associated with a reduced risk of chronic diseases such as heart disease, diabetes, and certain types of cancer (Chakraborty et al. 2022; Hargis et al. 2021). Similarly, engaging in active transportation, such as walking or cycling, not only reduces carbon emissions but also promotes physical activity, leading to improved cardiovascular health and mental well-being. Additionally, reducing exposure to environmental toxins through sustainable practices like using eco-friendly cleaning products and avoiding harmful chemicals can protect individuals from respiratory issues, allergies, and other adverse health effects. The United Nations Sustainable Development Goals (SDGs) provide a comprehensive framework for countries and organizations to prioritize sustainable practices that promote human health and environmental sustainability. Within this framework, SDG 3 focuses on ensuring healthy lives and well-being for all, while SDG 12 targets sustainable consumption and production patterns (Caban et al. 2019; Rasiah and Ren 2023). Additionally, the Paris Agreement, signed by nearly 200 countries, aims to combat climate change by reducing greenhouse gas emissions and transitioning to renewable energy sources, thereby benefiting both public health and the environment. The World Health Organization (WHO) Healthy Cities Initiative promotes urban planning strategies that prioritize health and well-being, such as sustainable transportation and access to green spaces. Furthermore, initiatives like the Circular Economy model and the Global Alliance for Clean Cookstoves promote resource efficiency and clean technologies to reduce waste generation, mitigate pollution, and improve public health (Sinkovics et al. 2014; Khmara and Kronenberg 2023). Moreover, sustainable practices play a crucial role in environmental protection by mitigating environmental degradation and preserving natural resources. By minimizing the impact of human activities on ecosystems, sustainable practices help maintain biodiversity, preserve habitats, and ensure the availability of clean air, water, and food for future generations. This environmental stewardship is essential for safeguarding the planet's health and resilience, which in turn directly impacts human health and well-being.

Furthermore, adopting sustainable practices promotes social equity by addressing disparities in resource distribution and ensuring that vulnerable populations have access to essential services and resources. Sustainable initiatives such as renewable energy projects, green infrastructure, and urban greenspaces can benefit underserved communities by improving air quality, reducing urban heat island effects, and enhancing overall quality of life. By prioritizing social equity in sustainability efforts, communities can work towards creating more inclusive and resilient environments where everyone has the opportunity to thrive (Vogel et al. 2023). Lastly, investing in sustainability initiatives can stimulate economic growth, create new job opportunities, and reduce healthcare costs associated with environmental pollution and climate-related disasters. Transitioning to renewable energy sources, implementing energy-efficient technologies, and promoting sustainable agriculture practices not only reduce environmental impact but also generate economic benefits by saving money on energy bills, creating new markets for green products and services, and reducing reliance on costly healthcare interventions for environmental-related health problems.

3 Strategies for Effective Community Engagement

3.1 Tailored Approaches Considering Cultural, Socioeconomic, and Demographic Factors:

Tailored approaches are essential for effective community engagement because they recognize the diversity within communities and acknowledge that one-size-fits-all solutions may not be suitable. By considering cultural, socioeconomic, and demographic factors, tailored approaches can:

- Cultural Sensitivity: Tailored approaches take into account the cultural norms, values, and beliefs that shape individuals' attitudes and behaviors related to environmental health. By recognizing and respecting diverse cultural perspectives, interventions can be designed to be culturally relevant and acceptable within different cultural groups (Rose 2014; Pafi et al. 2020). This fosters trust, engagement, and cooperation, ultimately enhancing the effectiveness of community initiatives. For instance, indigenous environmental health programs often integrate traditional healing practices and ceremonies, collaborating closely with local elders and healers to incorporate indigenous knowledge into interventions. Similarly, in multicultural societies, outreach materials may be created in languages commonly spoken by immigrant or minority communities to ensure accessibility and understanding. Additionally, considering religious and spiritual beliefs, tailored approaches may involve working with religious leaders to promote environmental stewardship within faith communities. Cultural events and festivals provide another opportunity for engagement, where environmental health initiatives can set up booths or workshops to connect with communities in familiar settings (Irawan and Benius 2022). Furthermore, community-based participatory research allows for the co-creation of interventions that address community-specific needs and preferences, ensuring that initiatives are both effective and culturally sensitive.

- Socioeconomic Considerations: Tailored approaches in environmental health initiatives recognize the significant socioeconomic disparities that exist within communities and aim to address these barriers to engagement. In low-income communities, individuals often face economic challenges that hinder their ability to participate in environmental health initiatives. Tailored interventions seek to overcome these barriers by providing targeted resources, support, and opportunities that meet the unique needs of these communities. For example, initiatives may offer financial incentives or subsidies to make participation more accessible, such as providing discounts on energy-efficient appliances or offering grants for home weatherization projects (Wuyts 2022; Kamboj et al. 2023). Additionally, access to affordable resources, such as community gardens or green spaces, can enhance the quality of life and promote environmental well-being in economically disadvantaged areas.

Furthermore, tailored approaches may provide assistance with navigating bureaucratic processes related to environmental regulations or community initiatives. Many low-income communities lack access to information and resources to address environmental concerns effectively. By offering support and guidance, tailored interventions empower community members to advocate for their needs and participate in decision-making processes that affect their environmental health. By addressing the economic challenges faced by disadvantaged communities, tailored interventions strive to promote equity and inclusivity within environmental health efforts (Brons et al. 2022; Cano et al. 2023; Diesch and Massari 2024). By ensuring that all members of the community have the opportunity to participate and benefit from environmental initiatives, tailored approaches contribute to building more resilient and sustainable communities. Ultimately, by addressing socioeconomic disparities, tailored interventions promote social equity and advance environmental justice goals, creating healthier and more vibrant communities for all.

- Demographic Diversity: Tailored approaches in environmental health initiatives acknowledge the diverse demographic makeup of communities and the varying needs and preferences of different groups (An et al. 2024). Recognizing that age, gender, ethnicity, and education level can significantly influence attitudes and behaviors towards environmental health, interventions are customized to cater to these differences. This involves adapting communication strategies, outreach methods, and programmatic interventions to effectively engage and resonate with specific demographic groups. For instance, initiatives targeting children may incorporate interactive educational materials, hands-on activities, and school-based programs to make learning about environmental health fun and engaging. By integrating environmental education into the curriculum and using age-appropriate teaching methods, tailored interventions can instill lifelong habits of environmental stewardship and promote a sense of responsibility towards the planet among young learners.

On the other hand, outreach efforts directed at older adults may focus on accessible community events, workshops, or gatherings held in senior centers or retirement communities. These initiatives may emphasize the importance of environmental health in relation to aging populations, such as promoting strategies for maintaining a healthy home environment or adapting to climate-related challenges. Furthermore, tailored approaches may consider gender-specific perspectives and preferences in designing interventions (Belal and Cooper 2011; Paillé et al. 2018). For example, initiatives aimed at women may address issues related to maternal and child health, household air quality, or sustainable household practices. By acknowledging and addressing the unique needs and concerns of different demographic groups, tailored approaches maximize relevance, engagement, and impact within the community. Ultimately, by customizing interventions to meet the specific characteristics and preferences of each group, environmental health initiatives can effectively promote behavior change, foster community empowerment, and achieve meaningful and sustainable outcomes.

3.2 *Importance of Clear Communication and Collaboration with Community Leaders:*

Clear communication and collaboration with community leaders are critical for building trust, fostering meaningful relationships, and mobilizing community members toward shared goals. Key considerations include:

- Building Trust: Establishing trust is foundational to effective community engagement. Healthcare practitioners and other stakeholders must demonstrate transparency, integrity, and a genuine commitment to listening to and addressing community concerns. This involves actively engaging with community members, acknowledging their perspectives, and ensuring that communication is honest, respectful, and responsive (Batat 2020). By fostering an environment of trust and mutual respect, healthcare practitioners can create a solid foundation for collaborative partnerships and meaningful engagement.

- Fostering Partnerships: Collaboration with community leaders, organizations, and influencers is essential for enhancing the reach and impact of engagement efforts. By partnering with trusted community entities, such as local nonprofits, religious institutions, or civic groups, healthcare practitioners can leverage existing networks and resources to amplify their message and engage a broader segment of the community. These partnerships enable healthcare practitioners to tap into local expertise, cultural knowledge, and community assets, leading to the co-creation of solutions that are tailored to meet the specific needs and priorities of the community.
- Empowering Community Voices: Community leaders play a pivotal role in amplifying community voices, advocating for change, and mobilizing grassroots support (Jones et al. 2021; Yang and Arhonditsis 2022). By empowering community leaders to take ownership of engagement initiatives, healthcare practitioners can ensure that interventions are culturally appropriate, locally relevant, and sustainable over the long term. This may involve providing training, resources, and support to community leaders, enabling them to effectively represent the interests and concerns of their constituents and serve as effective advocates for positive change within their communities.

3.3 Utilization of Various Platforms for Engagement:

Engaging communities effectively requires a multifaceted approach that leverages a diverse range of platforms and communication channels to reach different audiences and address their varying needs and preferences. Some key platforms and communication channels include:

- Workshops and Seminars: In-person workshops and seminars provide valuable opportunities for interactive learning, skill-building, and community dialogue. These events offer a platform for sharing knowledge, exchanging ideas, and fostering connections among participants. Facilitating face-to-face interactions, workshops and seminars create space for meaningful engagement, collaboration, and relationship-building within the community.
- Online Resources: Digital platforms, including websites, blogs, and online forums, offer a convenient and accessible way to disseminate information, resources, and tools to a wide audience. Online resources are particularly valuable for reaching diverse communities, including those in remote or underserved areas, as they can be accessed anytime, anywhere, with an internet connection (Oers et al. 2018; Bokun and Nazarko 2023). Additionally, online resources are scalable and cost-effective, making them an efficient means of delivering educational content and fostering community engagement on environmental health issues.
- Social Media Campaigns: Social media platforms such as Facebook, Twitter, Instagram, and TikTok provide powerful tools for engaging communities, raising awareness, and mobilizing action. Social media campaigns can leverage compelling visuals, storytelling, and interactive content to spark conversations,

inspire behavior change, and build community support for environmental health initiatives (Lapinski et al. 2015). By harnessing the widespread reach and influence of social media, organizations and healthcare practitioners can effectively amplify their message and engage with diverse audiences in real time.

- Community Events and Outreach Programs: Community events, such as health fairs, festivals, and neighborhood gatherings, offer opportunities to engage with community members in a more informal and interactive setting. These events provide a platform for disseminating information, conducting screenings or demonstrations, and connecting individuals with local resources and services. Additionally, outreach programs, such as door-to-door canvassing or community health education sessions, enable organizations to directly engage with residents, build trust, and address specific community needs and concerns.

4 Integration of Environmental Awareness into Healthcare Curricula

4.1 Potential Benefits of Incorporating Environmental Education into Healthcare Training Programs:

- Increased Awareness: By integrating environmental education into healthcare training programs, future professionals gain a deeper understanding of the links between environmental factors and human health. This heightened awareness enables them to recognize and address environmental determinants of health in clinical practice.
- Improved Patient Care: Healthcare professionals who receive environmental education are better equipped to recognize and address environmental risks that may impact patient health. By understanding the link between environmental factors and health outcomes, they can conduct more comprehensive assessments, provide more accurate diagnoses, and develop personalized treatment plans tailored to individual patients' needs (Svyrydenko and Revin 2024; Connell 2011). This results in improved health outcomes and a higher quality of care for patients, as healthcare practitioners can effectively mitigate environmental risks and address underlying contributors to illness.
- Preventive Medicine: Environmental education empowers healthcare professionals to adopt a preventive approach to healthcare, focusing on addressing the root causes of disease rather than just treating symptoms. By emphasizing prevention and promoting healthy environmental behaviors, healthcare practitioners can reduce the incidence of environmentally related illnesses and lower healthcare costs over time (Axon et al. 2018; Bianchi and Gonzalez 2021; Ravindra et al. 2023). This proactive approach not only improves individual health but also contributes to the overall well-being of communities by reducing the burden of preventable diseases and promoting long-term health and sustainability.

- Interdisciplinary Collaboration: Environmental education fosters collaboration between healthcare professionals and experts from other disciplines, such as environmental science, public health, and policy. This interdisciplinary approach enables holistic solutions to complex environmental health challenges, as healthcare practitioners work alongside specialists to identify and implement effective interventions. By leveraging diverse perspectives and expertise, interdisciplinary collaboration promotes innovation in healthcare delivery and enhances the effectiveness of environmental health initiatives.
- Advocacy and Policy Influence: Healthcare professionals who receive environmental education are better equipped to advocate for policies and practices that promote environmental sustainability and public health. By understanding the scientific evidence linking environmental factors to health outcomes, they can effectively communicate the importance of environmental protection and advocate for legislative changes, community interventions, and institutional reforms that address environmental health disparities and protect vulnerable populations (Karduri 2023; Zhao and Abeysekera 2023; Reading 2019). Through their advocacy efforts, healthcare professionals can influence policy decisions, shape public discourse, and mobilize support for initiatives that promote a healthier and more sustainable future for all.

4.2 Empowering Future Professionals to Integrate Sustainability Principles into Their Practices:

Environmental education for future healthcare professionals extends beyond theoretical knowledge to promote practical skills and values that contribute to sustainable healthcare practices. Here's an elaboration on some of the key benefits:

- Behavior Change: Environmental education empowers future healthcare professionals to adopt sustainable practices in their clinical settings, such as reducing energy consumption, minimizing waste generation, and promoting eco-friendly alternatives. By modeling sustainable behaviors, they can inspire patients, colleagues, and community members to adopt similar practices, creating a ripple effect that extends beyond individual healthcare settings and contributes to broader environmental stewardship (Walsh 2018; Hameed et al. 2019).
- Innovation and Research: Integrating sustainability principles into healthcare training programs encourages innovation and research in environmentally sustainable healthcare practices. Future professionals are encouraged to explore new technologies, interventions, and best practices that minimize environmental impact while optimizing patient care. By fostering a culture of innovation and inquiry, environmental education stimulates advancements in green healthcare solutions that benefit both patients and the planet.
- Ethical Responsibility: Environmental education underscores the ethical responsibility of healthcare professionals to consider the environmental consequences of

their actions and decisions. By prioritizing sustainability in their practices, future professionals uphold ethical principles of beneficence, non-maleficence, and justice, ensuring the well-being of both patients and the planet. This ethical framework guides healthcare professionals to make informed decisions that balance the needs of individuals, communities, and the environment.

- Leadership and Advocacy: Environmental education cultivates leadership skills and advocacy capabilities among future healthcare professionals (Gutschow et al. 2021; Vedeld et al. 2019). By empowering them to be agents of change within their organizations and communities, environmental education enables future professionals to influence policies, implement sustainable practices, and drive systemic change toward more environmentally responsible healthcare systems. By serving as advocates for environmental health, future professionals amplify the importance of sustainability in healthcare and contribute to shaping a more sustainable future for all.

4.3 Examples of Successful Integration Initiatives and Their Impact:

- Green Hospitals: Healthcare facilities globally are recognizing the importance of sustainability and are implementing various practices to reduce their environmental footprint while promoting the well-being of patients and staff. One example of this is energy-efficient building design. Hospitals are investing in sustainable infrastructure by incorporating features such as solar panels, energy-efficient lighting, and smart building systems. For instance, the Sutter Health Eden Medical Center in California utilizes solar panels to generate renewable energy, significantly reducing its reliance on traditional energy sources and lowering its carbon emissions. Waste reduction and recycling programs are another essential aspect of sustainability in healthcare facilities (Nugroho 2015; Bryan-Kinns et al. 2022; Durand et al. 2023). Hospitals produce a significant amount of waste, including medical supplies, packaging materials, and food waste. By implementing recycling programs and initiatives to reduce single-use plastics, hospitals can minimize the amount of waste sent to landfills and promote a circular economy. For example, the University of California San Francisco Medical Center has implemented a comprehensive recycling program that diverts over 50% of its waste from landfills, significantly reducing its environmental impact.

Green procurement policies are also gaining traction in healthcare facilities, with hospitals increasingly prioritizing the purchase of environmentally friendly products and materials. This includes sourcing products made from recycled materials, using eco-friendly cleaning supplies, and selecting equipment with minimal environmental impact. For instance, the Cleveland Clinic has implemented a green purchasing policy that prioritizes environmentally preferable products and services, reducing the hospital's overall environmental footprint (Buerkle et al. 2023). In addition to environmental benefits, green hospitals prioritize patient and staff health by creating healing environments that enhance comfort and indoor air quality. For example, hospitals may incorporate natural lighting, indoor plants, and healing gardens to promote relaxation and stress reduction. The Royal Jubilee Hospital in British Columbia, Canada, features a rooftop healing garden that provides patients, staff, and visitors with access to nature, promoting healing and well-being (Ansari et al. 2024b; Peruzzo et al. 2022).

- Climate and Health Education Programs: Universities and medical schools are increasingly acknowledging the urgent need to prepare future healthcare professionals to address the complex health impacts of climate change. One notable trend is the integration of climate and health education into curricula, offering interdisciplinary training that spans environmental science, public health, and clinical medicine. For example, the Harvard T.H. Chan School of Public Health offers a course titled "Climate Change and Health: From Science to Action," which explores the scientific evidence linking climate change to health outcomes and examines strategies for adaptation and mitigation (Villalba-Eguiluz et al. 2023; Bronson 2019). These educational programs aim to equip students with the knowledge and skills necessary to navigate the intricate relationship between environmental factors and human health. For instance, students learn about the physiological effects of climate change, such as heat-related illnesses, respiratory conditions exacerbated by air pollution, and the spread of vector-borne diseases. By understanding these connections, future healthcare professionals can better assess climate-related health risks and develop targeted interventions to protect vulnerable populations.

Moreover, climate and health education programs emphasize the importance of collaboration and interdisciplinary approaches in addressing climate-related health challenges. Students engage in case studies and group projects that require them to work across disciplines to develop holistic solutions. For example, the University of California, San Francisco, offers a Climate Changes Health course that brings together students from medicine, nursing, public health, and environmental science to explore the intersections of climate change and health and develop advocacy strategies. By educating the next generation of healthcare professionals about the health impacts of climate change, these programs empower students to become advocates for climate-resilient healthcare systems and communities (Gu et al. 2023; Jaciow and Wolny 2021). Armed with knowledge and skills in climate science, public health, and clinical medicine, graduates are prepared to lead initiatives that promote climate adaptation, mitigation, and resilience in healthcare practice and policy. Ultimately, integrating climate and health education into curricula ensures that healthcare professionals are equipped to address one of the most pressing challenges of our time and safeguard the health and well-being of future generations.

- Environmental Health Clinics: Several healthcare institutions have recognized the importance of addressing environmental health concerns and have established specialized clinics dedicated to this purpose. One such example is the Environmental Health Clinic at Mount Sinai Hospital in New York City. This clinic offers personalized assessments and interventions for patients with environmentally related health concerns, providing a comprehensive approach to addressing the complex relationship between environmental exposures and health outcomes. At these clinics, patients undergo thorough evaluations that include environmental exposure assessments and risk factor analyses. For instance, patients may be asked about their living and work environments, potential exposures to toxins or pollutants, and any symptoms or health issues they may be experiencing (Al Mamun et al. 2023; Xu et al. 2024). Clinicians utilize this information to identify environmental triggers of illness and develop tailored treatment plans that address both the underlying health condition and the environmental factors contributing to it.

For example, if a patient presents with asthma symptoms, clinicians at an environmental health clinic may investigate potential indoor air quality issues in the patient's home, such as mold or allergen exposure. They may then recommend remediation measures or lifestyle modifications to reduce exposure and improve respiratory health. Additionally, environmental health clinics offer preventive guidance to patients, empowering them to minimize future exposures and protect their health. Furthermore, environmental health clinics serve as hubs for interdisciplinary collaboration, bringing together experts from various fields to provide holistic care. For instance, clinicians may collaborate with environmental scientists, toxicologists, industrial hygienists, and community health workers to address patients' environmental health concerns comprehensively (Alomari et al. 2023; McKinnon et al. 2022). This interdisciplinary approach allows for a more nuanced understanding of the complex interplay between environmental factors and human health and ensures that patients receive well-rounded care that considers all aspects of their health and environment.

- Community Engagement Initiatives: Healthcare organizations are recognizing the importance of community engagement in addressing environmental health issues and are increasingly forging partnerships with community groups, environmental organizations, and local governments. One example of this collaboration is the Community Health and Environmental Partnership (CHEP) program implemented by Kaiser Permanente in California. CHEP brings together healthcare professionals, community leaders, and environmental advocates to address environmental health disparities in underserved communities. Through initiatives like CHEP, healthcare organizations aim to raise awareness about environmental health risks and build community capacity to address these challenges (Shah et al. 2024). For instance, they may organize outreach events such as health fairs or community forums to educate residents about the links between environmental exposures and health outcomes. These events provide opportunities for community members to learn about potential hazards in their environment and access resources and support to mitigate risks. Educational workshops are another key component of community engagement initiatives led by healthcare organizations. These workshops cover topics such as indoor air quality, lead poisoning prevention, and safe drinking water, providing residents with practical information and tools to protect their health. For example, the Children's Environmental Health Network (CEHN) collaborates with healthcare providers to host workshops on childhood lead poisoning prevention in communities with high rates of lead exposure. In addition to raising awareness and providing education, healthcare organizations engage communities in collaborative projects aimed at addressing environmental health challenges. For instance, they may partner with local schools or community centers to implement environmental monitoring programs or green space initiatives (Chenavaz and Dimitrov 2024; Rapada et al. 2023). These projects not only improve environmental quality but also promote community empowerment and resilience. Furthermore, healthcare organizations advocate for policies and practices that protect public health and the environment at the local, state, and national levels. They work alongside community members to identify priority issues, develop evidence-based solutions, and advocate for policy changes. For example, the American Public Health Association (APHA) collaborates with community-based organizations to advocate for clean air and water regulations and promote environmental justice.

5 Challenges and Solutions in Promoting Eco-Conscious Health

Identifying barriers such as resistance to change and information overload poses significant challenges for healthcare professionals aiming to promote environmentally sustainable practices. Resistance to change often stems from inertia, perceived inconvenience, or cultural norms within healthcare settings, hindering the adoption of new practices (Li et al. 2024; Walker and Ngara Manyamba 2020). Additionally, information overload, fueled by the vast amount of available data on environmental issues, can overwhelm professionals, making it difficult to discern credible sources and prioritize actions effectively. Overcoming these obstacles requires targeted interventions, such as comprehensive education and training programs that equip professionals with the necessary skills and knowledge to navigate challenges and advocate for change. Moreover, stakeholder engagement plays a crucial role in building support for environmental initiatives, as involving colleagues, administrators, patients, and community members fosters collaboration and shared responsibility. Clear and transparent communication is essential for addressing information overload, ensuring that messages are accessible, concise, and tailored to the needs of diverse audiences (Lydia et al. 2022; Smith et al. 2021). By adopting a gradual and iterative approach to change and highlighting the benefits of sustainable practices, healthcare professionals can overcome resistance, inspire action, and drive long-term engagement.

Grassroots efforts play a pivotal role in fostering environmental responsibility and driving positive change within communities. By empowering individuals and mobilizing collective action, grassroots initiatives have the potential to create lasting impacts on environmental health and sustainability. Community empowerment lies at the heart of grassroots movements, as they enable individuals and communities to take ownership of environmental issues, drive local solutions, and advocate for change (Argüelles et al. 2017; Lorenzoni et al. 2007). Through community-led initiatives such as clean-up campaigns, educational workshops, and advocacy efforts, grassroots organizations inspire behavioral change and promote sustainable practices among community members. Moreover, grassroots movements have the power to influence policy decisions by mobilizing public support, raising awareness, and advocating for legislative and regulatory changes at various levels of government. By nurturing grassroots initiatives and supporting community-led solutions, healthcare professionals can contribute to building resilient, sustainable communities where everyone can thrive now and in the future (Amayi 2021; Wakkee et al. 2019).

6 Conclusion

The chapter underscores the pivotal role of community engagement and education in promoting eco-conscious health, emphasizing the importance of tailored approaches that consider cultural, socioeconomic, and demographic factors. It highlights healthcare practitioners as key advocates for environmental awareness within their communities, leveraging their expertise and influence to drive change and improve public health outcomes. Clear communication and collaboration with community leaders are identified as essential strategies for building trust, fostering partnerships, and mobilizing collective action. Integrating environmental education into healthcare training programs empowers future professionals to recognize and address environmental determinants of health, promoting preventive medicine and advocating for policy changes. A call to action urges healthcare practitioners to prioritize community engagement and education for eco-conscious health, empowering patients, collaborating with communities, and leading by example. Future directions for research and practice include evaluating the impact of community engagement initiatives, exploring technology and innovation, advocating for supportive policies, and fostering global collaboration to address environmental health challenges.

References

Ahmad M, Ahmed Z, Alvarado R, Hussain N, Khan SA (2024) Financial development, resource richness, eco-innovation, and sustainable development: does geopolitical risk matter? J Environ Manage 351:119824. https://doi.org/10.1016/j.jenvman.2023.119824

Akenji L (2014) Consumer scapegoatism and limits to green consumerism. J Clean Prod 63:13–23. https://doi.org/10.1016/j.jclepro.2013.05.022

Al Mamun A, Yang Q, Naznen F, Aziz NA, Masud MM (2023) Modelling the mass adoption potential of food waste composting among rural Chinese farmers. Heliyon 9:e18998. https://doi.org/10.1016/j.heliyon.2023.e18998

Alomari MM, EL-Kanj H, Topal A, Alshdaifat NI (2023) Exploring the impact of the COVID-19 pandemic on energy literacy and conservation behavior in academic buildings of Kuwait. Heliyon 9:e21474. https://doi.org/10.1016/j.heliyon.2023.e21474

Amayi MK (2021) Flower growing firms' contribution to community water use management in Naivasha Sub County, Kenya, Environ. Challenges 5:100330. https://doi.org/10.1016/j.envc.2021.100330

An S, Eck T, Woosnam KM, Jiang L (2024) Applying mindfulness theory to enhance voluntourism experiences. Tour Manag Perspect 50:101203. https://doi.org/10.1016/j.tmp.2023.101203

Ansari A, Dutt D, Kumar V (2024a) Catalyzing paradigm shifts in global waste management: A case study of Saharanpur Smart city. Waste Manag Bull 2:29–38. https://doi.org/10.1016/j.wmb.2023.12.003

Ansari A, Dutt D, Kumar V (2024b) Revolutionizing urban waste: eco-innovation for Saharanpur' s sustainability (Class-II urban Centers). Waste Manag Bull 2:184–190. https://doi.org/10.1016/j.wmb.2024.01.006

Argüelles L, Anguelovski I, Dinnie E (2017) Power and privilege in alternative civic practices: examining imaginaries of change and embedded rationalities in community economies. Geoforum 86:30–41. https://doi.org/10.1016/j.geoforum.2017.08.013

Axon S, Morrissey J, Aiesha R, Hillman J, Revez A, Lennon B, Salel M, Dunphy N, Boo E (2018) The human factor: classification of European community-based behaviour change initiatives. J Clean Prod 182:567–586. https://doi.org/10.1016/j.jclepro.2018.01.232

Batat W (2020) Pillars of sustainable food experiences in the luxury gastronomy sector: a qualitative exploration of Michelin-starred chefs' motivations. J Retail Consum Serv 57:102255. https://doi.org/10.1016/j.jretconser.2020.102255

Belal AR, Cooper S (2011) The absence of corporate social responsibility reporting in Bangladesh. Crit Perspect Account 22:654–667. https://doi.org/10.1016/j.cpa.2010.06.020

Bianchi C, Gonzalez M (2021) Exploring sustainable fashion consumption among eco-conscious women in Chile. Int Rev Retail Distrib Consum Res 31:375–392. https://doi.org/10.1080/09593969.2021.1903529

Biehl A, Chen Y, Sanabria-Véaz K, Uttal D, Stathopoulos A (2019) Where does active travel fit within local community narratives of mobility space and place? Transp Res Part A Policy Pract 123:269–287. https://doi.org/10.1016/j.tra.2018.10.023

Bokun K, Nazarko J (2023) Smart villages concept—a bibliometric analysis and state-of-the-art literature review. Prog Plann 175:100765. https://doi.org/10.1016/j.progress.2023.100765

Boyd C, Parr H, Philo C (2023) Climate anxiety as posthuman knowledge, Wellbeing. Sp Soc 4:100120. https://doi.org/10.1016/j.wss.2022.100120

Brink E, Wamsler C (2019) Citizen engagement in climate adaptation surveyed: the role of values, worldviews, gender and place. J Clean Prod 209:1342–1353. https://doi.org/10.1016/j.jclepro.2018.10.164

Brons A, van der Gaast K, Awuh H, Jansma JE, Segreto C, Wertheim-Heck S (2022) A tale of two labs: rethinking urban living labs for advancing citizen engagement in food system transformations. Cities 123:103552. https://doi.org/10.1016/j.cities.2021.103552

Bronson K (2019) Looking through a responsible innovation lens at uneven engagements with digital farming. NJAS - Wageningen J Life Sci 90–91:100294. https://doi.org/10.1016/j.njas.2019.03.001

Bryan-Kinns N, Wang W, Ji T (2022) Qi2He: a co-design framework inspired by eastern epistemology. Int J Hum Comput Stud 160:102773. https://doi.org/10.1016/j.ijhcs.2022.102773

Buerkle A, O'Dell A, Matharu H, Buerkle L, Ferreira P (2023) Recommendations to align higher education teaching with the UN sustainability goals – a scoping survey. Int J Educ Res Open 5:100280. https://doi.org/10.1016/j.ijedro.2023.100280

Caban J, Vrábel J, Šarkan B, Ignaciuk P (2019) About eco-driving, genesis, challenges and benefits, application possibilities. Transp Res Procedia 40:1281–1288. https://doi.org/10.1016/j.trpro.2019.07.178

Cano JA, Londoño-Pineda AA, Campo EA, Fernández SA (2023) Sustainable business models of e-marketplaces: An analysis from the consumer perspective. J Open Innov Technol Mark Compl 9:100121. https://doi.org/10.1016/j.joitmc.2023.100121
Chakraborty D, Siddiqui A, Siddiqui M, Mohmmad F, Alatawi H (2022) Exploring consumer purchase intentions and behavior of buying ayurveda products using SOBC framework. J Retail Consum Serv 65:102889. https://doi.org/10.1016/j.jretconser.2021.102889
Chan HL, Wei X, Guo S, Leung WH (2020) Corporate social responsibility (CSR) in fashion supply chains: a multi-methodological study. Transp Res Part E Logist Transp Rev 142:102063. https://doi.org/10.1016/j.tre.2020.102063
Chenavaz RY, Dimitrov S (2024) From waste to wealth: policies to promote the circular economy. J Clean Prod 443:141086. https://doi.org/10.1016/j.jclepro.2024.141086
Connell KYH (2011) Exploring consumers' perceptions of eco-conscious apparel acquisition behaviors. Soc Responsib J 7:61–73. https://doi.org/10.1108/17471111111114549
Dickinson TD (2005) Strengthening women's studies through applied activism: theoretical, classroom, regional, and cross-border strategies for participating in change. Womens Stud Int Forum 28:115–126. https://doi.org/10.1016/j.wsif.2005.04.002
Diesch A, Massari M (2024) Hidden gems: the potential of places and social innovation for circular territories in Bogotá. Cities 144:104645. https://doi.org/10.1016/j.cities.2023.104645
Durand C, Hofferberth E, Schmelzer M (2023) Planning beyond growth: the case for economic democracy within ecological limits. J Clean Prod 437:140351. https://doi.org/10.1016/j.jclepro.2023.140351
Fouillouze A, Lacoeuilhe A, Truong MXA (2023) A step towards a greener green? Investigating golfers' relationships with nature and attitudes about biodiversity conservation in golf courses. J Outdoor Recreat Tour 43. https://doi.org/10.1016/j.jort.2023.100659
González-Hidalgo M, Del Bene D, Iniesta-Arandia I, Piñeiro C (2022) Emotional healing as part of environmental and climate justice processes: Frameworks and community-based experiences in times of environmental suffering. Polit Geogr 98. https://doi.org/10.1016/j.polgeo.2022.102721
Gu MM, Huang CF, Lee CKJ (2023) Investigating university students' digital citizenship development through the lens of digital literacy practice: a Translingual and transemiotizing perspective. Linguist Educ 77. https://doi.org/10.1016/j.linged.2023.101226
Gutschow B, Gray B, Ragavan MI, Sheffield PE, Philipsborn RP, Jee SH (2021) The intersection of pediatrics, climate change, and structural racism: ensuring health equity through climate justice. Curr Probl Pediatr Adolesc Health Care 51:1–6. https://doi.org/10.1016/j.cppeds.2021.101028
Hameed I, Waris I, Amin ul Haq M (2019) Predicting eco-conscious consumer behavior using theory of planned behavior in Pakistan. Environ Sci Pollut Res, 15535–15547. https://doi.org/10.1007/s11356-019-04967-9
Hargis K, McKenzie M, LeVert-Chiasson I (2021) A whole institution approach to climate change education: preparing school systems to be climate proactive
Irawan A, Elia, Benius (2022) Interactive effects of citizen trust and cultural values on pro-environmental behaviors: a time-lag study from Indonesia. Heliyon 8:e09139. https://doi.org/10.1016/j.heliyon.2022.e09139
Jaciow M, Wolny R (2021) New technologies in the ecological behavior of generation Z. Procedia Comput Sci 192:4780–4789. https://doi.org/10.1016/j.procs.2021.09.256
Jones MS, Teel TL, Solomon J, Weiss J (2021) Evolving systems of pro-environmental behavior among wildscape gardeners, Landsc. Urban Plan 207:104018. https://doi.org/10.1016/j.landurbplan.2020.104018
Kamboj S, Matharu M, Gupta M (2023) Examining consumer purchase intention towards organic food: an empirical study. Clean Respon Consum 9:100121. https://doi.org/10.1016/j.clrc.2023.100121
Karduri RKR (2023) Incentives for green technologies and community engagement in decision-making. SSRN Elect J 10. https://doi.org/10.2139/ssrn.4636510

Khmara Y, Kronenberg J (2023) On the road to urban degrowth economics? Learning from the experience of C40 cities, doughnut cities, Transition Towns, and shrinking cities. Cities 136:104259. https://doi.org/10.1016/j.cities.2023.104259

Lam C, Law R (2019) Readiness of upscale and luxury-branded hotels for digital transformation. Int J Hosp Manag 79:60–69. https://doi.org/10.1016/j.ijhm.2018.12.015

Lapinski MK, Funk JA, Moccia LT (2015) Recommendations for the role of social science research in One Health. Soc Sci Med 129:51–60. https://doi.org/10.1016/j.socscimed.2014.09.048

Li Y, Liu Y, Solangi YA (2024) Analysis of factors and strategies for the implementation of sustainable tourism in a green economic structure in China. J Clean Prod 434:140011. https://doi.org/10.1016/j.jclepro.2023.140011

Liu M, Luo X, Lu WZ (2023) Public perceptions of environmental, social, and governance (ESG) based on social media data: evidence from China. J Clean Prod 387:135840. https://doi.org/10.1016/j.jclepro.2022.135840

Lorenzoni I, Nicholson-Cole S, Whitmarsh L (2007) Barriers perceived to engaging with climate change among the UK public and their policy implications. Glob Environ Chang 17:445–459. https://doi.org/10.1016/j.gloenvcha.2007.01.004

Lozano R (2012) Towards better embedding sustainability into companies' systems: an analysis of voluntary corporate initiatives. J Clean Prod 25:14–26. https://doi.org/10.1016/j.jclepro.2011.11.060

Lydia O, Godwin A, Isaac L (2022) We have done nothing wrong": Youth miners' perceptions of the environmental consequences of artisanal and small-scale mining (ASM) in Ghana. Extr Ind Soc 12:101179. https://doi.org/10.1016/j.exis.2022.101179

McKinnon S, Breakey S, Fanuele JR, Kelly DE, Eddy EZ, Tarbet A, Nicholas PK, Ros AMV (2022) Roles of health professionals in addressing health consequences of climate change in interprofessional education: a scoping review. J Clim Chang Heal 5:100086. https://doi.org/10.1016/j.joclim.2021.100086

Moro A (2022) Co-design of public spaces for pedestrian use and soft-mobility in the perspective of communities reappropriation and activation. Transp Res Procedia 60:36–43. https://doi.org/10.1016/j.trpro.2021.12.006

Nandavar S, Kaye SA, Senserrick T, Oviedo-Trespalacios O (2023) Exploring the factors influencing acquisition and learning experiences of cars fitted with advanced driver assistance systems (ADAS). Transp Res Part F Traffic Psychol Behav 94:341–352. https://doi.org/10.1016/j.trf.2023.02.006

Nugroho RL (2015) Public engagement through Ecopreneurship: moving from 'Recyclable to Marketable', Procedia - Soc. Behav Sci 184:223–231. https://doi.org/10.1016/j.sbspro.2015.05.083

Van Oers LM, Boon WPC, Moors EHM (2018) The creation of legitimacy in grassroots organisations: a study of Dutch community-supported agriculture. Environ Innov Soc Trans 29:55–67. https://doi.org/10.1016/j.eist.2018.04.002

Pafi M, Flannery W, Murtagh B (2020) Coastal tourism, market segmentation and contested landscapes. Mar Policy 121:104189. https://doi.org/10.1016/j.marpol.2020.104189

Paillé P, Amara N, Halilem N (2018) Greening the workplace through social sustainability among co-workers. J Bus Res 89:305–312. https://doi.org/10.1016/j.jbusres.2017.12.044

Parreño-Rodriguez A, Ramallo-González AP, Chinchilla-Sánchez M, Molina-García A (2023) Community energy solutions for addressing energy poverty: a local case study in Spain. Energy Build 296. https://doi.org/10.1016/j.enbuild.2023.113418

Peiró-Signes Á, Segarra-Oña M, Trull-Domínguez Ó, Sánchez-Planelles J (2022) Exposing the ideal combination of endogenous–exogenous drivers for companies' ecoinnovative orientation: results from machine-learning methods. Socioecon Plann Sci 79. https://doi.org/10.1016/j.seps.2021.101145

Peruzzo F, Ball SJ, Grimaldi E (2022) Peopling the crowded education state: heterarchical spaces, EdTech markets and new modes of governing during the COVID-19 pandemic. Int J Educ Res 114:102006. https://doi.org/10.1016/j.ijer.2022.102006

Rahman MH, Tanchangya T, Rahman J, Aktar MA, Majumder SC (2024) Corporate social responsibility and green financing behavior in Bangladesh: towards sustainable tourism. Innov Green Dev 3:100133. https://doi.org/10.1016/j.igd.2024.100133
Rapada MZ, Yu DE, Janairo JI, Yu KD (2023) Impact of online-based information and interaction to proenvironmental behavior on plastic pollution. Clean Responsible Consum 9:100126. https://doi.org/10.1016/j.clrc.2023.100126
Rasiah R, Ren Y (2023) Sustainable management of a leading Chinese telecommunication multinational: a case study of company X in host country Malaysia. Clean Responsible Consum 8:100092. https://doi.org/10.1016/j.clrc.2022.100092
Ravindra K, Goyal A, Mor S (2023) Lifestyle for Environment (LiFE): a global initiative to fight against climate change through community engagement and lifestyle modification. Lancet Reg Heal Southeast Asia 15:100238. https://doi.org/10.1016/j.lansea.2023.100238
Reading M (2019) The anuvrat movement: a case study of jain-inspired ethical and eco-conscious living. Religions 10:1–19. https://doi.org/10.3390/rel10110636
Rose K (2014) Adopting industrial organizational psychology for eco sustainability, procedia. Environ Sci 20:533–542. https://doi.org/10.1016/j.proenv.2014.03.066
Shah P, Mehta N, Shah S (2024) Exploring the factors that drive millet consumption: insights from regular and occasional consumers. J Retail Consum Serv 76:103598. https://doi.org/10.1016/j.jretconser.2023.103598
Shimul AS, Faroque AR, Teah K, Azim SMF, Teah M (2024) Enhancing consumers' intention to stay in an eco-resort via climate change anxiety and connectedness to nature. J Clean Prod 442:141096. https://doi.org/10.1016/j.jclepro.2024.141096
Sinkovics N, Sinkovics RR, Yamin M (2014) The role of social value creation in business model formulation at the bottom of the pyramid – Implications for MNEs? Int Bus Rev 23:692–707. https://doi.org/10.1016/j.ibusrev.2013.12.004
Smith TSJ, Baranowski M, Schmid B (2021) Intentional degrowth and its unintended consequences: Uneven journeys towards post-growth transformations. Ecol Econ 190:107215. https://doi.org/10.1016/j.ecolecon.2021.107215
Svyrydenko D, Revin F (2024) Embracing the philosophy of responsible citizenship: on a path to Ukraine's societal transformation. Stud Warm 60:119–134. https://doi.org/10.31648/sw.9567
Tien ND, Lam Duyen TN, Thanh Huyen NT, Anh PQ, Oanh NT, Van Tich V, Dat DT, Hong Hanh NT, Trang VH (2024) Community-based ecotourism for sustainability: an evaluative analysis of Binh Son district. Quang Ngai province in Vietnam. Soc Sci Humanit Open 9:100807. https://doi.org/10.1016/j.ssaho.2024.100807
Trung Thanh H, Tschakert P, Hipsey MR (2021) Moving up or going under? Differential livelihood trajectories in coastal communities in Vietnam, World Dev 138:105219. https://doi.org/10.1016/j.worlddev.2020.105219
Vedeld T, Mathur M, Bharti N (2019) How can co-creation improve the engagement of farmers in weather and climate services (WCS) in India. Clim Serv 15:100103. https://doi.org/10.1016/j.cliser.2019.100103
Villalba-Eguiluz U, Sahakian M, González-Jamett C, Etxezarreta E (2023) Social and solidarity economy insights for the circular economy: limited-profit and sufficiency. J Clean Prod 418. https://doi.org/10.1016/j.jclepro.2023.138050
Virginie A, Roberta D, Jean-Loup L, Gun R, Barbara T, Matthew G, Beatrice B, Davide M, Duboys de Labarre M, Jelena F, Kathrin M, Áron T, Saša V, Pierre W, Vilde H, Péter C, Johannes S, Galjina O (2022) European food quality schemes in everyday food consumption: an exploration of sayings and doings through pragmatic regimes of engagement. J Rural Stud 95:336–349. https://doi.org/10.1016/j.jrurstud.2022.09.009
Vogel SM, Vasudev D, Ogutu JO, Taek P, Berti E, Goswami VR, Kaelo M, Buitenwerf R, Munk M, Li W, Wall J, Chala D, Amoke I, Odingo A, Svenning JC (2023) Identifying sustainable coexistence potential by integrating willingness-to-coexist with habitat suitability assessments. Biol Conserv 279:109935. https://doi.org/10.1016/j.biocon.2023.109935

Wakkee I, van der Sijde P, Vaupell C, Ghuman K (2019) The university's role in sustainable development: activating entrepreneurial scholars as agents of change. Technol Forecast Soc Change 141:195–205. https://doi.org/10.1016/j.techfore.2018.10.013

Walker J, Ngara Manyamba V (2020) Towards an emotion-focused, discomfort-embracing transformative tourism education. J Hosp Leis Sport Tour Educ 26:100213. https://doi.org/10.1016/j.jhlste.2019.100213

Walsh EA (2018) White fragility as an obstacle to anti-racist resilience planning: opportunities for equity-conscious partnerships. J Urban Manag 7:181–189. https://doi.org/10.1016/j.jum.2018.12.005

Wu Z, Ma J (2024) Chinese rural residents' identity construction with tourism intervention. Tour Manag Perspect 51:101218. https://doi.org/10.1016/j.tmp.2024.101218

Wuyts W (2022) An autoethnography about writing an eco-fiction on the Flemish circular economy. Futures 142:103000. https://doi.org/10.1016/j.futures.2022.103000

Xu Y, Nassani AA, Qazi Abro MM, Naseem I, Zaman K (2024) FinTech revolution in mineral management: exploring the nexus between technology adoption and sustainable Resource utilization in an industry 4.0 context. Heliyon 10:e24641. https://doi.org/10.1016/j.heliyon.2024.e24641

Yang C, Arhonditsis GB (2022) What are the primary covariates of environmental attitudes and behaviours in Canada? A national-scale analysis of socioeconomic, political, and demographic factors. Ecol Inform 69:101661. https://doi.org/10.1016/j.ecoinf.2022.101661

Zhao M, Abeysekera I (2023) The behaviour of Fintech users in the Alipay ant forest platform towards environmental protection. J Open Innov Technol Mark Complex 10:100201. https://doi.org/10.1016/j.joitmc.2023.100201

Zhou Y, Yuen KF (2024) A Bayesian network model for container shipping companies' organisational sustainability risk management. Transp Res Part D Transp Environ 126:103999. https://doi.org/10.1016/j.trd.2023.103999

Human Health in the Face of Extreme Weather: An Analysis of Impacts and Implications

Geeta Arora, Vandna Chhabra, and K. Akshit

Abstract Extreme weather events, such as droughts, floods, heatwaves, and storms, are becoming increasingly frequent and severe due to climate change. These events have profound implications for food security, human health and a sustainable environment. This present work is a comprehensive analysis of the impact of extreme weather events on human health. The study investigates the various dimensions of extreme weather events and their cascading effects. It examines the vulnerabilities and risks associated with different types of extreme weather events and evaluates their specific impacts on vector-borne diseases, malnutrition, depression, and lack of interest in personal hygiene. Furthermore, the study explores the socio-economic consequences of these events, including market disruptions, and increased food and health insecurity among vulnerable populations with low income. The findings and recommendations presented in this analysis provide valuable insights for policymakers, researchers, and practitioners in designing effective strategies to mitigate the adverse effects of extreme weather events on human health and build resilient health systems in expected climatic aberrations. Moreover, this is an effort to achieve sustainable goals naming good health (SDG 3), climate change (SDG 13) and life on land (SDG 15).

Keywords Climate · Extreme events · Human · Health · Sustainability

G. Arora (✉)
Department of Mathematics, Lovely Professional University, Phagwara, India
e-mail: geetadma@gmail.com

V. Chhabra
Department of Agriculture, Lovely Professional University, Phagwara, India

K. Akshit
Chintpurni Medical College, Pathankot, India

P. K. Prabhakar and W. Leal Filho (eds.), *Preserving Health, Preserving Earth*,
World Sustainability Series, https://doi.org/10.1007/978-3-031-60545-1_6

1 Introduction

The greatest threat to human health is climate change, and medical professionals are already taking action to mitigate its negative effects on health around the world. According to the Intergovernmental Panel on Climate Change (IPCC), the global temperature rise must be kept to 1.5 °C to avoid catastrophic health effects and millions of fatalities brought on by climate change. A certain amount of global temperature rises, and other climatic impacts are already unavoidable due to past emissions. Even 1.5 °C of global warming is not seen to be safe, and every additional tenth of a degree of warming will have a significant negative impact on people's lives and health. Everyone faces these risks, but those in low-income and underdeveloped nations and communities are those whose health is being harmed by the climate crisis first and worst. They also contribute the least to its causes and are least able to defend themselves and their families from it.

Although climate change influences human health, it is still difficult to predict the scope and severity of many climate-sensitive health hazards. But as science progresses, we can increasingly link an uptick in sickness and mortality to human-caused global warming and more precisely assess the severity of these health problems.

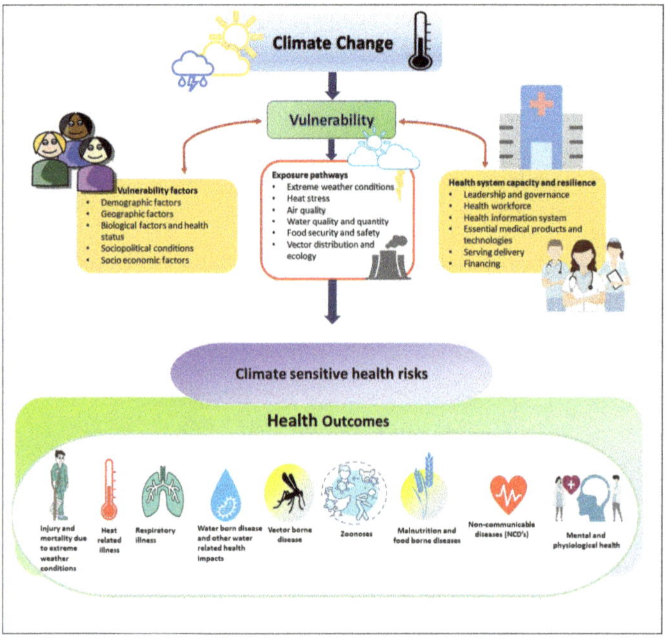

An overview of climate change's health concerns, exposure mechanisms, and risk factors.

2 Climate Change and Health

Epidemiologists were not particularly interested in climate-health relationships before the possibility of anthropogenic climate change surfaced. Contemporary epidemiology has mostly studied risk factors for noncommunicable illnesses in individuals rather than populations. In the interim, there have been sporadic studies looking at heat-related deaths, some epidemiological studies of air pollution that include temperature as a covariate, and a continuation of the longer-standing research interest in meteorological effects on microbes, vectors, and infectious disease transmission. The health hazards associated with thermal stress brought on by climate change, floods, and infectious diseases have generally been the most amenable to traditional epidemiological investigations.

3 Extreme Weather Events and Diseases

Human health is impacted by climate change both directly and indirectly (Ebi et al. 2018). A heat wave, an increase in sea level, a drought or flood, or an intense hurricane can directly damage, injure, or even kill people. Through environmental changes, the consequences of climate change can also have an indirect impact on health. For instance, worsening air pollution levels may have detrimental effects on cardiovascular and respiratory disorders. By adapting physiologically, behaviorally, culturally, and technologically to the local prevailing climate, regional populations achieve long-term survival. There are, however, times when extreme events stress populations beyond their ability to adapt. Understanding the health concerns associated with these occurrences is crucial since it is anticipated that as climatic variability and mechanisms change, so will the frequency and intensity of severe events in the future.

3.1 Heat Stress

All people are impacted by increasing global temperatures. However, some people are more susceptible to physiological stress, aggravated sickness, and a higher risk of passing away from exposure to excessive heat, either due to exposure or because they are more physiologically or socioeconomically fragile. Heat-related illnesses (such as heat exhaustion, heat syncope, and heat stroke) and death are directly linked to rising temperatures, and each person's risk varies greatly depending on their exposure, location, and sensitivity. Populations frequently exhibit a perfect temperature where the death rate (daily or weekly) is lowest. Since the 1950s, the number of hot days and nights has increased globally, while the number of cold days and nights has declined. Colder temperatures have a greater impact on individuals in hotter cities, and warmer temperatures have a greater impact on people in colder cities. Seasonal fatality rates and sickness occurrences are higher in the winter than in the summer in the UK and certain other northern high-latitude countries.

According to research, heat waves are positively correlated with mortality, among elderly persons (who have a reduced physiological capacity for thermoregulation, particularly women (being the most affected. According to additional studies, those who are mentally ill, children, persons who work in physically demanding environments, or people who have pre-existing illnesses are also at risk. People with chronic respiratory problems or pre-existing cardiovascular disease (heart attack and stroke) typically die during heat waves. Urban residents are more vulnerable than those who live in non-urban areas. Temperature increases are amplified and prolonged (especially overnight) by thermally inefficient housing and the so-called urban heat island effect, which occurs when inner metropolitan settings with high thermal mass and limited ventilation absorb and retain heat.

Changes in public health preparedness as well as physiological and behavioral adjustments might lessen morbidity and mortality during heatwaves. The overall decline in heatwave-related deaths over a recent three-decade span in US cities demonstrates how weather and mortality relationships can change over time. This decrease demonstrates the potential for risk reduction via climate change adaptations, such as air conditioning, better healthcare, increased public knowledge, and adjustments to underlying health status.

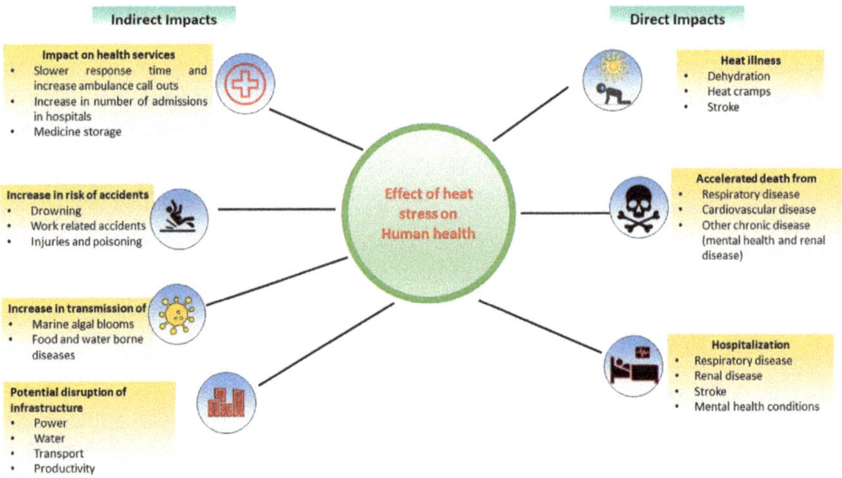

3.2 Floods

Before 2011, flooding was the most prevalent type of disaster on the planet, accounting for about half of the all-natural disaster victims and nearly US $185 billion in economic losses. Rainfall, surface runoff, evaporation, wind, sea level, and local terrain interact to cause floods. Flood patterns in inland places vary greatly according to catchment size, geography, and climate. Due to increasing sea levels and more frequent and severe precipitation events, flooding events are predicted to become more frequent and intense. Additionally, as urbanization levels rise, more individuals will be vulnerable to flooding incidents.

Floods are predicted to raise the burden of disease, morbidity, mortality, and social and economic disruptions on a worldwide scale. They are also predicted to continue to strain health services, particularly in low-resource nations. These nations are the most vulnerable and where most big floods occur. Flooding's effects on public health are influenced by geographic, socioeconomic, and population baseline vulnerabilities.

3.2.1 Health Consequences of Floods

Flood-related health effects can be broadly divided into direct and indirect effects. Drowning, injuries from debris, chemical poisoning, and hypothermia are just a few examples of the direct effects of flooding, which are those that arise from direct contact with the water and the surrounding area. Indirect effects include infectious diseases, starvation, diseases linked to poverty, and illnesses linked to displaced populations. They include dangers related to the harm caused by the water to the natural and built environments.

a. **Drowning**

Drowning is the primary cause of fatality from floods, and flash flooding, as opposed to longer riverine flooding, is more frequently to blame. Drowning frequently happens because of people acting inappropriately, leaving it too late to evacuate, or underestimating the strength of the current or depth of the water.

b. **Injuries**

In the early aftermath of floods, nonfatal injuries and the deterioration of existing illnesses are the main causes of morbidity among impacted populations and aid personnel. Injuries can happen before, during, and after a flood, as well as during the cleanup and repopulation phases. The most frequent causes of flood-related nonfatal injuries are cuts, falls, being struck by objects falling into floodwater or moving swiftly, being bitten, or being stung.

More than 7500 nonfatal injuries among locals and aid personnel were documented during post-Katrina/Rita surveillance in Greater New Orleans. Young to middle-aged males were most at risk in both groups, in part because they were actively involved in the relief and cleanup activities. Although there were differences in the injury patterns between residents and aid workers, cuts were the most common mechanism of injury in both groups (27% and 19%, respectively. During the repopulation era, more residents than rescue personnel were hurt, mostly because of falls (23% against 10%) and car accidents (9% vs. 4%).

Flooding might result in electrical injuries. A possible electrical hazard exists anywhere there is standing water near electrical lines, circuits, or equipment. As floodwaters disturb propane and natural gas lines, tanks, electricity lines and chemical storage tanks, burns and explosions may result. Floodwaters' surfaces may catch fire due to the presence of oil and other flammable, non-polar, low-density liquids. Floods can happen at any time of year, and hypothermia can happen with or without submersion.

c. **Water contamination**

Contact with flooding without drowning does not present a significant risk to health on its own (Centre for Disease Control and Prevention 1983). Floodwaters, however, may contaminate the area's water supply, food supply, and sewage system, leading to pollution and raising the risk of infectious diseases.

Waterborne diseases like hepatitis A virus, Shigella, Escherichia coli, and Salmonella are spread through contaminated water sources. According to several studies, heavy metals, dioxin, pesticides, and carbon monoxide have all been shown to be present in floodwaters and land. According to research, exposure to such toxins is known to increase the risk of developing cancer, cardiovascular, gastrointestinal, kidney, liver, and neurological problems.

d. **Communicable diseases**

According to the World Health Organization's fact sheet on flooding and communicable illnesses, there is a higher risk of water- and vector-borne diseases during floods. According to World Health Organisation, the danger rises when infrastructure is severely damaged, populations are uprooted, and water supply systems are affected, resulting in the contamination of drinking water facilities. Common illnesses brought on by tainted water include cholera, diarrheal illness, hepatitis A and E, leptospirosis, parasitic illnesses, rotavirus, shigellosis, and typhoid fever. Conjunctivitis, dermatitis, ear, nose, and throat infections, as well as other water-borne illnesses, are specifically linked to floods.

e. **Disability**

Disability brought on by flood stress is a common source of morbidity in the long run, especially if complications like infection make it worse. Asthma flare-ups and other chronic illnesses of the nose, throat, and lungs may worsen in people with disabilities.

f. **Mental health problems**

The aftermath of floods frequently includes mental health issues. Major life stresses, like natural disasters, make people more vulnerable to both physical and mental illness. It has been found that persons who have experienced flooding are four times more likely than those who must feel psychological distress, and their suicide rates are 13.8% greater than they were before the disaster. Physical health issues may contribute to mental health issues, as well as individual losses, societal disruptions, and financial difficulties. It has been demonstrated that young people, particularly females, are especially susceptible in developing societies.There should be sensitivity to the socio-economic status, culture, tradition, language, and pattern of local livelihoods by disaster mental health teams.

4 Vector-Borne Diseases and Climate Change

Due to their high disease burden, broad frequency, and great susceptibility to climatic influences, vector-borne diseases are among the conditions linked to climate change that have received the most research. The influence of climatic conditions is less direct and more varied, both within and between different diseases, in comparison to some other climate-sensitive health risks, such as heat stress or exposure to storms and floods.

Climate and weather conditions also exert a range of more indirect effects, through wider effects on the natural environment and human systems, For example, drought may affect water storage, land-use and irrigation practices, and population movement, in turn, affecting vector ecology, and human exposure to infection. In a recent WHO report summarizing the significance of vector-borne diseases, for instance, it is stated that previously relatively stable geographic distributions are now changing because of several factors, such as "climate change, intensive farming, dams, irrigation, deforestation, population movements, rapid unplanned urbanization, and phenomenal increases in international travel and trade" (WHO 2014). More than 80% of the world's population lives in high-risk locations where one or more forms of VBDs constitute a hazard, and each year, over 700,000 people die from them (WHO 2017). The survival and reproduction of arboviral diseases are suitable in 86% (218) of the world's 250 countries.

According to the World Health Organisation, the eight primary vector-borne diseases in the world include lymphatic filariasis, schistosomiasis, onchocerciasis, Chagas disease, leishmaniasis, and Japanese encephalitis. African trypanosomiasis, Lyme disease, tick-borne encephalitis, and West Nile fever are further vector-borne illnesses of regional significance.

Unquestionably, a favorable climate is required for the survival or emergence of a vector-borne disease. Because of the unfavorable climatic circumstances, neither the Aedes mosquito nor the virus have established themselves in Sweden, even though travel-related cases of dengue do occur there. When the temperature does not reach the ideal level, the chance of VBD transmission increases with the temperature rise; conversely, when the temperature rises over the ideal temperature's maximum, the risk of VBD transmission falls. Unsuitable temperature conditions may result in lower vector survival, which in turn reduces the ability of VBDs to transmit disease, which also has an immediate impact on the growth of vector-dependent diseases. Regarding diseases spread by mosquitoes, temperature can affect mosquito growth and survival, and there is a thermal optimum that will be repressed at either extremes of heat or cold. However, in South Africa, warming conditions may make some tick species' habitats less suitable, which will reduce the prevalence of diseases associated with ticks.

Additionally, rainfall is generally helpful to mosquito oviposition and reproduction, while rainfall and malaria incidence increase with increasing precipitation.. The population of ticks can be decreased by prolonged and heavy precipitation, which can wash away ticks, their eggs, and their larvae. The spread of vector-borne diseases is significantly impacted by climate change, which will generally result in an expansion of the climate-adaptive transmission zone for these diseases. The expansion of VBDs is impacted differently by several extreme climate events, including heat waves, cold waves, floods, and droughts. Heat waves affect mosquito populations differently depending on when they start and how long they last; they typically encourage population expansion in the early stages of development but frequently repress the entire life cycle. In wetlands that occasionally endure droughts, mosquito populations unexpectedly increase as their competitors and predators are removed.

4.1 Climate Change and Emergence of Coronaviruses

Climate change has altered environments in addition to causing humans and animals to migrate in different ways and modify their behaviours, which has brought them into closer contact with the sources of new illnesses.The increase in temperature and relative humidity, which creates conditions that favour the long-term survival of infectious particles in the environment, may be one of the more significant effects of climate change that influences transmission efficacy. It has been demonstrated that temperature and humidity have an impact on how long particulate matter is suspended in the air and how viruses spread.

The loss of wild animal habitats over time may weaken the natural barrier between people from animals that carry harmful microbes, resulting in new contact zones (forest patches close to human settlements) and the potential for spillover. This is especially important at this time in history because of the severe coronavirus pandemic known as severe acute respiratory syndrome coronavirus (SARS-CoV2) that started in the Chinese area of Hubei and quickly spread over the entire world. In addition to the coronaviruses, there are probably some of the common cold-causing as well as the Nipah and Ebola viruses.

Some researchers have hypothesised that continuous climate changes may have contributed to the migration of pangolins, a possible intermediate host of SARS-CoV2, along with changes in bat populations. If true, coronaviruses may have evolved in pangolins to become more pathogenic for humans. For instance, a rise in humidity increases the amount of dangerous chemicals in the air, which also creates a favourable environment for the spread of new coronaviruses and other viral organisms that cause respiratory disorders.

5 Air Quality and Climate Change

Changing climate is likely to alter patterns of air pollution concentrations having a substantial and well-established influence on air quality. The chemical processes that produce ozone and secondary particle production are accelerated by higher temperatures. Additionally, higher temperatures and possibly higher CO_2 concentrations cause vegetation to emit more ozone-relevant VOC precursors.

All contaminants in the atmosphere are affected by weather patterns through the impact of winds, vertical mixing, and rainfall. The vertical and horizontal dispersion of air pollutants can be limited by atmospheric conditions. For instance, during winter's morning rush hour, the dispersion of traffic emissions is limited by calm winds and cool air above. Power plant emissions significantly rise when there are heat waves since air conditioner consumption peaks then. In July, August, and September of 2004, nitrogen oxide (NOx) emissions from a few Californian power plants more than doubled on days when the daily maximum temperature increased from 75 to 95 °F.

Higher ozone concentrations in future decades are associated with greater relative increases due to climate change. A study in Germany (Forkel and Knoche et al. 2006) that used a scaled-down integrated modeling approach to assess the ozone in the 2030s and the 1990s supported the discovery of greater climatic impacts on extreme ozone readings. There was an increase of 2–6 parts per billion (6–10%) in the maximum ozone concentrations daily across the study site. In contrast, there were nearly four times as many cases of daily maximum ozone exceeding 90 parts per billion, from 99 to 384.

It is possible that climate change could increase the frequency and area impacted by wildfires because the risk of their origin and spread is increased by higher temperatures, decreasing soil moisture, and protracted periods of drought. A major concern near fires as well as in areas downwind are the resulting increases in fine particulate air pollution, which have numerous health and economic impacts.

6 Social Determinants Affected Due to Climate Change

A variety of health effects caused by climate change now exist and are anticipated to continue to exist across different population groups. Any group's susceptibility depends on how sensitive it is to health hazards associated with climate change, how exposed it is to those risks, and how well it can adapt to and deal with climate unpredictability and change. Depending on social, political, and economic factors—collectively known as social determinants of health—individuals or communities may be vulnerable to health risks.

Social determinants of health, such as living conditions and access to healthcare, disproportionately disadvantage some groups by limiting resources and opportunities for health-promoting behaviors and conditions. Health risks and vulnerability may rise in areas or situations where certain combinations of social determinants of health that heighten health concerns happen at the same time, close together, or both.

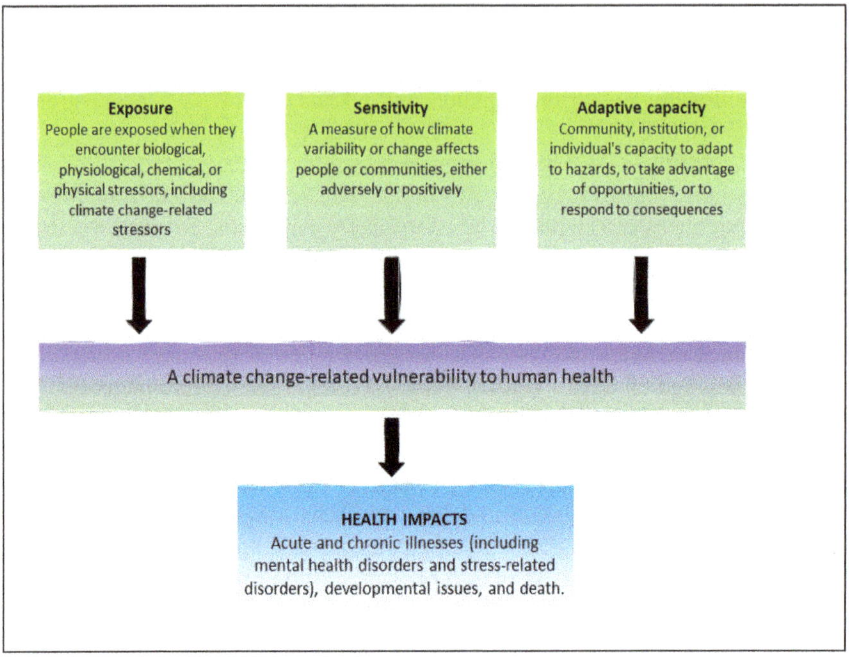

The determinants of vulnerability to health impacts associated with climate change.

6.1 *Population of Concern*

Many of the variables that make communities of color, low-income groups, and certain immigrant groups—especially those who are undocumented—vulnerable to the health effects of climate change are present in these groups. These populations are more likely to reside in risky areas (such as urban heat islands, remote rural areas, or coastal and other flood-prone areas), in locations with outdated or poorly maintained infrastructure, or locations with higher air pollution burdens, increasing their risk of exposure. Inequalities in health and vulnerability are exacerbated by the negative effects of climate change, which disproportionately affect low-income communities (such as some communities of colour, immigrant groups, and indigenous people), children and pregnant women, older adults, vulnerable occupational groups (such as outdoor workers), and people with pre-existing health conditions or disabilities.

6.1.1 Children

Children are particularly susceptible to the negative health impacts of environmental exposures because of their immature physiology and metabolism, specific exposure pathways, biological sensitivity, and limited capacity for adaptation. Beginning in the womb and lasting into their second decade of life, children experience a series of vulnerable windows. As a proportion of their body weight, children consume more air, food, and water than adults. Children will be exposed to more health risks because of climate change due to its interaction with factors like economic position, diet, living circumstances, and developmental stage. When determining how children will experience the effects of climate change, the effect of poverty on their health must be considered. A healthy child's growth depends on fundamental needs being met, which can be difficult for low-income and poor households to do.

Children, especially infants and toddlers, are particularly vulnerable to climate change's effects on water contamination, excessive heat, and air pollution. Children are more vulnerable to the effects of climate change on air pollution, extreme heat, and contaminated water. This is especially true of newborns and toddlers.

6.1.2 Pregnant Women and Babies

Low birth weight (less than 5.5 pounds), preterm birth (birth before 37 weeks of pregnancy), increased neonatal mortality, dehydration and associated renal failure, malnutrition, diarrhea, and respiratory disease are just a few of the negative pregnancy and newborn health outcomes that may result from exposure to climate-related factors (Sheffield and Landrigan 2011; Rylander et al. 2013). Congenital cataracts (VanZutphen et al. 2012) and low birth weight, preterm birth, and other unfavorable birth outcomes (Strand et al. 2012) have all been linked to extreme heat occurrences.

In addition, exposure to inhaled particulate matter while pregnant is linked to poor delivery outcomes. Floods are linked to a higher risk of environmental contaminants and mold exposure for pregnant women, decreased access to clean water and food, psychological stress, and disruptions in healthcare. Maternal risk of anemia (a disorder associated with low red blood cell counts occasionally induced by poor iron intake), eclampsia (a condition that can cause convulsions in pregnant women), and spontaneous abortion are additional flood-related health effects for mothers and babies.

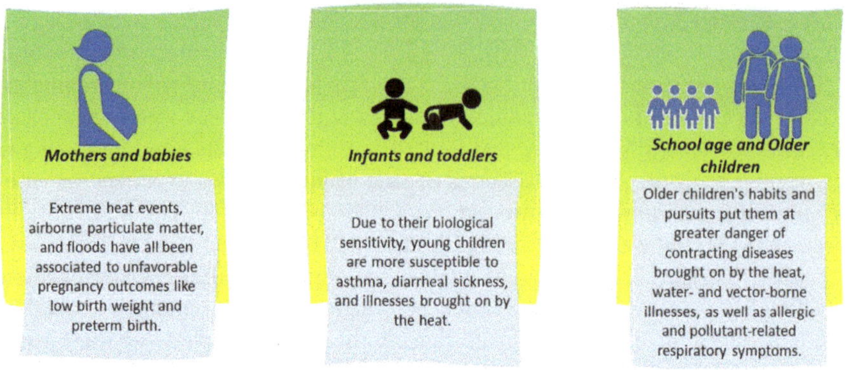

The vulnerability to the health impacts of climate change at different life stages.

6.1.3 Older Adults

The health effects of climate change and weather extremes are particularly harmful to older folks, who are commonly considered to be those who are 65 years of age and older. Extreme heat exposure in older people can have a variety of negative impacts. Extreme heat events are expected to increase in frequency, intensity, and length during the future decades, especially at higher latitudes and major metropolitan centers. Extremely high temperatures in New York City were linked to a rise in hospital admissions for cardiovascular and respiratory conditions, with the elderly being among those most negatively impacted. Elderly patients had the highest hospital admission rates for respiratory illnesses, increasing by 4.7% for every degree Celsius (Lin et al. 2009). Physical, mental, or emotional trauma occurs before, during, and after hurricanes and other extreme weather occurrences. Even in adults in good health, exposure to ground-level ozone can harm lung function, increase trips to hospital. In older persons, air pollution can aggravate asthma and raise their risk of having a heart attack, particularly if they have diabetes or are obese.

Older adult populations may become more exposed to infections because of the projected changes in the distribution of disease vectors like ticks and mosquitoes. Certain vector-borne illnesses, such as the West Nile and St. Louis encephalitis viruses spread by mosquitoes, provide a higher risk of illness for vulnerable older persons with weakened immune systems. Additionally, aging can affect the processes that control body temperature, especially for people who use drugs that affect this regulation, such as psychotropic drugs used to treat a range of mental diseases like depression, anxiety, and psychosis. Increased exposure to outdoor air pollutants, including ozone and fine particulate matter, aeroallergens, and wildfire smoke- all of which may be made worse by climate change will exacerbate the respiratory deficits that older persons currently face.

7 Climate Change and Human Mental Health

Extreme weather due to climate change affects people's mental health directly and immediately. Climate change can have a variety of effects on mental health, ranging from minor stress and distress symptoms to clinical disorders, such as anxiety and sleep difficulties (Blanc et al. 2019) to depression, post-traumatic stress disorder, and suicidal ideation (WHO 2013; Ursano et al. 2017). Other effects could be how people and communities see and experience their daily lives, as well as the need to deal with and effectively adapt to climate change and its Posttraumatic stress disorder (PTSD), depression, general anxiety, increased substance use or abuse, and suicidal thoughts are notorious reactions to extreme events that cause life disruption, such as the loss of life, resources, social networks, or extensive relocation.

People who frequently experience disasters should pay particular attention to these mental health effects since they might have a cumulative psychological impact. When exposed to Hurricane Katrina, veterans with prior mental illnesses were 6.8 times more likely to experience any subsequent mental illnesses than soldiers without such conditions (Sullivan et al. 2013). Extreme disasters (such as storms and floods) that result in the loss of life, property, resources, social networks, or events that cause substantial relocation and life disruption are also frequently followed by depression and general anxiety.

Long-term drought affects people's lives and livelihoods as well as how individuals, families, and communities operate over time as it interacts with many environmental and social pressures. In rural areas, a protracted drought has been accompanied by cascading and interrelated economic, social, and everyday living problems. In Australian locations where a drought has been proclaimed, anxiety and psychological distress have grown, especially for individuals who are enduring a loss of livelihood and industry. In Australia, men farmers have a higher frequency of suicide when there has been a prolonged drought.

Extreme heat's negative effects on mental health are linked to an increase in illness and fatalities, aggressive behavior, violence, and suicide, as well as an increase in people with mental health or psychiatric issues being admitted to hospitals and emergency rooms. People's outdoor activities are restricted by rising temperatures and deteriorating air quality. Many people find that cutting back on outdoor exercise and other stress-relieving activities results in worsened physical and mental health as well as increased stress.

8 Adaptive Strategies and Public Health Responses

By preserving or boosting adaptive capability and resilience, climate change adaptation is defined as "any activity that intends to reduce the vulnerability of human or natural systems to the impacts of climate change and climate-related risks" (OECD

2020). This includes a wide range of tasks, such as gathering information and knowledge, building capacity, planning, and putting into action proactive interventions to stop or lessen the effects of climate change. Five phases of adaptation exist awareness, appraisal, planning, execution, monitoring, and evaluation (Lempert et al. 2018). In general, adaptation techniques aim to strengthen systems' resistance to climate change and lessen the impact of upcoming climate change events.

Adaptation techniques are necessary to lessen the negative consequences of climate change events by making systems more resilient to future negative effects of climate change. Building blocks of such systems include (a) a skilled health workforce with the resources needed to support climate resilience; (b) health information systems that support efficient management of the health risks of extreme events; (c) effective service delivery, including emergency planning; and (d) adequate funding (Ebi et al. 2019).

To guide adaptation planning, health systems must be prepared for climate extremes and disasters with solid knowledge of present impacts and projected risks. To guide adaptation planning, health systems must be prepared for climate extremes and disasters with solid knowledge of present impacts and projected risks (WHO 2015). Creating a health national adaptation plan (HNAP) or subnational plan, which includes carrying out a vulnerability and adaptability assessment, is a crucial step. To decrease vulnerabilities and develop the necessary capacity and resilience, HNAP aids in ensuring that the health hazards of climate change are given priority from local to national levels of decision-making.

Stress tests can help officials prepare for and manage progressively severe climate change-related shocks and stressors by providing information. Vulnerability and adaptation assessments should include examinations of the capacity of health systems and services to endure extreme events and disasters. When an extreme event's health effects approach a tipping point, the system is significantly altered, and population health outcomes become more serious. This was seen in 2017, when Hurricane Maria caused an estimated 4,645 more fatalities in Puerto Rico, with a third of those fatalities resulting from delayed or interrupted medical care (Kim et al. 2018).

Given the significant carbon footprint of health-related activities connected to energy, food, anesthetic gases, and transportation, health-sector leaders have a wide range of options to assist in lowering the future amount and pattern of extreme weather and climate events through the reduction of GHGs (Karliner et al. 2020). In addition, well-planned GHG mitigation activities can result in considerable health advantages that contribute to resilience, such as enhanced respiratory and cardiovascular health due to better air quality.

9 Mitigation Strategies

"Strategies, policies, and measures to improve disaster risk understanding, foster disaster risk reduction and transfer, and promote continuous improvement in disaster preparedness, response, and recovery practice" are all included in mitigation strategies (UN 2020). If health-emergency management plans and activities are based on information and concerns about climate change, they can help with adaptation and resiliency-building initiatives. Aligning finance methods and procedures as well as developing dual- or multipurpose government policies and strategies are examples of opportunities to integrate catastrophe risk management with climate change adaptation; Integration of disaster risk management and disaster adaptation into national and international institutions, programs, and organizations; use of widely used participatory hazard assessment techniques for disaster risk management; The encouragement of locally driven initiatives or plans that fully handle environmental issues and disasters from all hazards.

To address community-specific hazards, incorporate priorities for vulnerable groups, and acknowledge community assets and coping mechanisms, community-based disaster risk management techniques are required. Local public health organizations are particularly positioned to foster resilience since it is crucial to put adaptation strategies in place at the local level. Due to their duties, experience, and knowledge in the field, public health professionals are well-suited to handle climate change. Additionally, if the frequency of climate-related hazards rises, public health initiatives carried out in the context of disasters—such as quick needs assessments, surveillance, and epidemiologic studies—will be more crucial.

Integrating health into disaster risk management plans and strategies, as well as including disaster risk management into all health policies, is a crucial step in closing the health adaptation gap related to heat and weather extremes (UNEP 2018). For state and municipal health departments to participate in climate change adaptation efforts, more collaboration opportunities, partnerships, communication (within and across sectors), resources, authority, training, and capacity are needed.

10 Conclusion

Globally, extreme weather and climate events are likely to increase in frequency and intensity in the coming decades, with significant consequences for populations and healthcare systems. Thus, the complex relationship between climate change and human health emphasizes the need for swift, all-encompassing action to reduce risks and prepare for new challenges. The ramifications for human health are significant and far-reaching as the world experiences changes in temperature, precipitation patterns, and the frequency and intensity of extreme weather events. Human health is affected by climate change now more than ever before; the consequences are a present reality that demands our attention and concerted action. A fraction of the health risks individuals and communities face include heat-related illnesses, vector-borne diseases, air quality deterioration, waterborne infections, malnutrition, mental health impacts, and population displacement. For people who are already socially and economically poor and marginalized, these dangers are heightened.

With a strong focus on inclusivity, equity, and community involvement, adaptation measures must be customized to the vulnerabilities of regions and populations. Additionally, limiting the severity of future health impacts will depend heavily on mitigation measures to lower greenhouse gas emissions.

Collaborating is paramount in navigating this complex landscape. Climate change demands innovative solutions that involve governments, international organizations, non-governmental organizations, and private companies. We can build resilient communities that are better prepared to address the coming health issues by fusing scientific knowledge, technology innovation, policy creation, and grassroots activism.

References

Blanc J, Spruill T, Butler M, Casimir G, Jean-Louis G (2019) 0885 is resilience a protective factor for sleep disturbances among earthquake survivors? Sleep 42:A356

Centers for Disease Control and Prevention (CDC, 2000) Morbidity and mortality associated with Hurricane Floyd--North Carolina, September-October 1999. MMWR. Morbidity Mortality Weekly Report 49(17):369-372

Ebi KL, Berry P, Hayes K, Boyer C, Sellers S, Enright PM, Hess JJ (2018) Stress testing the capacity of health systems to manage climate change-related shocks and stresses. Int J Environ Res Public Health 15(11):2370

Ebi KL, Berry P, Campbell-Lendrum D, Cissé G, Hess J, Ogden N, Schnitter R (2019) Health system adaptation to climate variability and change. Global Center on Adaptation, Washington, DC

Forkel R, Knoche R (2006) Regional climate change and its impact on photooxidant concentrations in southern Germany: simulations with a coupled regional climate-chemistry model. J Geophy Res Atmosp 111(D12)

Karliner J, Slotterback S, Boyd R, Ashby B, Steele K, Wang J (2020) Health care's climate footprint: the health sector contribution and opportunities for action. Eur J Public Health 30(Supplement_5):ckaa165–843

Kim Y, Knowles S, Manley J, Radoias V (2018) Long-run health consequences of air pollution: evidence from Indonesia's forest fires of 1997. Econ Hum Biol 26:186–198

Lempert R, Arnold J, Pulwarty R, Lempert R, Gordon K, Greig K, Hoffman CH, Sands D, Werrell C, Lazarus MA (2018) Reducing risks through adaptation actions 28. Adaptation Response. Impacts, Risks, and Adaptation in the United States: The Fourth National Climate Assessment, Volume II. Intergovermental Panel on Climate Change; Washington, DC, USA

Lin S, Luo M, Walker RJ, Liu X, Hwang SA, Chinery R (2009) Extreme high temperatures and hospital admissions for respiratory and cardiovascular diseases. Epidemiology, 738–746

OECD D (2011) Handbook on the OECD-DAC Climate markers. Paris: Organisation for Economic Co-operation and Development's Development Assistance Committee (DAC). https://www.oecd.org/dac/stats/48785310. pdf

Rylander C, Øyvind Odland J, Manning Sandanger T (2013) Climate change and the potential effects on maternal and pregnancy outcomes: an assessment of the most vulnerable–the mother, fetus, and newborn child. Glob Health Action 6(1):19538

Sheffield PE, Landrigan PJ (2011) Global climate change and children's health: threats and strategies for prevention. Environ Health Perspect 119(3):291–298

Strand LB, Barnett AG, Tong S (2012) Maternal exposure to ambient temperature and the risks of preterm birth and stillbirth in Brisbane, Australia. Am J Epidemiol 175(2):99–107

Sullivan G, Vasterling JJ, Han X, Tharp AT, Davis T, Deitch EA, Constans JI (2013) Preexisting mental illness and risk for developing a new disorder after hurricane Katrina. J Nerv Ment Dis 201(2):161–166

UN Environ. Progr. (UNEP) (2018) The adaptation gap report 2018 Rep., United Nations Environ. Progr., Nairobi, Kenya. https://wedocs.unep.org/bitstream/handle/20.500.11822/27114/AGR_2018.pdf

UN Off. Disaster Risk Reduct (2020) Disaster risk reduction & disaster risk management. PreventionWeb https://www.preventionweb.net/risk/drr-drm

Ursano RJ, Morganstein JC, Cooper R (2017) APA resource document: mental health and climate change. Mental-Health-Climate-Change. pdf

Van Zutphen AR, Lin S, Fletcher BA, Hwang SA (2012) A population-based case–control study of extreme summer temperature and birth defects. Environ Health Perspect 120(10):1443–1449

World Health Organization (2013) Global action plan for the prevention and control of noncommunicable diseases 2013–2020. World Health Organization

World Health Organization (2014) WHO guidance to protect health from climate change through health adaptation planning

World Health Organization (2014) A global brief on vector-borne diseases (No. WHO/DCO/WHD/2014.1). World Health Organization

World Health Organization (2015) Operational framework for building climate resilient health systems. World Health Organization

World Health Organization (2017) Global vector control response 2017–2030. *Global vector control response 2017–2030*

Exploring the Emergence of Sustainable Practices in Healthcare Research and Application as a Path to a Healthier Future

Harshita Jain

Abstract Beyond long-standing norms, there has been a substantial paradigm shift in favour of sustainability in the rapidly evolving healthcare landscape. This chapter examines the rising trend of incorporating environmentally friendly concepts into medical research and practise. Through an analysis of several studies, projects, and instances, it clarifies the rising significance of sustainability in healthcare. The chapter covers a growing range of sustainable therapies, such as energy-efficient medical equipment and hospital architecture that takes the environment into consideration. It is important to recognise the interconnectedness between human and environmental health, since sustainable healthcare methods have been demonstrated to have good benefits on both. This examination of the literature and expert opinions demonstrates that sustainable practises increase patient outcomes and community health resilience. As stakeholders come to understand the significance of sustainability in healthcare, this chapter serves as a road map, outlining the challenges, opportunities, and requirements for interdisciplinary collaboration, appropriate legislation, and innovative solutions. By breaking down the different components of sustainable healthcare, this chapter gives scholars, practitioners, and policymakers the tools they need to steer toward a healthier, more sustainable future.

Keywords Community health resilience · Multidisciplinary cooperation · Sustainable healthcare · Environmentally friendly design · Energy efficiency

1 Introduction

The need to reconcile healthcare practises with the requirement of sustainability is driving a major revolution in this crucial sector of the economy. Conventional healthcare models are useful for meeting medical demands, but they frequently fail to take society and the environment into consideration. The necessity of reassessing and

H. Jain (✉)
Amity Institute of Environmental Sciences, Amity University Noida, Uttar Pradesh 201301, India
e-mail: hjain@amity.edu

P. K. Prabhakar and W. Leal Filho (eds.), *Preserving Health, Preserving Earth*,
World Sustainability Series, https://doi.org/10.1007/978-3-031-60545-1_7

reshaping healthcare procedures has become imperative considering global sustainability trends. Historically, the development of healthcare practices has been characterized by an emphasis on therapeutic procedures, frequently overlooking the wider ecological environment (Proctor et al. 2015, Sherman et al. 2020). A paradigm shift is required due to the increasing rate of environmental deterioration and the rising knowledge of the connection between the planet's health and human health. To comprehend the historical trajectory, worldwide trends, and the urgent environmental concerns within the healthcare industry, this chapter explores the context and reasoning for the establishment of sustainable practises in the field. A thorough conceptual model detailing the many interactions necessary to achieve sustainability in healthcare is shown in Fig. 1. Fundamentally, the model represents the main goal of "Sustainable Healthcare," signifying the peaceful coexistence of social justice, environmental responsibility, and economic viability in healthcare procedures. Specific goals to understand the complexities of sustainable healthcare. This section outlines the goals of the chapter, which include an analysis of the state of the art, a discussion of important ideas, and a determination of the factors that are pushing the transition to sustainability. By outlining these goals, the chapter hopes to give a thorough and organised examination of sustainable healthcare, supplying knowledge that will benefit academics, influence healthcare procedures, and direct politicians. As a complex idea, sustainable healthcare has to have its boundaries carefully defined in order to allow for a more in-depth analysis (Sherman et al. 2020). The boundaries of geography, culture, and economy that the study will take place inside are described in this section. In addition, it highlights how important sustainable healthcare is to many stakeholders and how it may improve healthcare quality while simultaneously advancing larger social objectives. This chapter aims to go beyond conventional limits by providing a comprehensive viewpoint on the relationship between sustainability and healthcare. The goal of this investigation is to not only clarify the difficulties of sustainable healthcare, but also to provide stakeholders with a road map for navigating the transition to a future in which environmental sustainability and health are interdependent (Shelton et al. 2018). Understanding the history, justification, goals, extent, and importance of sustainable practises in healthcare is the first step in this journey.

2 The Paradigm Shift: Sustainability in Healthcare

A revolutionary paradigm shift that goes beyond the conventional boundaries of medical practise and embraces a comprehensive commitment to sustainability is taking place in the complex field of healthcare. In the second part of our investigation, the focus is on how the sustainability imperative is changing healthcare practises, regulations, and philosophies. It is a storey about breaking away from the norm and travelling toward a healthcare philosophy that not only treats the sick person but also promotes the health of the entire world.

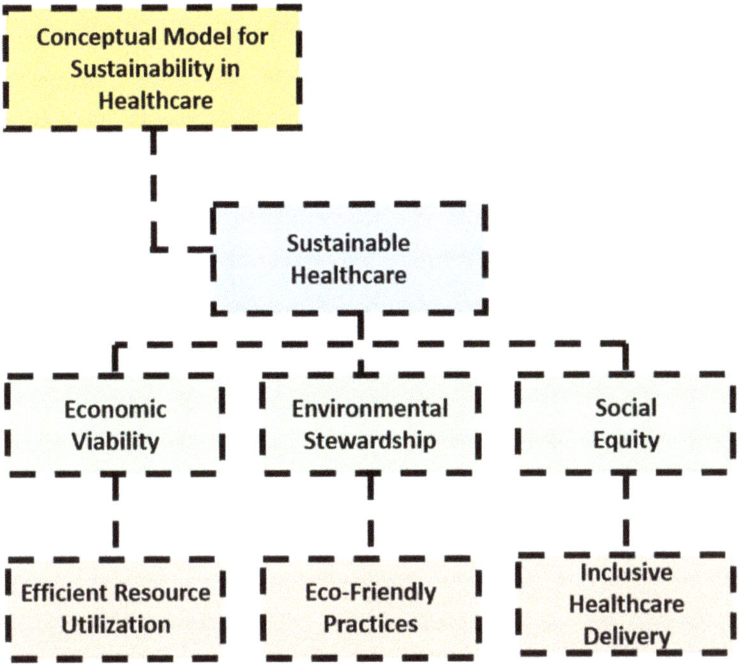

Fig. 1 Conceptual model for sustainability in healthcare

2.1 Evolution of Sustainable Practices

An important historical change in health care methods is at the heart of the current shift in healthcare. In the past, the primary focus of healthcare has been on treating patients' urgent medical needs, frequently ignoring the implications for larger ecological settings. This part explores the history of healthcare methods, exposing the flaws in the traditional method and highlighting the urgent need for change. Table 1 provides a thorough historical account of how healthcare practises have changed throughout the course of many eras and periods. Every section of the table outlines the distinctive features and focal points of particular eras, providing an overview of the ways in which medical care paradigms have changed. From antiquity to the present, the table deftly explains the unique characteristics of every period. It highlights the revolution in healthcare practises that has occurred, from localised and reactive methods to the current focus on eco-friendly and sustainable procedures (Molero et al. 2021).

The storey takes place in ancient times, when the medical system was mostly reactive and relied mainly on herbal cures and crude surgical procedures for treatment. As time goes on, the Medieval Ages see the establishment of organised medicine; nonetheless, healthcare is still entwined with a narrow understanding of ecological interdependencies and alchemical and herbal remedies. With little regard for the effects on the environment, the Industrial Revolution brings with

Table 1 Evolution of healthcare practices

Era/period	Healthcare practices	Characteristics
Ancient times	Predominantly reactive healthcare	Heavily reliant on herbal remedies and basic surgery
	Limited understanding of ecological interdependencies	Medical care centered on symptom alleviation
Middle ages	Emergence of formalized medical institutions	Increased reliance on alchemy and herbal treatments
	Limited awareness of environmental impacts	Healthcare primarily a localized community affair
Industrial revolution	Advancements in medical science and technology	Surge in urbanization; rise of industrial healthcare
	Escalating use of synthetic drugs and chemicals	Limited consideration for environmental repercussions
Late twentieth century	Rapid expansion of medical knowledge and technology	Accelerated pharmaceutical developments
	Surge in healthcare infrastructure and specialization	Limited focus on sustainable or eco-friendly practices
Present day	Growing awareness of environmental issues	Increasing emphasis on sustainable healthcare practices
	Integration of technology for sustainable healthcare solutions	Recognition of the need for an eco-conscious approach

it advances in medical science and technology, urbanisation, and industrial healthcare. Medical knowledge and technology are expanding at a tremendous pace as the storey moves into the late twentieth century. Infrastructure and expertise in healthcare are expanding, but there is less emphasis on environmentally friendly or sustainable techniques in this period. The contemporary era is characterised by an increased consciousness of environmental issues and a notable transition towards sustainable healthcare methodologies (Shelton et al. 2018). Healthcare solutions are easily integrated with technology, and the need for an environmentally responsible approach is acknowledged. Table 1 illustrates the progression from reactive healthcare approaches to a modern focus on sustainability, which reflects a better comprehension of ecological interdependencies. This path of transformation is a turning point in the philosophy of medical treatment. The table encapsulates the spirit of each age, underscoring the need for a comprehensive, ecologically responsible strategy in the healthcare industry and the ever-expanding understanding of the fundamental connection between healthcare and the environment (Mallee 2017).

2.2 Drivers of Change in Healthcare Sustainability

It is critical to identify the complex dynamics driving the transition toward sustainability in healthcare to better understand what is driving this change. This section makes its way through a complicated network of variables that both highlight the state of healthcare now and move it in the direction of a more sustainable future. Social expectations are at the forefront of this transformation process (Stoughton and Ludema 2012). Globally, societies are becoming more cognizant of one another, which calls for a more responsible and thoughtful approach to healthcare. Contemporary communities, motivated by heightened consciousness of ecological concerns and an aspiration for enhanced welfare, anticipate that healthcare systems will cater to more than just urgent medical need. Therefore, the change in society serves as a catalyst, forcing healthcare organizations and providers to reassess their operations using a sustainable framework. Another key element driving the transformation of healthcare sustainability is technological innovation. The unrelenting advancement of technology has not only transformed healthcare practices but also created new opportunities for long-term healthcare solutions. Technology, from digital health records to sophisticated diagnostics, not only increases the effectiveness of healthcare services but also offers resources for reducing environmental effects. In the quest for sustainability, integrating eco-friendly technologies such as smart infrastructure and energy-efficient medical devices becomes essential, not optional (Sherman et al. 2020).

Furthermore, the momentum propelling the change to sustainable healthcare is greatly enhanced by a deeper comprehension of environmental interdependencies. The complex interplay between environmental factors and human health is becoming more and more evident, necessitating a review of medical procedures. The understanding of how natural systems and human well-being are interrelated grows along with scientific knowledge. The understanding that human health and environmental health are inextricably linked has become a motivating factor for the healthcare industry to embrace preventative, preventive, and environmentally sound approaches. Analyzing these forces is important because it clarifies why sustainability is now a must rather than a choice in the modern healthcare environment. Setting the stage for a transformative journey is a worldwide appeal for responsible citizenship, which amplifies societal expectations. The ability of technological advancement to expand the parameters of what is possibly makes integrating sustainable practises into healthcare both necessary and possible (Karamat et al. 2019). Growing awareness of environmental interdependencies serves as a compass, steering the healthcare industry in the direction of a more ecologically conscious and holistic approach. As a result, the forces influencing healthcare sustainability are combining to create a powerful force that will lead the industry toward a time when providing quick medical attention is not as important as protecting the environment. Advancements in technology, societal expectations, and a deeper comprehension of environmental interdependencies combine to create a storey in which sustainability is viewed as a critical component of healthcare procedures. In addition to identifying these forces, this investigation aims to highlight how important it is for them all to work together to move the healthcare industry in the direction of a more resilient and sustainable future (Marimuthu et al. 2016).

2.3 Current State of Sustainability Integration in Healthcare

As the healthcare industry continues to change, it becomes increasingly important to integrate sustainability into its core values. This need necessitates a thorough evaluation of the level of sustainability integration in the healthcare industry today. It is based on the widespread understanding of the environmental difficulties we face. This evaluation process is essential because it reveals the patchwork of current programmes, laws, and inventions that either support or obstruct the adoption of sustainable healthcare practises globally. It is urgent to bring healthcare into line with sustainable principles since we are all aware of the environmental crisis and the fact that the healthcare industry has a substantial influence on the ecological balance due to its extensive reach and resource use. Therefore, assessing how sustainability integration is doing right now is an important technique to make sure you're on the right track and to consider areas that may use some improvement (Lennox et al. 2018).

This evaluation examines how sustainability has been assiduously knitted into the fundamental fabric of healthcare procedures worldwide, digging into the specifics of current programmes. It goes beyond a cursory analysis in an attempt to uncover the extent of healthcare systems' dedication to implementing sustainable practises. The adoption of a global lens guarantees a thorough examination, taking into account local differences and the many obstacles that healthcare systems encounter globally. The regulations controlling healthcare practises have a significant influence on how sustainable integration is shaped. This assessment examines the effectiveness of current policies, determining whether they facilitate or impede efforts to achieve sustainable healthcare. When policies are in line with sustainable principles, they can work as change agents, creating an atmosphere that is favourable to the adoption of eco-friendly activities and technologies. On the other hand, deficiencies in policy frameworks might provide obstacles that hinder the smooth integration of sustainability into the healthcare model (Lennox et al. 2020).

The assessment also looks at current advancements in the healthcare industry. It looks for and honours ground-breaking projects that best represent a dedication to sustainability. Innovations come in many forms, from improvements in environmentally friendly medical gadgets to trash reduction tactics and energy-efficient infrastructure. By highlighting these advancements, the evaluation not only acknowledges progress but also lays the groundwork for visualising a time when sustainability will be a basic feature of healthcare procedures rather than only an ideal. The next investigation of environmentally friendly advances in medical research and practise is built around this assessment snapshot, which is ingrained in the present condition of sustainability integration. It serves as an introduction to a more thorough examination of the revolutionary possibilities of sustainable healthcare methods. As the evaluation progresses, it creates the foundation for comprehending the difficulties encountered, the accomplishments attained, and the directions still to be explored in the direction of a more resilient and sustainable healthcare future (Hester et al. 2015).

3 Eco-Friendly Innovations in Medical Study and Practice

The goal of sustainability is evident in the dynamic field of healthcare, not only in broad ideas but also in concrete developments that influence medical research and practise. This section delves into the world of environmentally friendly advancements, where innovative discoveries are changing the face of healthcare and education. The investigation of these developments offers a window into a future in which healthcare and sustainability are inextricably linked, from ecologically friendly medical teaching methods to energy-efficient medical equipment.

3.1 Environmentally Sensitive Hospital Architecture

The concept of environmentally sensitive hospital architecture goes beyond the aesthetic appeal, delving into holistic design strategies that prioritize ecological responsibility. This approach considers the entire life cycle of the hospital facility, from construction to daily operations, aiming to minimize resource consumption, energy use, and environmental emissions. Utilizing eco-friendly materials in hospital construction is a fundamental aspect of environmentally sensitive architecture. These materials, sourced sustainably and often recycled, reduce the carbon footprint associated with the building's construction while promoting resource conservation (Rad et al. 2012). Incorporating energy-efficient design principles is paramount in environmentally sensitive hospital architecture. This involves optimizing natural light, implementing energy-efficient HVAC systems, and utilizing renewable energy sources such as solar panels. These strategies not only reduce energy consumption but also contribute to long-term cost savings. Sustainable hospital architecture emphasizes waste reduction and efficient recycling systems. From construction waste management to daily operational practices, hospitals are integrating robust recycling programs. This not only aligns with environmental goals but also instills a culture of responsibility among staff and patients. Integrating nature into the hospital environment is a cornerstone of environmentally sensitive architecture. Healing gardens, green roofs, and biophilic design principles are incorporated to enhance patient well-being. These elements not only contribute to a visually pleasing environment but also have therapeutic effects, promoting faster recovery and reducing stress. Sustainable hospital architecture places a strong emphasis on water conservation. Implementation of water-efficient fixtures, rainwater harvesting systems, and innovative landscaping practices helps reduce water consumption and minimizes the hospital's impact on local water resources (Abdelaal and Soebarto 2018).

3.2 *Energy-Efficient Medical Devices*

With its wide range of medical instruments, the healthcare industry has always been a significant energy user. These technologies, which range from life support systems to diagnostic imaging scanners, are essential to patient care yet frequently have substantial energy requirements. In order to overcome this difficulty, the idea of energy efficiency in medical equipment optimises energy utilisation without sacrificing the therapeutic effectiveness of the devices. Intelligent power management systems are given priority in energy-efficient medical devices. These systems are made to optimise energy usage by modifying power levels according to the needs of the device's operation. For example, diagnostic equipment that dynamically modifies its power consumption based on the imaging complexity not only consumes less energy but also lasts longer. Advancements in technology have paved the way for the development of energy-efficient medical devices that leverage cutting-edge innovations (Daniluk and Niewiadomska-Szynkiewicz 2012). Smart automation, machine learning algorithms, and sophisticated sensors work together to minimise energy use and enable precise operation of these devices. For instance, smart infusion pumps can use energy-saving features while not in use and dispense medication with the highest precision. The integration of renewable energy sources into medical device operations is a growing example of the synergy between sustainable energy practises and healthcare. Healthcare institutions like hospitals are looking at ways to power medical devices using renewable energy sources like solar, wind, or other energy. This lessens dependency on conventional energy grids and strengthens the foundation for a sustainable and resilient healthcare system. Energy-efficient medical equipment provide advantages for the environment, but they also save a lot of money and improve overall operational effectiveness in healthcare facilities. The long-term financial gains are particularly attractive to healthcare providers facing budgetary constraints and seeking avenues for sustainable cost reduction (Nia et al. 2015).

Energy-efficient medical device integration is a natural fit with the larger objectives of supporting sustainable healthcare practises. It displays a diligent attempt to strike a balance between providing patients with top-notch treatment and environmental stewardship. Healthcare providers are progressively integrating these technologies into their operational frameworks due to their dedication to patient well-being and ecological stewardship. The adoption of energy-efficient medical devices contributes significantly to reducing the overall environmental impact of healthcare operations (Jia and Zhao 2023). Reduced carbon emissions from lower energy use put healthcare facilities in line with international efforts to slow down climate change. The benefits reach patient outcomes in addition to the surroundings. Energy-efficient equipment, which is frequently fitted with cutting-edge technology, can help to improve patient care and recovery by increasing the precision of diagnosis and treatment. Energy-efficient medical device integration offers a potential path toward sustainable healthcare, but obstacles still need to be overcome. The upfront expenses linked with obtaining this cutting-edge technology may be a challenge for certain healthcare establishments. However, as the demand for energy-efficient solutions grows, economies of scale are likely to drive down costs, making these innovations more accessible (Sodhro et al. 2021).

3.3 Case Studies and Exemplary Projects

Table 2 summarises two notable case studies in healthcare sustainability and provides a brief overview of the innovative projects that Parkland Hospital and the Cleveland Clinic have been involved in. Within the framework of the Cleveland Clinic's initiative to "Green the Operating Room," the organisation coordinated a multimodal strategy. Surgical waste was significantly reduced because of a thorough waste reduction programme that was implemented in operating rooms. The programme placed a strong emphasis on material sorting and recycling. Energy-efficient LED lighting technologies have also been adopted, which has improved the working atmosphere in addition to reducing its environmental impact. The strategic transition to reusable surgical textiles not only demonstrated environmental responsibility but also yielded tangible financial benefits, exemplifying a harmonious blend of sustainability and cost-effectiveness (Weismann et al. 2020).

Conversely, Parkland Hospital made a great step toward "Sustainable Infrastructure Design," establishing itself as a model of environmentally friendly hospital design. With the attainment of LEED Gold accreditation, the hospital demonstrated its unwavering dedication to incorporating green construction methods. The hospital's design became replete with green roofs and ecological landscaping, which not only improved energy efficiency but also created healing outdoor areas for staff and patients. In addition, the hospital adopted the integration of renewable energy and installed solar panels to generate clean energy and lessen reliance on conventional energy sources. Parkland Hospital's outcomes underscored the potential for healthcare facilities to achieve sustainability goals while creating healing environments and fortifying operational resilience (Rich et al. 2021).

4 Positive Impacts on Human and Environmental Health

The adoption of sustainable practises in healthcare ushers in a revolutionary era when the health of people and the earth are intertwined, with numerous positive effects on both human and environmental health. This mutually beneficial link between sustainable healthcare and improved health outcomes is essential for paving the way for a day when patient welfare and environmental responsibility coexist. The

Table 2 Case studies in healthcare sustainability

Case study	Initiatives	Outcomes
Cleveland Clinic - Greening the Operating Room (Weismann et al. 2020)	- Waste Reduction: Comprehensive program for segregating and recycling materials - Energy-Efficient Lighting: Adoption of LED technologies in operating rooms - Reusable Surgical Textiles: Transition to reusable gowns and drapes	- Significant reduction in surgical waste, diverting a substantial amount from landfills - Lowered energy consumption through LED lighting, improving the working environment - Financial benefits from the use of reusable surgical textiles
Parkland Hospital - Sustainable Infrastructure Design (Rich et al. 2021)	- LEED Certification: Integration of green building practices for resource efficiency - Green Roof and Landscaping: Inclusion of green roofs and sustainable landscaping - Renewable Energy Integration: Use of solar panels for clean energy	- Achieved LEED Gold certification, showcasing a commitment to sustainable building practices - Improved energy efficiency and reduced stormwater runoff through green roofs and landscaping - Integration of renewable energy sources for a more sustainable and resilient energy supply

primary benefit of implementing eco-friendly technology and practises in healthcare institutions is the decrease in environmental damage. From energy-efficient medical devices to sustainable infrastructure designs, the healthcare sector is mitigating its ecological footprint (Salonen et al. 2012). Communities that surround healthcare institutions benefit from this decrease in environmental impact since it results in better air and water quality. By adopting sustainable energy sources and reducing waste, the healthcare sector contributes to a more robust and ecologically balanced world while also protecting ecosystems.

Furthermore, the benefits penetrate people's health and wellbeing in addition to the immediate environmental domain. Improved community health resilience and better patient outcomes have been associated with sustainable healthcare practises. For example, environmentally conscious hospital design frequently includes features that enhance patient comfort and recovery. Natural light, green spaces, and improved air quality in these facilities not only promote a positive patient experience but also facilitate faster recovery times (Reis et al. 2015). Positive impacts on public health have also been shown when sustainable techniques are included in medical research and practise. There is a push to lessen the health hazards connected with environmental pollution at the same time that the healthcare industry is aggressively working to reduce its carbon footprint. Healthcare facilities help to minimise incidence of respiratory disorders and other environmentally related diseases by switching to greener energy sources and minimising hazardous waste. This preventive strategy is in line with the fundamental principles of public health, which place a strong emphasis on treating underlying issues and fostering wellbeing both within the person and the society. A salient feature of the beneficial effects on human health is the comprehensive outlook that sustainable healthcare practises embrace. Beyond the advantages for physical health, the relationship between environmental and mental well-being is acknowledged. Stress reduction among patients and healthcare personnel has been linked to healthcare facilities that include green areas, natural surroundings, and aesthetically pleasant locations. This acknowledgment of the healing power of nature contributes to a more comprehensive and patient-centered approach to healthcare delivery (Gutierrez et al. 2022).

Furthermore, tackling global health issues—particularly those made worse by climate change requires the implementation of sustainable healthcare practises. Environmental variables are linked to the rising incidence of vector-borne illnesses, harsh weather, and interruptions to the healthcare system. The healthcare industry actively seeks to reduce the effect of these difficulties in addition to adapting to them by promoting sustainability. By taking a proactive approach, healthcare organisations may position themselves as major contributors to the worldwide endeavour to protect human health in the face of climate change by strengthening their resilience against new health risks. The positive impacts on human and environmental health also extend to the realm of social equity (Yazdi et al. 2014). Considering the needs of disadvantaged and vulnerable groups is a common component of sustainable healthcare practises. Sustainable practises address health inequities and lead to more equitable health outcomes by encouraging inclusion in healthcare delivery. This focus on social justice is in line with the larger objectives of public health, which include making sure that everyone has access to high-quality treatment and lives in surroundings that promote health and wellbeing. Adopting sustainable healthcare practises has real health advantages, but it also encourages innovation and a culture of accountability within the healthcare industry. Institutions foster a continual improvement attitude as they make commitments to waste reduction, resource conservation, and environmental effect minimization. This ethos of innovation and adaptability is crucial in addressing emerging health challenges and ensuring the long-term resilience of healthcare systems (Qureshi et al. 2015).

5 Enhancing Patient Outcomes and Community Health Resilience

The incorporation of sustainable practises into healthcare systems is a key factor in improving patient outcomes and strengthening the resilience of community health. This ground-breaking project acknowledges the fundamental link between patient treatment, environmental awareness, and communities' overall resilience. From energy-efficient medical technologies to eco-centric hospital designs, sustainable healthcare practices usher in a paradigm shift that places the well-being of individuals and communities at its core (Abrash Walton et al. 2021). Patient well-being is positively impacted by sustainable healthcare practises, and this is a key finding. A great patient experience is given priority in hospitals and other healthcare facilities that are constructed with sustainability in mind. Natural light, greenery, and aesthetically beautiful surroundings are examples of elements that have been shown to lower stress levels, improve mood, and hasten the healing process. This patient-centric approach echoes the principles of holistic healthcare, acknowledging that well-being extends beyond the absence of illness to encompass positive physical, mental, and social states (Nicholls et al. 2017). Sustainable healthcare practices contribute significantly to the resilience of healthcare infrastructure, especially in the face of emerging challenges like climate change and public health crises. Hospitals designed with energy efficiency in mind, coupled with the integration of renewable energy sources, exhibit a heightened ability to endure disruptions to traditional energy grids. This resilience is paramount during extreme weather events or other emergencies, ensuring the continuous delivery of healthcare and safeguarding the well-being of patients and communities (Chandra et al. 2010).

The paradigm of preventative healthcare is well aligned with the use of sustainable techniques in medical research and practise. Sustainable initiatives help to foster a proactive approach that not only enhances individual health outcomes but also lays the groundwork for community health resilience by actively addressing the underlying causes of health difficulties. The healthcare industry plays a crucial role in preventative health initiatives, contributing to the long-term well-being of communities by minimising the harm to the environment and the health risks connected with pollution. Social equality is promoted in the provision of healthcare services via sustainable healthcare practises. Making inclusion a top priority guarantees that healthcare facilities are built with cost and accessibility in mind, bridging the gap in health that frequently affects marginalised populations. The foundation of community health resilience is this emphasis on social equality, which acknowledges the close relationship between each member's health outcomes and the community's overall well-being. Initiatives involving communities in sustainable projects, educational programs, and collaborative efforts not only raise awareness about environmental and health issues but also empower individuals to actively participate in their own well-being (DeMello et al. 2020). By fostering a feeling of shared accountability for health outcomes, this community-centric approach lays the groundwork for resilience and long-term well-being. Public health policies that aim to improve overall community health and resilience align with the emphasis on preventative interventions and the development of a healthy environment. The relationship between health and economic well-being emphasises how important sustainable healthcare practises are to building comprehensive community resilience. The adoption of sustainable practises in healthcare is emerging as a paradigm-shifting force that puts the health of people and communities first in place of established methods. Sustainable healthcare practises are critical to a resilient and prosperous future because of their favourable effects on patient outcomes and community health resilience. This path represents a break from traditional healthcare paradigms, adopting a holistic perspective that acknowledges the complex relationships among environmental, personal, and community health. In embracing sustainability, the healthcare sector becomes a pivotal contributor to the enduring positive health outcomes of individuals and the collective resilience of communities (Rubin et al. 2013).

6 Overcoming Challenges: The Road to Sustainability

Setting off on the road to sustainability in healthcare requires careful evaluation of the obstacles that might stand in the way of advancement. This chapter explores the many challenges that the healthcare industry has in implementing sustainable practises. By examining these challenges with a critical lens, the chapter aims to illuminate potential solutions, strategies, and collaborative efforts required to navigate the complexities on the road to achieving a sustainable healthcare system (Zajac et al. 2021). Financial demands and resource limitations are two of the main obstacles to implementing sustainability in the healthcare industry. Healthcare organisations may find it difficult to pay for the up-front expenses of adopting sustainable practises, eco-friendly technology, and infrastructure changes, especially if they are already struggling with financial limitations. In order to help healthcare organisations make sustainability a viable and long-term commitment, this section examines creative funding models, economical fixes, and various incentives that might lessen financial pressures. The smooth integration of sustainable projects might be hampered by institutional cultures that are firmly embedded, resistance to change, and a preference for traditional methods. Strategic interventions are necessary to change attitudes and promote an environmentally conscious culture in healthcare facilities. The chapter examines case studies of successful cultural transformations, emphasizing the importance of leadership commitment, stakeholder engagement, and educational programs to overcome resistance and instill a culture that embraces sustainability (Crisp 2013).

Sustainability in healthcare demands interdisciplinary collaboration, yet siloed approaches often hinder effective communication and cooperation between diverse departments. Breaking down these silos and fostering collaboration between medical professionals, environmental experts, administrators, and policymakers are essential components of a sustainable healthcare ecosystem. Strategies for promoting cross-disciplinary partnerships and communication channels are explored in this section to bridge the gap and ensure a holistic approach to sustainability. Navigating the complex landscape of regulations and policies poses significant challenges for healthcare institutions aiming to adopt sustainable practices (Couturier et al. 2012). Achieving these goals will need an understanding of the regulatory environment, active participation in policy creation, and advocacy for favourable policies. The chapter explores instances of effective policy lobbying and cooperative initiatives between regulatory agencies and the healthcare industry to promote a more seamless shift to sustainable healthcare practises. When new technologies are integrated, especially ones that prioritise sustainability, worries about data security and operational disturbances could surface. Healthcare institutions have to manage the complexities of technology changes while maintaining patient data integrity and confidentiality.

This section examines solutions and best practises for a safe and smooth technology transition, highlighting the significance of risk assessments and strong cybersecurity defences. A sustainable healthcare model must involve patients in the transition to sustainability and raise public understanding of how healthcare practises affect the environment. Creating efficient communication plans, educational campaigns, and interactive projects are necessary to overcome obstacles pertaining to patient education and involvement. This section examines effective patient involvement models, showing how knowledgeable and involved patients support the implementation of sustainable healthcare practises. The healthcare industry's intricate worldwide supply networks make it difficult to get drugs, medical equipment, and sustainable materials. This chapter looks at methods for encouraging suppliers to use sustainable practises, increasing transparency, and optimising supply chains. Case studies highlighting successful initiatives to simplify and green healthcare supply chains offer insights into overcoming the challenges posed by global interconnectedness (Ippolito et al. 2023).

7 Opportunities and Solutions

Increasing funding for research and development of sustainable healthcare solutions is one important possibility. Fostering research and development (R&D) activities may stimulate innovation and build a solid foundation for sustainable practises, ranging from eco-friendly medical technology to green infrastructure. The chapter highlights the potential for breakthroughs in sustainable healthcare solutions by examining successful cases where R&D expenditures have produced game-changing outcomes. There are chances for cooperation between political agencies, environmental groups, and healthcare institutions as collaboration emerges as a potent way to tackle difficulties (Rowan and Galanakis 2020). By fostering collaborative relationships, institutions can share knowledge, resources, and best practices. Case studies illustrating successful collaborations underscore the importance of building networks to collectively address sustainability challenges in healthcare. The integration of technology offers a myriad of opportunities to enhance efficiency and sustainability. Technological innovations have the potential to improve patient care by streamlining procedures, reducing resource consumption, and improving smart healthcare systems and telemedicine solutions. The chapter explores the transformational power of technology, presenting examples of creative solutions that have improved healthcare delivery and advanced sustainability objectives. One of the most important opportunities to foster a sustainable culture in the healthcare industry is to empower stakeholders via educational efforts. Conscientious stakeholders, ranging from administrators and patients to healthcare experts, take up the cause of sustainable practises. This section examines effective educational initiatives, highlighting their contribution to raising consciousness, altering behaviour, and promoting a shared commitment to sustainability (Wu et al. 2023).

By placing a focus on resource efficiency, waste reduction, and product life extension, adopting the concepts of the circular economy gives a comprehensive approach to sustainability. There is discussion of how waste management, product design, and healthcare supply chains may all benefit from using circular economy principles. Case studies highlight situations where the circular economy has been effectively incorporated, offering insightful information to healthcare organisations looking for environmentally friendly solutions. Healthcare facilities have the chance to actively participate in policy lobbying in order to get regulatory support for sustainable practices (Ranjbari et al. 2022). Institutions may help create frameworks that require and promote sustainable healthcare by working with legislators. The chapter examines effective policy advocacy cases, emphasising the role that regulatory assistance plays in easing the shift to sustainable healthcare models. Building solid relationships with constituents is made possible for healthcare organisations through community engagement and public awareness campaigns. Institutions can get support for sustainable practises by implementing participatory projects, awareness campaigns, and community outreach activities. One effective way to encourage the adoption of sustainable practises is to offer both monetary and non-monetary incentives. We look at ways to take use of official awards, certification schemes, and rewards for long-term accomplishments. Case studies show how these incentives benefit organisations, which promotes a wider adoption of environmentally friendly methods in the healthcare industry (Daú et al. 2019; Fanta et al. 2021).

8 Conclusion

Through navigating the transformational terrain of sustainable healthcare, the chapters present an engaging storey of development, obstacles, chances, and perseverance. A paradigm change from traditional healthcare methods to a holistic, environmentally sensitive approach that acknowledges the inextricable relationship between environmental health and human health may be seen in the historical trajectory. The investigation of obstacles, ranging from lack of resources to opposition to change, acts as a lighthouse for creativity, and the solutions it offers include cooperative alliances, technology integration, and policy lobbying. Prospects for sustainable healthcare surface as rays of hope, enticing funding for R&D, adopting the concepts of the circular economy, and encouraging community involvement. The call to action, imploring stakeholders to commit to a sustainable future where healthcare institutions become stewards of both patient well-being and environmental preservation, is loud as we stand at the intersection of difficulties and possibilities. This chapter not only marks a turning point in the trip, but it also lays out a plan for healthcare organisations to leave a legacy in which sustainability is a constant commitment to a resilient and healthy world rather than merely an objective.

References

Abdelaal M, Soebarto V (2018) The death of modern hospitals: Towards a comprehensive approach for restorative healthcare environmental design

Abrash Walton A, Marr J, Cahillane MJ, Bush KJS (2021) Building community resilience to disasters: a review of interventions to improve and measure public health outcomes in the northeastern united states 13(21):11699

Chandra A, Acosta J, Meredith LS, Sanches K, Stern S, Uscher-Pines L, Williams M, Yeung DJSM, CA: RAND Corporation (2010) Understanding community resilience in the context of national health security

Couturier J, Sola D, Borioli G, Raiciu CJC and Strategies (2012) How can the internet of things help to overcome current healthcare challenges (87):67–81

Crisp HJJORC (2013) Overcoming challenges in improvement work 39(S2):30–34

Daniluk K, Niewiadomska-Szynkiewicz E (2012) Energy-efficient security in implantable medical devices. 2012 federated conference on computer science and information systems (FedCSIS), IEEE

Daú G, Scavarda A, Scavarda LF, Portugal VJTJS (2019) The healthcare sustainable supply chain 4.0: the circular economy transition conceptual framework with the corporate social responsibility mirror 11(12):3259

DeMello A, Egan R, Drew JJJOTRSONZ (2020) Resilience-building by community health organizations: a guiding model for practice 50(4):552–571

Fanta GB, Pretorius L, Nunes B (2021) Enabling circular economy in healthcare using industry 4.0 digital technologies. Proceedings of the 30th International Conference of the International Association for Management of Technology, IAMOT

Gutierrez AM, Frazar EM, Klaus MVX, Paul P, Hilt JZJAHM (2022) Hydrogels and hydrogel nanocomposites: enhancing healthcare through human and environmental treatment 11(7):2101820

Hester J, Stange PV, Seeff LC, Davis J, Craft CA (2015) Toward sustainable improvements in population health: overview of community integration structures and emerging innovations in financing

Ippolito A, Sorrentino M, Capalbo F, Di Pietro AJIJOP, Management P (2023) How technological innovations in performance measurement systems overcome management challenges in healthcare 72(9):2584–2604

Jia Y, Zhao LJIS-SCM (2023) Implantable medical devices for wireless optical neuromodulation and neural recording: energy-efficient integrated circuit and system design 15(4):30–40

Karamat J, Shurong T, Ahmad N, Afridi S, Khan S, Khan NJS (2019) Developing sustainable healthcare systems in developing countries: examining the role of barriers, enablers and drivers on knowledge management adoption 11(4):954

Lennox L, Linwood-Amor A, Maher L, Reed JJHRP and systems (2020) Making change last? Exploring the value of sustainability approaches in healthcare: a scoping review 18(1):1–24

Lennox L, Maher L, Reed JJIS (2018) Navigating the sustainability landscape: a systematic review of sustainability approaches in healthcare 13(1):1–17

Mallee HJCOIES (2017) The evolution of health as an ecological concept 25:28–32

Marimuthu M, Paulose HJP-S, Sciences B (2016) Emergence of sustainability based approaches in healthcare: expanding research and practice 224:554–561

Molero A, Calabrò M, Vignes M, Gouget B, Gruson DJAOLM (2021) Sustainability in healthcare: perspectives and reflections regarding laboratory medicine 41(2):139–144

Nia AM, Mozaffari-Kermani M, Sur-Kolay S, Raghunathan A, Jha NKJITOM-SCS (2015) Energy-efficient long-term continuous personal health monitoring 1(2):85–98

Nicholls K, Picou SJ, McCord SCJJOPHM and practice (2017) Training community health workers to enhance disaster resilience 23:S78-S84

Proctor E, Luke D, Calhoun A, McMillen C, Brownson R, McCrary S, Padek MJIS (2015) Sustainability of evidence-based healthcare: research agenda, methodological advances, and infrastructure support 10:1–13

Qureshi MI, Khan NU, Rasli AM, Zaman KJES, Research P (2015) The battle of health with environmental evils of Asian countries: promises to keep 22:11708–11715

Rad NB, Ghasemzadeh B, Kolour YJAC (2012) Hospitals adaption to principles of sustainable architecture (Case study in 10 hospitals) 2:187–195

Ranjbari M, Esfandabadi ZS, Shevchenko T, Chassagnon-Haned N, Peng W, Tabatabaei M, Aghbashlo MJJOHM (2022) Mapping healthcare waste management research: past evolution, current challenges, and future perspectives towards a circular economy transition 422:126724

Reis S, Morris G, Fleming LE, Beck S, Taylor T, White M, Depledge MH, Steinle S, Sabel C, Cowie HJPH (2015) Integrating health and environmental impact analysis 129(10):1383–1389

Rich RK, Jimenez FE, Puumala SE, DePaola S, Harper K, Roy L, Brittin JJHHER, Journal D (2021) From fable to reality at Parkland Hospital: the impact of evidence-based design strategies on patient safety, healing, and satisfaction in an adult inpatient environment 14(3):65–82

Rowan NJ, Galanakis CMJSOTTE (2020) Unlocking challenges and opportunities presented by COVID-19 pandemic for cross-cutting disruption in agri-food and green deal innovations: Quo Vadis? 748:141362

Rubin SE, Roszak ARJJOPHM, Practice (2013) Improving community resilience through public health preparedness partnerships 19(4):388–390

Salonen H, Lappalainen S, Lahtinen M, Knibbs L, Morawska L (2012) Positive impacts of environmental characteristics on health and wellbeing in health-care facilities: a review. Healthy Buildings International Conference (10th)

Shelton RC, Cooper BR, Stirman SWJAROPH (2018) The sustainability of evidence-based interventions and practices in public health and health care. 39:55–76

Sherman JD, Thiel C, MacNeill A, Eckelman MJ, Dubrow R, Hopf H, Lagasse R, Bialowitz J, Costello A, Forbes MJR (2020) Conservation and recycling. The green print: advancement of environmental sustainability in healthcare 161:104882

Sodhro AH, Al-Rakhami MS, Wang L, Magsi H, Zahid N, Pirbhulal S, Nisar K, Ahmad A (2021) Decentralized energy efficient model for data transmission in IoT-based healthcare system. 2021 IEEE 93rd Vehicular Technology Conference (VTC2021-Spring), IEEE

Stoughton AM, Ludema JJJOOCM (2012) The driving forces of sustainability 25(4):501–517

Weismann M, Ganske S, Delgado OJTCJ (2020) Cleveland Clinic Florida "pay-for performance" reimbursement: why the best care does not always make the happiest patients 16(5):625–663

Wu K, Wang XJES, Research P (2023) Studying financial aspect of green credit and regional heterogeneity on technology innovation in China 30(41):93708–93721

Yazdi S, Zahra T, Nikos MJRAIAE (2014) Public healthcare expenditure and environmental quality in Iran 1:126–134

Zajac S, Woods A, Tannenbaum A, Salas E, Holladay CLJFIC (2021) Overcoming challenges to teamwork in healthcare: a team effectiveness framework and evidence-based guidance 6:606445

Sustaining Sustainable: Investigating the Full Spectrum of Food Waste, from Production Through Consumption to Disposal

Leena N. Fukey and Kedarnath Rajaram

Abstract **Purpose:** The purpose of this research is to explore the primary factors that contribute to food waste. Additionally, it creates practical strategies to cope with food waste and encourages to perform sustainable practices to improve the environment. Additionally, the study presents an analytical framework for supply chain problems as well as the methods that are environmentally friendly.

Methodology: The research begins by defining how sustainable development should be incorporated in the hospitality sector and by briefly outlining its attributes. Next, it discusses the expanding interest in supply chain management and outlines an overview of the breadth of academic research on sustainability in the literature related to the hotel industry.

Findings: The paper examines the enormous ecological and economic effects of food waste, including how it contributes to the adverse global warming, the dwindling of natural resources, and also the loss of worthwhile financial investments. Additionally, it highlights the social effects of food wastage, such as how it contributes to gaps in access to nourishing foods and food insecurity.

Research Limitations: It attempts to shed light on the scope of food loss, identify major contributing variables, and suggest methods to reduce food loss along the whole supply chain through an examination of current literature and data.

Practical Implication: The practical application of this research is to offer evidence-based insights and practical recommendations to policymakers, organizations, and people with an aim to decrease food waste and enhance the effectiveness of the food system.

Keywords Sustainability · Supply chain management · Food waste · Resource management

---

Sustaining Sustainability: Investigating the Full Spectrum of Food Waste, from Production through Consumption to Disposal.

L. N. Fukey (✉) · K. Rajaram
Department of Hotel Management, School of Business and Management, Christ (Deemed To Be University), Bangalore, Karnataka, India
e-mail: leena.n.fukey@christuniversity.in

P. K. Prabhakar and W. Leal Filho (eds.), *Preserving Health, Preserving Earth*, World Sustainability Series, https://doi.org/10.1007/978-3-031-60545-1_8

1 Introduction

Food waste management continues to be one of the most difficult issues to address in the hospitality business. Out of the 7.5 billion people that inhabit this planet, 795 million go hungry, which indicates that one amongst nine individuals worldwide have a serious condition of chronic malnutrition (Hunger News 2016). The paradox in this situation is that despite a notable quantity of food being discarded each day, a very huge number of individuals still sleep hungry. The estimation given by the Food and Agriculture Organisation of the United Nations (FAO 2017) states that more than a billion tonnes of food produced worldwide for human consumption are discarded. The United Nations (UN) proposes the Sustainable Development Goals (SDG) to recognise the hyperlink between food waste and food insecurity. The condition of emerging economies shows an even greater problem with food waste in the food service industry (Papargyropoulou et al. 2019). For instance, China's catering industry accounts for half of the nation's overall food waste (Wen et al. 2018). This lopsided proportion is mostly related to Native individual's higher eating out frequency as a result of rising living standards (Filimonau et al. 2019). The absence of comprehensive research pertaining to food waste within emerging economies' food service industries poses a difficulty in comprehending its fundamental causes and devising practical strategies for mitigation. This challenge arises due to the scarcity of empirical analyses tailored to specific market conditions in these economies (Principato et al. 2018) (Table 1).

Table 1 Systematic representation of total food waste *Source* Classification of food and drink waste relating to the possibility of mitigating their production (Culgin et al. 2017)

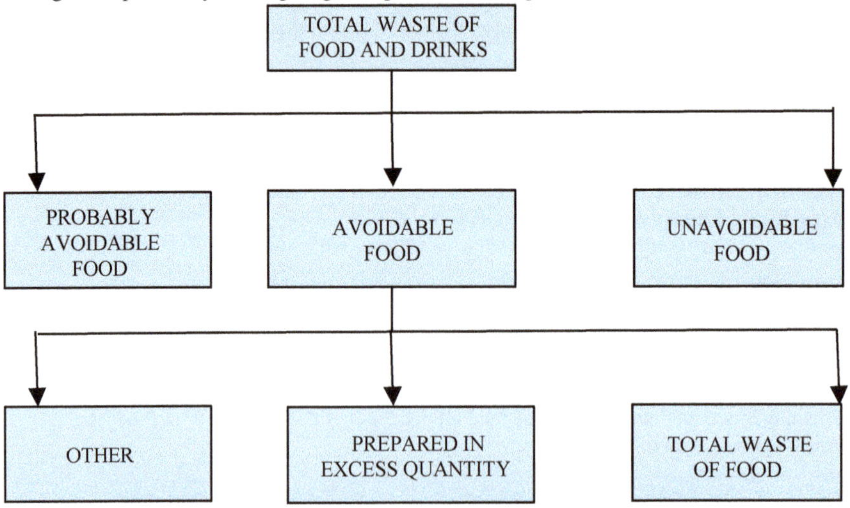

The above table provided discusses the concept of "avoidable food waste" and highlights its components. It also emphasizes that most avoidable food waste is not hazardous and should be composted or sent to different waste treatment programs.

Avoidable food waste typically includes the following categories:

- Excessive Food Preparation: Food that is prepared or served in excessive amounts, leading to leftover food that may go to waste.
- Cooking-Related Waste: Food that has been damaged during cooking, such as burnt or overcooked food.
- Mismatched Volume or Recipe: Food that is not used because it was prepared in the wrong quantity or doesn't match the intended recipe.
- Expired Food: Food that has not been consumed before its expiration date.

It's important to note that these types of avoidable food waste are not inherently hazardous to health, unlike some other types of food waste. Therefore, they can be managed and disposed of in more environmentally friendly ways, such as composting. Composting involves breaking down organic matter (food scraps, in this case) into nutrient-rich soil or compost that can be used to enrich the soil in gardens or agriculture. Proper waste separation and recycling programs can help divert these avoidable food wastes away from landfills, reducing environmental impact and promoting sustainability (Cuglin et al. 2017).

Sustainability has gained significant prominence within the hotel industry in recent years, with numerous studies focusing on this subject (Barber 2014; Barber and Deale 2013; Kim et al. 2012). The growing awareness of rapidly emerging environmental issues that can profoundly impact business operations has led society to prioritize sustainability over other environmental concerns (Singjai et al. 2018). Sustainability, encompassing environmental, economic, and socio-cultural aspects, is commonly defined as "development that meets the present generation's needs without jeopardizing the ability of future generations to fulfill their own requirements" (United Nations Economic Commission for Europe 2017). The implementation of sustainable practices is vital for ensuring a high quality of life, both now and in the future. In academic circles, many researchers emphasize the significance of establishing a positive connection between sustainability and industry operations (Han et al. 2010; Lu and Leung 2003; Wijesinghe 2013).

Environmental sustainability has been a well-researched topic for many years, encompassing a wide range of content (Singjai et al. 2018). In this research, as well as in the analysis of previous researches and future research suggestions, we define a sustainable hotel as an accommodation business that conducts its operations while taking into account their environmental impact (e.g., conserving water and energy, reducing emissions and waste production), economic impact (e.g., encouraging suppliers to use less packaging), and social impact (e.g., involvement in local initiatives related to education, poverty alleviation, and charitable contributions) (Fairmont 2012; Graci and Kuehnel 2011). In this context, managers implement sustainable enhancements.

(International Hotels Environment Initiative [IHEI] 1993) and work to minimize adverse environmental, economic, and social consequences (Bohdanowicz 2005; Chan and Lam 2003, 2007).

The hospitality sector within the tourism industry faces considerable environmental challenges, including escalating energy demand, heightened pressure on solid waste management, and the pollution of water bodies, soil, and the air. These negative environmental impacts create significant uncertainty, especially regarding their long-term implications, such as their role in contributing to global climate change. Consequently, there is a growing imperative to make the hospitality sector more environmentally sustainable and establish carbon–neutral accommodations in various destinations.

In light of the recognized risks of environmental degradation, governments, the environmental advocacy within the hospitality industry, and travelers are all increasingly aware of the necessity for more effective environmental protection measures. To achieve substantial improvements, it is essential for managers and operators in the hospitality industry to adopt environmentally responsible practices. Implementing recycling and waste reduction measures can significantly reduce greenhouse gas emissions, mitigate further ozone layer depletion, and thereby reduce global warming and pollution levels. The anaerobic digestion (AD) of food waste is considered an effective approach for energy production and environmental enhancement. Additionally, the byproduct of the AD process, known as digestate, can serve as a natural fertilizer, replacing synthetic alternatives.

India holds the position of the world's second-largest fruits and vegetables producer, trailing only behind China (National Horticulture Database 2014). During the year 2014, India's fruit production accounted for 88.97 million metric tonnes, while vegetable production stood at 162.98 million metric tonnes (Handbook on Horticulture Statistics 2014). These figures indicate 12.6% and 14% correspondingly, of the global fruit and vegetable output. Despite this substantial production, only a meager 1 to 2% is earmarked for export. Throughout the journey from harvest to consumption, approximately 18% of the total fruits and vegetables produced face wastage (Food and Agribusiness Strategic Advisory & Research 2014). Urban regions have evolved into hubs of intensive resource consumption, including water, energy, and food (WEF), driven by population growth and urbanization trends. Addressing the issue of scarce WEF resources necessitates equal emphasis on managing food consumption alongside production. Food wastage is defined by the Food and Agriculture Organisation (FAO) as "wholesome edible material intended for human consumption, arising at any point in the food supply chain (FSC) that is instead discarded, lost, or degraded" (Gustavson et al. 2011). The loss or wastage of edible materials along various stages of the food supply chain, including harvesting, post-harvest handling, processing, distribution, and consumption, contributes to this. The energy and resources used in food production and delivery are considerable. Food losses in the supply chain can occur due to factors such as mechanical damage, spillage, and degradation caused by improper handling, packaging, and threshing. These losses have direct implications for food quality, security, safety, economic development, and the environment (Gustavsson et al. 2011a, b).

They also have a direct negative impact on both farmers and consumers, as valuable natural resources used in food production go to waste. Moreover, such losses lead to unnecessary CO2 emissions, further exacerbating existing environmental challenges.

The research aims to address inquiries concerning the reasons and repercussions of food wastage. It also seeks to investigate the management of food waste within the hospitality sector and explore potential enhancements to the operational system for improved overall management outcomes. The research introduces an Internet of Things (IoT)-based system for managing restaurant food waste (RFW), encompassing all stages from waste generation to disposal, with a focus on creative techniques to prevent, reuse, or recycle food waste in restaurant settings. Additionally, the study aims to identify and rank the factors contributing to food wastage in the fruits and vegetables supply chain.

2 Review of Literature

2.1 Supply Chain

One of the major issues facing a developing country like India is food loss along the supply chain. With the recent rise in public health, food safety, and food quality concerns, demand and price volatility, and the short shelf life of these products, managing fresh food supply chains is particularly complicated. Inventory management is crucial for agri-food supply chains because of the short shelf life and perishability of the products (Beshara et al. 2012). Traditionally, supply chain management has been seen as a process where raw resources are transformed into finished goods before being shipped to customers. But during the last few decades, the trash and emissions generated by the supply chain have emerged as one of the primary causes of major environmental issues including acid rain and global warming. So, among the common business practices, the necessity for environmental sustainability has been gaining ground. In light of the aforementioned changes, "Green Supply Chain Management" (GSCM) has been able to pique the growing interest of operations and Individuals engaged in supply chain management research and practice explore the realm of Green Supply Chain Management (GSCM). This domain encompasses the complete lifecycle of a product, spanning its conception, manufacturing, distribution, utilization by end users, and eventual disposal (Borade and Bansod 2007). Supply chain management (SCM) is being made more environmentally conscious because of GSCM (Gilbert 2000). The green supply chain within the context of hotels constitutes an interlinked network of diverse operations and processes responsible for delivering environmentally conscious services and products offered by hotels. This network spans from suppliers to the hotel's clientele, ultimately elevating the ecological value of the hotel's environmental offerings (Zhang et al. 2009). This enhancement is a direct consequence of efforts to improve the environmental value of hotel services (Al-Aomar and Hussain 2017; Schwartz et al. 2008).

For the effective delivery of daily hospitality services, hotels and restaurants, which are significant providers of hospitality, collaborate with suppliers to acquire beverages, food, culinary supplies, laundry linen, and manage inbound logistics (Fantazy et al. 2010; Pullman and Rodgers 2010; Smith and Xiao 2008; Song 2011). In this scenario, substantial effort goes into integrating and coordinating with multiple suppliers. The ongoing aim is to offer top-quality services through partnerships with different business counterparts on projects that complement each other. As a result, managing the relationships between hospitality businesses and their suppliers becomes a noteworthy challenge (Fantazy et al. 2010; Ku et al. 2011; Wang et al. 2013; Zhang et al. 2009). Central to the establishment and maintenance of these interorganizational relationships is collaborative teamwork between these entities (Aldrich and Herker 1977; Schopler 1987; Soltani and Wilkinson 2010). Joint teams are formed, involving a minimum of two representatives from each organization participating in the supply chain. These team members act as ambassadors for their respective organizations, ensuring the smooth flow of information, materials, and products. This approach bolsters interorganizational collaboration, enhances relationship quality, and maximizes the value derived from these interorganizational bonds (Ferguson et al. 2011; Narus and Anderson 1995; Stock 2006).

Significant shifts in society, such as industrialization, globalization, changing lifestyles, demographic variations, and technological progress, have largely contributed to the substantial evolution of our food cultivation, preparation, and retail practices over the last century (Guthrie et al. 2006). The economic influence of food-related activities, including the hospitality sector, which heavily relies on food and its components, is estimated at nearly 20% of the nation's GDP (Connolly and McGing 2007). The manner in which we produce, distribute, and sell our food has been under scrutiny and reexamination due to the growing consumer demand for healthier food options (Guthrie 2006). A mounting conversation surrounds the extent to which our pursuit of efficiency through large-scale operations and industrialized food production techniques has distanced us from our agrarian origins. This dialogue is substantiated by a movement that has led decision-makers, advocacy groups, and specialized interests to come together, recognizing the necessity to adapt our food practices to align with shifting consumer preferences, societal changes, and sustainability imperatives.

2.2 *Sustainability Approaches*

The primary aim of sustainable practices in the hospitality sector, particularly in "green" hotels, is to mitigate their adverse social and environmental impacts. Among American businesses, the lodging industry stood out as the fourth-largest energy consumer, as identified by the alliance to preserve energy (Fauziah et al. 2017). Activities linked to tourism can significantly harm the environment, culture, delicate ecosystems, and lead to pollution, conflicts between visitors and locals, and overcrowding.

Consequently, the hospitality industry's endeavors to promote conservation, protection of these resources, and raising visitor awareness about their importance should be endorsed. The United Nations has dedicated its sustainable goals to inform and guide global leaders in making sustainability-focused decisions, considering the role of human activity in global warming.

The Leadership in Energy and Environmental Design (LEED) standards guide builders in constructing certified green structures. LEED certification evaluates a hotel's sustainability based on five criteria: sustainable site design, water conservation, energy efficiency, material selection, and indoor environmental quality (Fukey and Issac 2014). Since its introduction in 2000, approximately 60 hotels have sought LEED certification (Butler 2008). Some operators are hesitant to invest in the future due to the perception that clients are primarily concerned about recycling (Butler 2008).

As the world's significant service sector, tourism generates over $2 trillion annually and employs 15% of the global workforce (Martinez 2013). Properly managed, tourism can contribute to economic growth, generate employment, income, and foreign exchange, develop infrastructure, aid in environmental preservation, and enhance local well-being (Dodds and Kuehnel 2010). However, irresponsible tourism can lead to negative consequences like leakage, low wages, environmental degradation, cultural dilution, and more (Agarwal 2002).

Sustainability remains a subject of ongoing conceptual discussions, tied to broader socio-economic, ideological, and political trends. These debates necessitate further exploration in sustainable tourism research (Bramwell and Lane 2005). The notion of sustainability should be seamlessly integrated into all forms of tourism activities and environments (Saarinen 2006). The concept of sustainable tourism, which originally emerged as a reaction to change and fear, is now evolving to identify positive benefits (Bramwell and Lane 2012).

2.3 *Energy Conversion*

The concern of food waste (FW) stems from its potential to contaminate water and land and contribute to greenhouse gas emissions. According to the FAO, roughly one-third of the global food supply is lost or wasted, with some estimates suggesting this could be as high as 50% (Lundqvist et al. 2008). In the EU alone, there is an astonishing quantity of more than 180 million tons of FW, resulting in significant economic costs amounting to approximately 143 billion euros. The primary source of FW occurs within the distribution chain, with a major contribution coming from the hospitality industry.

Restaurants, hospitals, cafeterias, and catering services collectively form the hospitality sector, which generates a substantial portion of global FW. Among these, restaurants are the largest contributors, accounting for around 21% of FW (Principato et al. 2018). FW within the hospitality sector encompasses various stages of the food supply chain, starting from challenges in storage and transportation and extending to issues during cooking and the disposal of uneaten food. In urban waste systems, FW from the hospitality sector represents a significant proportion, comprising approximately 30–60% of solid urban residue (SUR) (Kaur et al. 2019). Consequently, local services are often responsible for its collection and management. Adapting governmental regulations to facilitate the use of FW as a resource in industrial processes is crucial. An example is the European Waste Framework Directive, which outlines technical specifications for composting and digestion processes involving FW. Additionally, standards like RSB-STD-01–010 are in place to regulate the production of biofuels and bioenergy using end-of-life products, detailing parameters such as hygiene, impurity levels, odor, pH, and greenhouse gas emissions. To effectively leverage FW in bioprocesses, a comprehensive understanding of its composition and degradation is essential (Neves et al. 2008).

2.4 Food Waste Causes

In the realm of food and beverage, when a diner or staff member in a restaurant or food service establishment doesn't consume a food product, it falls under the category of waste. This encompasses unconsumed food and beverages meant for customers as well as kitchen-related refuse like eggshells, unused portions of fresh ingredients, oils, and spoiled or expired items. Pre-consumer food waste, as identified (Baldwin and Shakman 2012), often arises due to several factors such as unforeseen fluctuations in demand, excessive inventory, ineffective production practices, communication gaps, behaviors of the staff, and inadequate trimming techniques. Unidentified demand is the primary and most frequent reason why food is wasted. Sometimes it's difficult for kitchen staff to predict how many visitors they'll serve and what menu will be the most appropriate and economical. Studying consumer preferences as well as other elements like, weather trends, and local rivalry is essential for proving the prognosis (Baldwin and Shakman 2012). Another factor that causes food service locations to waste food is overstocking. Some restaurants prepare more food than they actually need because they don't like to tell their clients when they can't meet an order. This margin of error strategy may be effective in a variety of circumstances, however excessive margin prediction can result in wastage (Baldwin and Shakman 2012). Additionally, ineffective communication among restaurant staff members might result in the production of food waste. Due to a variety of factors, including the physical structure of the operation, linguistic or cultural differences, or a lack of time, miscommunication between the front and back of the house can lead to failure in controlling food production (Baldwin and Shakman 2012).

Overproduction is the primary cause of food waste in restaurants. Chefs and restaurant managers desire not to run out of any food items, thus they frequently stock up and produce in enormous quantities. Despite the fact that the goal is a good one, this generally results in food waste. Overproduction is regarded as a risk management strategy in the food service sector, yet we pay a high price for it in the form of food waste (Hogan 2016).

2.5 Materials and Methods

This particular study conducted a systematic review of the literature, exploring several key themes: (a) Supply Chain, (b) Sustainable Approaches, (c) Sustainability Audits, (d) Energy Conversion, and (e) Food Waste Causes, all within the context of the hospitality industry. The study collected data from well-known databases including Emerald Full Text, EBSCO, Elsevier's ScienceDirect, JSTOR, Taylor & Francis, and Springer-Verlag. The researchers employed search terms such as "Green Hotels," "Sustainable development," and "Resource management" to gather information. The main objective was to synthesize and organize knowledge disseminated between 2009 and 2023, aiming to enhance the visibility of sustainable development within the hospitality sector and contribute to further investigations in the field. The study's intention was to identify gaps and limitations in existing literature and suggest potential research directions. The research process was divided into four stages: data identification, initial data screening, eligibility selection, and data inclusion. Google Scholar was chosen as the primary database due to its extensive content and broad scope, which facilitated access to numerous publications spanning many years.

In the third stage, all articles published between 2009 and 2023 were identified. Downloading all papers that complied with the requirements for the research mentioned in the previous phase was the fourth step. The titles and abstracts of this information were kept in a folder for subsequent review. During the fifth and sixth stages, the titles and abstracts of the materials were examined to classify them. Irrelevant papers were excluded, and relevant information from the retained papers was documented in a spreadsheet for subsequent review.

The research design is the comprehensive framework that guides the entire research process, encompassing the planning, execution, and data analysis. In this particular investigation, the survey method was chosen as the research design to examine sustainable practices in the hotel industry and to identify the factors contributing to food waste. The study focuses on hotels and restaurants across Karnataka as the target sample from the broader population. This research design provides the structure for data collection, measurement, and analysis, enabling a systematic approach to achieve the research objectives.

The chapter employed a data-driven approach called bibliometrics to analyze research trends regarding environmental sustainability from 2005 to 2018. This method, as supported by several references, involved examining bibliographic materials to identify patterns and themes within the literature. By using bibliometrics, the chapter effectively categorized and organized the data, allowing for the analysis of organizational trends surrounding the perception, management concepts, and tools related to environmental sustainability over the 11-year period.

Years	Themes				Total
	Supply chain	Sustainability approaches	Energy conversion	Food waste causes	
2009–10	1		1	1	3
2010–11		2	1		3
2011–12	2			2	4
2012–13	2		2		4
2013–14		3		1	4
2014–15	3	3	3		9
2015–16		1		3	4
2016–17	2	1	1		4
2017–18				2	2
2018–19		2	2		4
2019–20	1			3	4
2020–21					
2021–22		3	3	4	10
2022–23	4		3		7
Total	15	15	16	16	62

Source Authors own illustration

TABLE ILLUSTRATION

The table that is being shown provides an insightful overview of several issues that range from 2009–2023 while illuminating trends and priorities throughout time. The themes cover "Supply Chain," "Sustainability Approaches," "Energy Conversion," and "Food Waste Causes," offering a thorough canvas of important areas to take into account.

The total number of responses for each theme over the years emphasizes their importance in the data even more. The contributions from supply chain and sustainability strategies are roughly equal at 24.19% each, demonstrating their continuity in importance through time. The sources of energy conversion and food waste each have a somewhat greater percentage, or about 25.81%, which reflects the various levels of attention they receive across the given period of time.

3 Findings and Discussion

The hotel industry is a dynamic and diverse sector encompassing a wide range of accommodation services, from budget-friendly motels to luxurious resorts. It plays a vital role in the global economy, providing travelers with a home away from home and contributing significantly to local and national economies. However, the industry is also known for its resource-intensive operations, which can have adverse environmental and social impacts. This is where sustainable practices come into play.

Hospitality and Sustainability are something which should go hand in hand to make this earth a better space to live in. The concepts under each of them brighten the perspective of sustainable development. In reality it is very difficult to inculcate sustainable practices in the ever changing customer centric industry. It might be considered "unfair" to claim that "nothing" has occurred. Adopting green and economically friendly practices is being considered as a bon by many of the researchers and practitioners. On the contrary, the hospitality businesses have not failed to show "the turning point in mankind's recognition of the biodiversity of creation," (Hegarty 1992). A system based view which addresses the environmental, economical, and social imperatives of sustainable development, suggest that sustainable hospitality can be said to be "infancy," with notable "silent spaces" (Chung and Parker 2010). The state of sustainable hospitality is nothing but the area which is left behind within the huge context of sustainable tourism (Bramwell and Lane 2012). The social-ecological-economic linkages that necessitate a systems-level approach is nothing but sustainable development. This viewpoint is best defined as "a way of understanding reality that emphasizes the relationships between a system's parts rather than the parts themselves" (Hjorth et al. 2006). The key question is how sustainable tourism might be pursued in a way mindful of wider complex systems, considering not only destination-level management issues but also the linkages formed throughout the tourism value chain (Pomering et al. 2011).

Food waste is a complex and pervasive issue that significantly challenges the overarching goal of sustainability. From the moment food is produced and harvested, through its distribution and consumption, to its ultimate disposal, this waste represents an inefficient and unsustainable use of resources. Food waste and sustainable practices play a pivotal role in the hotel industry's ongoing efforts to reduce its environmental footprint while enhancing operational efficiency. Hotels, being hubs of culinary activity, generate significant food waste in their daily operations. Addressing this challenge is essential not only for minimizing economic losses but also for promoting environmental responsibility. Sustainable practices in the hotel industry involve measures such as efficient inventory management, portion control, and innovative solutions for repurposing surplus food. In addition to reducing food waste, hotels are increasingly embracing sustainability through local sourcing of ingredients, energy-efficient technologies, water conservation strategies, and community engagement. These practices not only contribute to reducing the environmental impact but also resonate with environmentally conscious guests, making sustainability a key differentiator for the modern hotel industry. As hotels strive to find the delicate balance between culinary indulgence and responsible stewardship, their commitment to sustainable practices is central to the broader quest for a greener and more environmentally friendly hospitality sector.

After a comprehensive examination of numerous papers in this particular research domain, a notable research gap emerges, which pertains to the absence of a standardized sustainability audit framework. This framework is needed not only to evaluate and quantify the waste generated by hotels but also to assess the effectiveness of their sustainable practices. The lack of a universally recognized and established sustainability audit tool poses a significant challenge in the hospitality industry.

Without such a framework, hotels struggle to systematically review and measure their waste management processes and sustainable initiatives. This deficiency hinders their ability to accurately gauge their environmental impact and their progress toward sustainability goals. In essence, the identified research gap underscores the pressing need for the development and implementation of a structured sustainability audit system tailored to the unique needs of the hotel industry. Such a system would provide a standardized approach to assess, monitor, and enhance sustainability practices, thereby addressing a crucial void in the field.

The benefits of using sustainable practices in the hotel industry are numerous. Firstly, embracing sustainability can lead to reduced operational costs through energy and water conservation, waste reduction, and more efficient resource management. Sustainability initiatives also enhance a hotel's reputation, attracting eco-conscious travelers and potentially increasing occupancy rates. Moreover, by reducing their carbon footprint and environmental impact, hotels can contribute to a healthier planet and a cleaner local community, aligning with societal expectations and regulatory requirements. Sustainable practices can also lead to positive employee morale, as staff members often take pride in working for environmentally responsible establishments. Overall, integrating sustainability into the hotel industry not only fosters economic advantages but also promotes social and environmental well-being.

The research conducted for sustainability and food waste management in the hospitality industry is of paramount significance. It serves as a vital catalyst for reducing the environmental impact of the industry by identifying and implementing eco-friendly practices, such as minimizing waste, conserving resources, and adopting sustainable sourcing. These efforts not only contribute to a healthier planet but also result in cost savings and compliance with stringent environmental regulations.

Furthermore, the research bolsters the reputation of hospitality establishments, making them more attractive to eco-conscious consumers. By engaging with local communities and addressing their concerns, businesses can foster goodwill and support. Additionally, it fuels innovation and adaptability, positioning businesses for long-term success in a changing world.

Beyond its environmental implications, research in this field can have socio-economic benefits by potentially reducing food insecurity and facilitating food donations to those in need. In essence, research in sustainability and food waste management is an integral and forward-looking endeavor, with wide-reaching implications for both businesses and society, enhancing their long-term viability and contributing to a more responsible and eco-friendly hospitality industry.

4 Conclusion

The research study digs into the diverse world of sustainable practices in the hospitality industry, with a particular emphasis on solving the critical issue of food waste. The study intends to shed light on the reasons contributing to food waste, explore sustainability approaches, and outline viable solutions through a comprehensive assessment of literature. The study provides a unified narrative by analyzing data from many sources and exploring numerous perspectives, emphasizing the relevance of sustainable solutions and the difficulty of implementing them in the hospitality sector. Around 62 papers were chosen, downloaded, and analyzed in order to fulfill the goals of the current study. Sustainability is the main theme of these articles, and energy conversion, Supply chain, resource management and sustainable development are major topics of discussion.

The chosen publications that had been published within the last fifteen years as our time period because it was easy to see an increase in scholars' interest in this subject year after year. As opportunities on this subject, it was possible to comprehend the major challenges the hospitality industry faces in relation to food waste, whether in the delivery of sustainable practices, putting sustainability goals into reality, having employees adopt it, or due to poor management or internal understanding. The counter arguments opposed are the application of active communication strategy through sustainable reports, the promulgation of sustainable culture by opting for a Chief Sustainability Officer (CSO), speculation in research and development areas. Instead of establishing the CSO position, hotels might employ the team members' leadership to foster a supportive and cooperative atmosphere.

Suggesting that fuel, from manufacturing and distribution to consumption and disposal. Food waste has far-reaching repercussions that go beyond monetary losses, negatively hurting the environment, increasing climate change, and exacerbating social disparities. The preceding study is a valuable tool for comprehending the complex relationship between sustainability and food waste in the hospitality business. It illustrates their research opportunities take into account all options for sustainable development in the hospitality industry. However, if we were to pick only one, we would advise a deeper study of issues with Sustainability Reports or Audits. Numerous studies have emphasized how these documents are not standardized and how accountability for sustainable practices suffers from poor communication. How implementing sustainability reports/ Audits are beneficial to the hospitality industry might be a good starting point for queries on this topic. Or, how much would a sustainability report motivate everyone to follow it.

The insights of this research generates numerous crucial pieces of evidence that underscore the gravity of the problem at hand. It emphasizes the contradictory coexistence of massive food waste and global food insecurity, highlighting the critical need for effective food management systems. The research shows a widespread issue that affects the entire supply chain. Opportunities for waste exist at every level: from manufacturing and distribution to consumption and disposal. The problem of food waste is inextricably linked to wider sustainability issues, which include environmental, economic, and social elements. The report presents a path for the industry to move to more sustainable practices. It does, however, emphasize the difficulty of implementing such changes due to inherent operational challenges and the necessity for systemic shifts. Finally, the report emphasizes the importance of joint action and innovative solutions to guarantee the hospitality sector has a more sustainable and responsible future.

References

Abrizah A, Inuwa S, Afiqah-Izzati N (2016) Systematic literature review informing LIS professionals on embedding librarianship roles. J Acad Librariansh 42(6):636–643. https://doi.org/10.1016/j.acalib.2016.08.010

Agarwal S (2002) Restructuring seaside tourism. Ann Tour Res 29(1):25–55. https://doi.org/10.1016/s0160-7383(01)00002-0

Al-Aomar R, Hussain M (2017) An assessment of green practices in a hotel supply chain: a study of UAE hotels. J Hosp Tour Manag 32:71–81. https://doi.org/10.1016/j.jhtm.2017.04.002

Baldwin CJ, Shakman A (2012) Food waste management. In Greening Food and Beverage Services: A Green Seal Guide to Transforming the Industry (pp. 57–58). Green Seal Inc

Beshara Balaji M, Kaur A (2012) Modeling the causes of food wastage in Indian perishable food supply chain. Res Conserv Recycl 114:153–167. https://doi.org/10.1016/j.resconrec.2016.07.016

Borade AB, Bansod SV (2007) Domain of supply chain management - a state of art. DOAJ (DOAJ: Directory of Open Access Journals). https://doaj.org/article/be226e7020cd4858a3e0330d23c26f99

Bramwell B, Lane B (2005) From niche to general relevance?: Sustainable tourism, research and the role of tourism journals. J Tour Stud 16(2):52–62. https://search.informit.com.au/documentSummary;dn=200603952;res=IELAPA;type=pdf

Bramwell B, Lane B (2012) Towards innovation in sustainable tourism research? J Sustain Tour 20(1):1–7. https://doi.org/10.1080/09669582.2011.641559

Butler J (2008) The compelling "Hard case" for "Green" hotel development. Cornell Hospit Quart 49(3):234–244. https://doi.org/10.1177/1938965508322174

Chung LH, Parker LD (2010) Managing social and environmental action and accountability in the hospitality industry: a Singapore perspective. Account Forum 34(1):46–53. https://doi.org/10.1016/j.accfor.2009.10.003

Connolly P, McGing G (2007) High performance work practices and competitive advantage in the Irish hospitality sector. Int J Contemp Hosp Manag. https://doi.org/10.1108/09596110710739903

Corrado S, Sala S (2018) Food waste accounting along global and European food supply chains: state of the art and outlook. Waste Manage 79:120–131. https://doi.org/10.1016/j.wasman.2018.07.032

Dodds R, Kuehnel J (2010) CSR among Canadian mass tour operators: good awareness but little action. Int J Contemp Hosp Manag 22(2):221–244. https://doi.org/10.1108/09596111011018205

Fauziah Deraman, Ismail N, Izzat M, Izzuan AMM (2017) Green practices in hotel industry: factors influencing the implementation. J Tour Hosp 9(2):305–316. https://www.cabdirect.org/cabdirect/abstract/20183073491

Fantazy K, Kumar V, Kumar U (2010) Supply management practices and performance in the Canadian hospitality industry. Int J Hosp Manag 29(4):685–693. https://doi.org/10.1016/j.ijhm.2010.02.001

Ferguson MA, Branscombe NR, Reynolds KJ (2011) The effect of intergroup comparison on willingness to perform sustainable behavior. J Environ Psychol 31(4):275–281. https://doi.org/10.1016/j.jenvp.2011.04.001

Filimonau V, Fidan H, Alexieva I, Dragoev S, Marinova DD (2019) Restaurant food waste and the determinants of its effective management in Bulgaria: an exploratory case study of restaurants in Plovdiv. Tourism Manag Perspect 32:100577. https://doi.org/10.1016/j.tmp.2019.100577

Frame B, Brown J (2008) Developing post-normal technologies for sustainability. Ecol Econ 65(2):225–241. https://doi.org/10.1016/j.ecolecon.2007.11.010

Fukey L, Issac SS (2014) Connect among green sustainability and hotel industry: a prospective simulation study. Zenodo (CERN European Organization for Nuclear Research). https://doi.org/10.5281/zenodo.1090835

Gilbert Ballou RHSM, Mukherjee A (2000) New managerial challenges from supply chain opportunities. Indust Mark Manag 29(1):7–18. https://doi.org/10.1016/s0019-8501(99)00107-8
Grober 2012 (2012) Sustainability: a cultural history. https://researchgate.net
Gustavsson J, Cederberg C, Sonesson U, Van Otterdijk R, Meybeck A (2011) Global food losses and food waste. Gothenburg, SIK & Rome, FAO. (2011). Global food losses and food waste
Gustavsson J, Cederberg C, Sonesson U, Van Otterdijk R, Meybeck A (2013) Estimates of European food waste levels
Gustavsson J, Cederberg C, Sonesson U, Van Otterdijk R, Meybeck A (2011) Global food losses and food waste. Gothenburg, SIK & Rome, FAO
Hegarty JA (1992) Towards establishing a new paradigm for tourism and hospitality development. Int J Hosp Manag 11(4):309–317. https://doi.org/10.1016/0278-4319(92)90048-Z
Hjorth P, Bagheri A (2006) Navigating towards sustainable development: a system dynamics approach. Futures 38(1):74–92. https://doi.org/10.1016/j.futures.2005.04.005
Hogan and L., Oza, S., D., Chu, Y., Perin, J., Zhu, J., Lawn, J. E., Cousens, S., Mathers, C., & Black, R. E., 2016 Hogan, Liu L, Oza SD, Chu Y, Perin J, Zhu J, Lawn JE, Cousens S, Mathers C, Black RE (2016) Global, regional, and national causes of under-5 mortality in 2000–15: an updated systematic analysis with implications for the Sustainable Development Goals. The Lancet 388(10063):3027–3035. https://doi.org/10.1016/s0140-6736(16)31593-8
Hu Wen Z, De Clercq D, Beck M, Zhang H, Hua-Nan Z, Fan, Liu J (2018b) Design, implementation, and evaluation of an Internet of Things (IoT) network system for restaurant food waste management. Waste Manag 73:26–38. https://doi.org/10.1016/j.wasman.2017.11.054
Jones BA, Madden GJ, Wengreen HJ (2014) The FIT Game: preliminary evaluation of a gamification approach to increasing fruit and vegetable consumption in school. Prev Med 68:76–79. https://doi.org/10.1016/j.ypmed.2014.04.015
Kaur, Khanna A, Kaur S (2019) Evolution of Internet of Things (IoT) and its significant impact in the field of precision agriculture. Comput Electr Agricult 157:218–231. https://doi.org/10.1016/j.compag.2018.12.039
Ku and C., Harrold, T., Armour, S., Krikidis, I., Videv, S., Grant, P., Haas, H., Thompson, J., I., Wang, C., Le, T. A., Nakhai, M. R., Zhang, J., & Hanzo, L., 2011 Ku, Han C, Harrold T, Armour S, Krikidis I, Videv S, Grant P, Haas H, Thompson JI, Wang C, Le TA, Nakhai MR, Zhang J, Hanzo L (2011) Green radio: radio techniques to enable energy-efficient wireless networks. IEEE Commun Magaz 49(6):46–54. https://doi.org/10.1109/mcom.2011.5783984
Kumar CG, Anand SK (1998) Significance of microbial biofilms in food industry: a review. Int J Food Microbiol 42(1–2):9–27. https://doi.org/10.1016/s0168-1605(98)00060-9
Lundqvist J, De Fraiture C, Molden D (2008) Saving water: from field to fork. Curbing losses and wastage in the food chain. Indiana University. https://agris.fao.org/agris-search/search.do?recordID=GB2013203081
Martínez-Ferrero J, Frías-Aceituno JV (2013) Relationship between sustainable development and financial performance: international empirical research. Bus Strateg Environ 24(1):20–39. https://doi.org/10.1002/bse.1803
Moscardo Budeanu A, Miller GG, Ooi C (2016) Sustainable tourism, progress, challenges and opportunities: an introduction. J Clean Product 111:285–294. https://doi.org/10.1016/j.jclepro.2015.10.027
Narus JA, Anderson JC (1995) Using teams to manage collaborative relationships in business markets. J Bus Bus Mark 2(3):17–46. https://doi.org/10.1300/j033v02n03_02
Neves L, Gonçalo E, Oliveira R, Alves MM (2008) Influence of composition on the biomethanation potential of restaurant waste at mesophilic temperatures. Waste Manage 28(6):965–972. https://doi.org/10.1016/j.wasman.2007.03.031
Papargyropoulou E, Steinberger J, Wright N, Lozano R, Padfield R, Ujang Z (2019) Patterns and causes of food waste in the hospitality and food service sector: food waste prevention insights from Malaysia. Sustainability 11(21):6016. https://doi.org/10.3390/su11216016
Principato L, Pratesi CA, Secondi L (2018) Towards zero waste: an exploratory study on restaurant managers. Int J Hosp Manag 74:130–137. https://doi.org/10.1016/j.ijhm.2018.02.022

Pullman ME, Rodgers S (2010) Capacity management for hospitality and tourism: a review of current approaches. Int J Hosp Manag 29(1):177–187. https://doi.org/10.1016/j.ijhm.2009.03.014

Saarinen J (2006) Traditions of sustainability in tourism studies. Ann Tour Res 33(4):1121–1140. https://doi.org/10.1016/j.annals.2006.06.007

Schopler JH (1987) Interorganizational groups: origins, structure, and outcomes. Acad Manag Rev 12(4):702–718. https://doi.org/10.5465/amr.1987.4306723

Schwartz K, Font X, Tapper R, Kornilaki M (2008) Sustainable supply chain management in tourism. Bus Strateg Environ 17(4):260–271. https://doi.org/10.1002/bse.527

Seuring S, Müller M, Westhaus M, Morana R (2005) Conducting a literature review—the example of sustainability in supply chains. In Physica-Verlag eBooks (pp 91–106). https://doi.org/10.1007/3-7908-1636-1_7

Smith SL, Xiao H (2008) Culinary tourism supply chains: a preliminary examination. J Travel Res 46(3):289–299. https://doi.org/10.1177/0047287506303981

Soltani E, Wilkinson A (2010) What is happening to flexible workers in the supply chain partnerships between hotel housekeeping departments and their partner employment agencies? Int J Hosp Manag 29(1):108–119. https://doi.org/10.1016/j.ijhm.2009.06.006

Song L, Liang D (2011) Lean construction implementation and its implication on sustainability: a contractor's case study. Can J Civ Eng 38(3):350–359. https://doi.org/10.1139/l11-005

Song H, Huang GQ (2009) Tourism supply chain management: a new research agenda. Tour Manage 30(3):345–358. https://doi.org/10.1016/j.tourman.2008.12.010

Stenmarck Å, Jensen C, Quested T, Moates GK, Buksti M, Cseh B, Juul S, Parry AD, Politano A, Redlingshofer B, Scherhaufer S, Silvennoinen K, Soethoudt J, ZüBert C, Östergren K (2016) Estimates of European food waste levels

Stock, Marcellino MJH, Watson MW (2006) A comparison of direct and iterated multistep AR methods for forecasting macroeconomic time series. J Econom 135(1–2):499–526. https://doi.org/10.1016/j.jeconom.2005.07.020

Zhang X, Song H, Huang GQ (2009) Tourism supply chain management: a new research agenda. Tour Manage 30(3):345–358. https://doi.org/10.1016/j.tourman.2008.12.010

Navigating Health Challenges: Approaches for Sustainable Development in a Changing Climate

Amir Ahmad Dar, Mehak Malhotra, and Akshat Jian

Abstract The role of healthcare and climate change in sustainable development is crucial for ensuring a healthy and prosperous future for people and the planet. Addressing the health impacts of climate change requires a multi-faceted approach that includes reducing greenhouse gas emissions to mitigate the drivers of climate change, enhancing adaptation strategies to build resilience in communities, and implementing public health measures to protect vulnerable populations. Collaboration between governments, international organizations, healthcare providers, and communities is essential in confronting the health challenges posed by climate change. Healthcare and climate change are interconnected and play pivotal roles in sustainable development. A holistic approach that considers the health impacts of climate change promotes sustainable healthcare practices, and addresses health disparities is vital for achieving a sustainable and resilient future for all. In this chapter, we are going to discuss some strategies for climate change and health policies that can reduce the burden of illness related to climate change and work towards a healthier and more sustainable future for all.

Keywords Heath · Sustainable development · Climate change · Challenges

1 Introduction

The interconnected problems of healthcare and climate change have become critical issues in today's world that require immediate response. The effects of climate change are deeply rooted in human health and go well beyond issues related to the environment. The purpose of this introduction is to clarify the current status of climate change, highlight its significant effects on health, and emphasize the necessity of sustainable development as a keystone for ensuring the welfare of people and the environment. The globe is currently dealing with the obvious consequences

A. A. Dar (✉) · M. Malhotra · A. Jian
School of Chemical Engineering and Physical Sciences, Lovely Professional University, Punjab, India
e-mail: sagaramir200@gmail.com

P. K. Prabhakar and W. Leal Filho (eds.), *Preserving Health, Preserving Earth*, World Sustainability Series, https://doi.org/10.1007/978-3-031-60545-1_9

of climate change, which include rising temperatures, worse weather, and ecological disturbances. Without swift and significant action, these climate changes will worsen, increasing already existing weaknesses and posing previously unheard-of health risks, according to the Intergovernmental Panel on Climate Change (IPCC). The health effects of climate change are extensive and need coordinated efforts for adaptation and mitigation, ranging from the rise in vector-borne disease prevalence to the intensity of heat waves (Change 2013).

According to predictions from the World Health Organisation (WHO), the mortality toll from climate change is already predicted to exceed 150,000 per year in the next several decades. Changes in climatic patterns are making it easier for disease vectors to transmit infectious illnesses like dengue and malaria to new areas. Extreme heat waves can provide a direct risk as they worsen pre-existing medical issues and lead to heat-related illnesses (World Health Organization 2018).

The interconnected effects of climate change on ecosystems, economies, and, most importantly, human lives highlight how urgent it is to address this relationship. Vulnerable people suffer significantly as a result of climate change's disturbance of long-standing disease patterns, threats to the security of water and food supplies, and escalation of extreme weather events. In order to reduce these effects and improve resilience, a thorough and prompt response is required to safeguard public health (Watts et al. 2018).

Let me introduce sustainable development, which is the core of a strategy framework that aims to promote long-term well-being by coordinating economic, social, and environmental issues. According to the World Commission on Environment and Development (WCED), sustainable development is a practical strategy for guaranteeing a future in which the needs of the present are satisfied without harming the ability of future generations to satisfy their own needs. It is not just an abstract idea. Understanding how the environmental, social, and economic spheres are intertwined, sustainable development becomes a vital paradigm for addressing the multifaceted issues raised by climate change and preserving the health of people everywhere (Lousley 2015).

Effective approaches to the dual concerns of healthcare and climate change revolve upon sustainable development. Sustainable development offers a comprehensive framework for tackling the underlying causes of climate change while also strengthening public health infrastructure by combining environmental stewardship, social equality, and economic success. This strategy is crucial for promoting adaptive measures that strengthen communities' resilience to changing climate circumstances, going beyond simple mitigation initiatives. We explore a variety of approaches to address the health effects of climate change in the upcoming chapters. This entails closely examining methods to lower greenhouse gas emissions, strengthen adaption plans, put public health initiatives into action, and encourage cooperation between many stakeholders. We hope to shed light on a route towards a resilient and sustainable future through these talks—one in which the complex interactions between healthcare, sustainable development, and climate change come together for the benefit of both the earth and humankind (Field et al. 2014; World Health Organization 2021a; Costello et al. 2009).

2 The Health Impacts of Climate Change

Examining particular and alarming expressions becomes essential when exploring the complex web of health effects caused by climate change. The most prominent of them is the increased transmission of infectious illnesses, a result closely associated with changing climatic patterns. Changes in temperature and precipitation patterns provide favourable conditions for mosquitoes and ticks, which in turn expands the geographic range of illnesses, including dengue fever, malaria, and Lyme disease. The global health burden is increased as a result of this growth, which presents a greater hazard to communities that were previously immune to these diseases (Patz et al. 2005).

A further noticeable aspect of the health damage caused by climate change is heat-related disorders. Vulnerable people are more prone to heat exhaustion, heatstroke, and other heat-related illnesses as temperatures rise above historical averages. Long-term exposure to intense heat can cause respiratory and cardiovascular problems, with the elderly, young children, and people with pre-existing medical disorders being especially vulnerable. Significantly, hundreds of extra casualties were caused by the heatwave that struck Europe in 2003, highlighting the real and immediate effects of increasing temperatures on human health (Watts et al. 2019; Robine et al. 2008).

The health narrative around climate change is further punctuated by the effects of severe storms. Climate change worsens natural disasters such as hurricanes, floods, wildfires, and droughts that lead to a series of health risks. After such incidents, there is a chance of harm, relocation, and interruption of vital medical services. As an illustration of the far-reaching effects that persist long after the initial climatic disturbance, the aftermath of Hurricane Katrina in 2005 saw a rise in waterborne illnesses, mental health problems, and a stressed healthcare system (Finlay et al. 2012).

These observations are supported by scientific facts, which gives the discourse an empirical foundation. In its publications, the Intergovernmental Panel on Climate Change (IPCC) categorically declares that infectious illnesses are threatened by climate change, which increases their frequency, spread, and seasonality (IPCC). The World Health Organisation (WHO) and Lancet Countdown have both conducted studies on the health effects of heatwaves, and they both highlight how rising global temperatures directly contribute to an increase in the frequency and severity of heat-related illnesses. Additionally, the Lancet Countdown highlights the complex interactions between the rise in infectious illnesses that are vulnerable to climate change, clarifying the complex character of the current health issues (Masson-Delmotte 2018).

The recent instances highlight how urgent and immediate the health threat linked to climate change is. One striking example of how climate change may quickly materialize into serious public health risks is the Zika virus outbreak that occurred in South America in 2015–2016. This outbreak was connected to changing climate conditions that favor the development of Aedes mosquitoes. These incidents demonstrate that the effects of climate change on health are not ghosts but rather present realities that call for swift action and coordinated international efforts (Kraemer et al. 2015).

The effects of climate change on health are extensive and diverse, going beyond academic discussion to pose real risks to the general public's health. The increase in heat-related illnesses, the spread of infectious diseases, and the heightened health hazards resulting from extreme weather events necessitate a comprehensive and prompt response. Recent incidents and scientific data highlight how urgent it is to solve these issues to protect human health in the face of a constantly shifting environment (Patz et al. 2005; Watts et al. 2019; Kraemer et al. 2015).

3 Reducing Greenhouse Gas Emissions

Reducing greenhouse gas emissions in the healthcare industry requires a multimodal approach that combines environmental awareness with high-quality patient care. Using renewable energy sources to power healthcare facilities is one important direction. Healthcare practices are brought into line with a cleaner, more sustainable energy model by switching to solar, wind, or other sustainable energy sources, which also reduces the industry's carbon impact. Furthermore, a key component of lowering greenhouse gas emissions in healthcare institutions is the use of energy-efficient techniques. By putting energy-saving lighting, heating, ventilation, and air conditioning (HVAC) systems into place, for example, resources may be redirected toward patient care efforts while simultaneously reducing energy usage. In addition, the healthcare industry's sustainable mobility measures are essential for cutting emissions. These actions support the industry's commitment to greener, more ecologically friendly practices, such as using electric cars for patient transportation and encouraging employees to take public transit (Guenther and Vittori 2008; Hoban et al. 2021).

Several success stories highlight healthcare institutions that have considerably reduced their carbon footprint. In this sense, the American healthcare system run by Kaiser Permanente is a model. By putting money into on-site wind and solar energy generation, Kaiser Permanente was able to achieve carbon neutrality for all of its facilities in the United States by 2020. This historic accomplishment encourages others to follow suit by demonstrating the viability of large-scale healthcare systems switching to renewable energy. Furthermore, the well-known healthcare facility Cleveland Clinic has put energy-saving measures into place, such as updating its LED lighting and streamlining its building management system. A remarkable 30% decrease in energy use was the outcome of these efforts, demonstrating the practical effects of focused energy-efficient practices in healthcare institutions. These case studies not only highlight the potential for considerable emissions reductions in the healthcare industry but also provide other institutions with a model to follow (Senay et al. 2023).

Apart from healthcare institutions operating independently, cooperative efforts have surfaced as potent stimulants for diminishing greenhouse gas emissions. This cooperative approach is best demonstrated by the "Health Care Climate Challenge," which is spearheaded by the Global Green and Healthy Hospitals network. This effort encourages sustainable practices and the development of an international network dedicated to climate change by involving healthcare institutions globally and mobilizing the industry to lower its carbon footprint. The accomplishment of these cooperative projects highlights the significance of shared accountability and group effort in attaining significant emissions reductions in the healthcare sector (Armstrong et al. 2018).

It is morally and environmentally required for the healthcare industry to significantly decrease its greenhouse gas emissions. Adopting energy-efficient procedures, sustainable transportation, and renewable energy are strategies that not only help healthcare organizations meet their environmental targets but also save money and provide better care for patients. Effective case studies, like the ones from Cleveland Clinic and Kaiser Permanente, highlight the possibility of significant emissions reductions in the healthcare industry. Collaborative projects such as the Health Care Climate Challenge also emphasize how important it is to work together to bring about significant and long-lasting change in the healthcare sector (Guenther and Vittori 2008; Senay et al. 2023; Armstrong et al. 2018).

4 Adaptation and Resilience

In the context of adaptation and resilience to climate change, it becomes necessary for communities and healthcare systems to critically evaluate various options. Improving adaptation requires concentrating on developing resilience to successfully address the complex issues brought about by climate change. The foundation of this effort is sound infrastructure design, which includes creating community infrastructure and healthcare facilities that are robust to the effects of climate change and can adapt to them. One way to ensure that healthcare facilities are prepared to withstand extreme weather events and continue delivering important services without compromising patient and staff safety is to incorporate climate-resilient architectural designs (Ebi 2011).

The development of early warning systems designed to foresee and lessen the effects of climate change on health is equally important. By acting as sentinel guardians and informing communities and medical professionals promptly, these systems enable communities to take preventative measures against new health hazards. Early warning systems are more effective when they combine cutting-edge technology with data-driven models. This allows for accurate forecasts and practical insights for making strategic decisions in the face of shifting climate trends (World Health Organization 2015).

It becomes clear that community involvement is essential to promoting resilience and adaptability. Giving communities the information and tools they need to take an active role in health efforts connected to climate change encourages a feeling of accountability and ownership. Community-based adaptation measures, including the formation of neighbourhood health task groups or educational initiatives, raise awareness while guaranteeing that treatments are tailored to the unique requirements of various communities and are culturally sensitive. This community-based engagement helps to co-create resilient communities that can navigate the changing terrain of health issues associated with climate change (Campbell-Lendrum et al. 2015).

Furthermore, strengthening resilience in healthcare systems requires strong public health initiatives. This entails creating and putting into effect comprehensive health strategies that specifically address the vulnerabilities brought about by climate change. A resilient healthcare response might include, for example, modifying immunization programs to combat the shifting patterns of infectious disease transmission or developing heat action plans to safeguard vulnerable groups during periods of excessive heat. Healthcare systems that establish adaptive strategies not only protect people's health but also strengthen their ability to withstand and adjust to changing climate conditions. Achieving climate change resistance and adaptation needs a thorough strategy that takes into account aspects of the community and the healthcare system. To establish resilience against health concerns associated with climate change, it is essential to have strong infrastructure planning, early warning systems, and community participation. The ability to not only reduce risks but also create resilient and sustainable societies becomes attainable as communities and healthcare systems work together to apply adaptive strategies (Ebi 2011).

5 Public Health Measures

Implementing specialized public health interventions is a vital precaution in the arena of climate change and its subsequent health implications, especially for vulnerable populations. A critical strategy is the implementation of focused interventions intended to protect the most vulnerable populations from the health risks associated with climate change. Proactive heat action plans, for example, those designed for the elderly, small children, and people with pre-existing medical disorders, can dramatically lower the chances of heat-related diseases during periods of excessive temperature. These actions frequently involve setting up cooling centres, issuing heat advisories, and engaging the community to make sure those who are susceptible are properly informed and assisted (Ebi 2011).

To strengthen the resilience of people confronting health concerns due to climate change, education is essential. Communities with more knowledge are better able to comprehend, foresee, and reduce the threats related to climate change. Providing advice on preventative measures, educating the public on the health effects of climate change, and teaching practical skills for handling disasters caused by climate change are a few examples of educational activities. Education becomes a vital component in enabling communities to make knowledgeable decisions about their health in the face of environmental changes by cultivating a culture of awareness and knowledge (World Health Organization 2021b).

In addition to teaching, awareness campaigns are effective means of reaching a variety of audiences and sharing important information. These campaigns can use a variety of media platforms to provide information about health hazards associated with climate change, preventative actions, and accessible services. A deeper understanding and resonance within communities is promoted by the purposeful use of imagery, narratives, and community testimonies to boost the impact of awareness campaigns. Additionally, these efforts may stress the value of community solidarity and inspire people to help one another in times of health emergencies brought on by climate change. The knowledge that populations at risk need specific, focused interventions to successfully address the health implications of climate change serves as the foundation for these public health initiatives. Public health methods can promote community resilience by integrating education, awareness campaigns, and community-based interventions. This will help communities adjust to the difficulties presented by changing climate conditions. By doing this, it becomes possible to achieve the goals of reducing health inequities and promoting a society that is stronger, more knowledgeable, and better equipped (Bayes et al. 2023).

6 Collaboration for Effective Solutions

A collaborative and coordinated effort, including a range of stakeholders, including governments, international organizations, healthcare providers, and local communities, is required to address the complex difficulties presented by climate change. To develop and implement practical solutions that address the complex interactions between healthcare, climate change, and overall social well-being, collaboration across these disparate institutions is essential. Enacting laws and policies that promote resilience and sustainability is primarily the responsibility of governments. Because of their global outlook, international organizations may help with international coordination, information exchange, and resource activation. As first responders to the health effects of climate change, healthcare professionals play a critical role in promoting sustainable healthcare practices and putting adaptive measures into place. Local communities provide invaluable insights, resilience, and grassroots activities since they are closely linked to the immediate effects of climate change (Smith et al. 2015).

Many instances of successful teamwork demonstrate the benefits that may come from working together. 2015 saw the adoption of the Paris Agreement, a worldwide agreement that represents international cooperation in which nations pledge to reduce global warming and increase adaptive capacity (United Nations Framework Convention on Climate Change (UNFCCC). The need for concerted efforts to address the underlying causes of climate change is highlighted by this historic accord. In the healthcare industry, programs such as the Global Green and Healthy Hospitals (GGHH) network unite medical facilities throughout the globe to lessen their environmental impact, build resilience, and advance sustainable practices. To achieve the sustainability goals of the healthcare industry, GGHH places a strong emphasis on cooperation across healthcare institutions, establishing a network where best practices, creative ideas, and lessons learned are exchanged (Agreement 2015; Duindam 2022).

In addition, community-based organizations and healthcare professionals work together locally to put creative solutions into practice. For example, the Healthy Air Campaign in Los Angeles brought together environmental groups, medical experts, and local communities to combat air pollution, a serious health risk associated with climate change. The program lowered respiratory health concerns in the area by improving air quality via collaborative advocacy, education, and policy activities. These cooperative initiatives show that small-scale projects may have a significant positive impact on healthcare outcomes and the mitigation of climate change when they are included in larger frameworks of collaboration (Kotcher et al. 2018).

Building long-term resilience and tackling current obstacles are equally important for joint endeavours to be successful. The One Health concept is a prime example of cross-sector collaboration that works by acknowledging the interdependence of environmental, animal, and human health. Integrating knowledge from environmental science, veterinary medicine, public health, and other fields, the One Health approach encourages comprehensive solutions that take into consideration the intricate network of variables affecting climate and health. Overcoming the complex difficulties posed by climate change requires cooperation between governments, international organizations, healthcare professionals, and local communities. Effective instances, including the Paris Agreement, the Global Green and Healthy Hospitals network, and neighbourhood-based programs like the Healthy Air Campaign, highlight the benefits of teamwork. In addition to addressing the current effects of climate change, we also provide the foundation for a resilient and sustainable future by encouraging collaboration amongst many stakeholders (Heerden et al. 2023).

7 Sustainable Healthcare Practices

Encouraging environmentally conscious and resilient healthcare systems requires integrating sustainability into healthcare practices. Waste reduction methods, which seek to reduce the environmental effects of hospital operations, are a critical component of sustainable healthcare practices. By putting policies in place such as waste segregation, recycling initiatives, and the adoption of environmentally friendly replacements for medical equipment and supplies, hazardous medical waste production is lessened, which lowers pollution and conserves resources. Adopting green building design concepts also improves sustainability in healthcare facilities by lowering carbon emissions, increasing energy efficiency, and creating a healthy interior atmosphere. Low operating costs, better patient outcomes, and a smaller environmental impact are all facilitated by features like natural ventilation systems, energy-efficient lighting, and sustainable building materials (Guenther and Vittori 2008; Molero et al. 2021).

Another essential component of sustainable healthcare practices is responsible resource management, which includes the effective use of energy, water, and other resources. Healthcare facilities may reduce resource consumption while upholding high standards of care delivery by using water-saving technology, energy-efficient equipment, and renewable energy sources. In addition, eco-friendly product purchasing and local supplier support are examples of sustainable procurement practices that support sustainable resource management and foster community economic resilience (Guenther and Vittori 2008).

Using sustainable practices in healthcare facilities may be a catalyst for larger social change in addition to being an act of environmental stewardship. As powerful organizations in their communities, healthcare facilities have a big say in what kinds of norms and behaviors are accepted. Healthcare organizations encourage and inform patients, employees, and the general public about the value of sustainable living by emphasizing sustainability and providing concrete examples of ecologically friendly operations. Healthcare institutions may cultivate a sustainable culture that extends beyond their walls by educating the public and raising awareness of the connection between human health and environmental health through outreach programs, educational initiatives, and public engagement activities (Lee and Lee 2022).

Furthermore, healthcare facilities operate as role models for outstanding practices, showcasing the viability and advantages of sustainability in action. Through the display of creative solutions like waste reduction programs, renewable energy installations, and green building designs, healthcare facilities encourage imitation and emulation in a variety of social sectors. The adoption of sustainable practices by corporations, government organizations, and individuals in their respective areas is influenced by this ripple effect, which transcends the healthcare industry. Therefore, healthcare organizations have a critical role to play in promoting systemic change that will lead to a more resilient and sustainable future by setting a good example. Including sustainability in healthcare procedures is critical to encouraging responsibility for the environment, building resilience, and driving social change. Reducing waste, designing green buildings, and managing resources responsibly are essential elements of sustainable healthcare practices that support resource efficiency and environmental preservation. Healthcare organizations may lessen their environmental impact and lead the way in society's transition to a more sustainable and just future by implementing sustainable practices (Osorio-González et al. 2020).

8 Addressing Health Disparities

With its wide-ranging effects, climate change amplifies already-existing health disparities, disproportionately harming disadvantaged communities and aggravating pre-existing imbalances. The complex relationship between health disparities and climate change is readily apparent, given that marginalized communities—which frequently lack resources and socioeconomic advantages—bear a disproportionate share of the health effects associated with climate change. These people are often found in ecologically vulnerable locations, making them more vulnerable to emergencies, excessive heat, and air pollution—all of which are made worse by climate change. These environmental stresses aggravate the health inequalities that disadvantaged groups face by causing a multitude of health issues, such as respiratory disorders, heat-related illnesses, and a rise in the incidence of vector-borne diseases (Hajat et al. 2014).

Beyond conventional healthcare measures, a comprehensive strategy is needed to address the health inequities made worse by climate change. Social justice and sustainability concepts must be included in policy initiatives targeted at closing these gaps. Making sure disadvantaged populations have access to healthcare facilities and services that can withstand climate change is a key tactic. This entails modernizing the infrastructure of hospitals in high-risk regions to survive severe weather, guaranteeing the provision of critical services even in the case of weather-related disruptions. Furthermore, measures that support fair resource distribution might lessen the unequal effects of climate change on health. Examples of these policies include financing transportation infrastructure and climate-resilient housing in underserved neighbourhoods (Cutter et al. 2008).

To reduce health inequalities, community-based interventions play a critical role. They emphasize active involvement with affected groups to customize solutions to meet their unique requirements. Communities may be made more resilient by actively participating in their development as well as being receivers of help when they are empowered via education, capacity-building, and participatory decision-making procedures. Outreach initiatives that consider cultural sensitivity might help vulnerable communities overcome obstacles to healthcare access by building trust and encouraging preventative actions. By recognizing the distinctive insights these groups have into their ecosystems, including traditional knowledge from indigenous communities into climate change adaptation methods, further improves the relevance and efficacy of interventions (Ariyaningsih & Shaw 2023).

Furthermore, tackling the socioeconomic determinants of health that support health inequalities should be the top priority for policies and treatments. Vulnerabilities to the health effects of climate change are made worse by prejudice, lack of access to education, and income inequality. Therefore, more comprehensive socioeconomic policies that support fairness, inclusion, and social justice must be included in initiatives for eliminating health inequalities. Policies that offer economic opportunities, reasonably priced housing, and education, for example, have the potential to elevate marginalized people and increase their ability to withstand the effects of climate change and health inequities. Tackling health inequities made worse by climate change requires an all-encompassing, inclusive strategy that acknowledges the complex interplay of social, economic, and environmental issues. In addition to concentrating on healthcare interventions, strategies and policies need to address the underlying causes of vulnerability. Society may aim to build a more resilient and just future where the negative effects of climate change on people's health are lessened for all by making sure that the advantages of sustainable development are distributed fairly (Hajat et al. 2014).

9 Conclusion

The complex interactions that exist between sustainable development, healthcare, and climate change are highlighted by this investigation. The thorough knowledge that results from this conversation highlights how various areas are interrelated and highlights the need for coordinated policies to create a resilient and sustainable future. Several important issues have been brought to light, including the serious effects of climate change on human health, such as the proliferation of infectious diseases and the escalation of heat-related ailments, and the necessity of government, international organizations, healthcare providers, and community cooperation in tackling these issues. A comprehensive strategy must include tactics like cutting greenhouse gas emissions, improving adaption measures, putting public health initiatives into action, and embracing sustainable healthcare practices. It has also been stressed how important it is to address health inequities, especially those that disproportionately affect disadvantaged communities as a result of climate change. A more resilient and ecologically sensitive healthcare industry has been emphasized by the importance of integrating sustainability into healthcare practices, such as waste reduction, green building design, and responsible resource management. Additionally, the investigation has explored the transformational power of education, awareness campaigns, and community participation in fostering health resilience.

A call to action becomes necessary as we consider the linked concerns of sustainable development, healthcare, and climate change. Governments, healthcare organizations, communities, and people are responsible for setting priorities and carrying out the plans that have been established. It is essential that we reduce global warming, protect public health, and promote sustainability. At the local, national, and international levels, action must be taken in the form of concrete measures like switching to renewable energy, strengthening the resilience of the healthcare system, and enacting democratic laws. For these programs to succeed, cooperation across borders, disciplines, and sectors is crucial. Adopting and supporting sustainable practices helps to enhance public health outcomes, advance social fairness, and safeguard the environment. It takes a team effort, creative thinking, and perseverance to fulfil our shared duty for a healthier and more sustainable future. We can ensure the welfare of present and future generations by navigating the complexity of our linked environment through these coordinated efforts.

References

Agreement P (2015) United nations framework convention on climate change (UNFCCC). Climate Change Secretariat: Bonn, Germany 4:471–474

Ariyaningsih, Shaw R (2023) Community-based approach for climate resilience and COVID-19: case study of a climate village (Kampung Iklim) in Balikpapan Indonesia. Land 12(3):650

Armstrong F, Cooke S, Rissik D, Tonmoy F (2018) Human health and wellbeing climate change adaptation plan for Queensland

Bayes R, Bolsen T, Druckman JN (2023) A research agenda for climate change communication and public opinion: the role of scientific consensus messaging and beyond. Environ Commun 17(1):16–34

Campbell-Lendrum D, Manga L, Bagayoko M, Sommerfeld J (2015) Climate change and vector-borne diseases: what are the implications for public health research and policy? Philos Trans R Soc B Biol Sci 370(1665):20130552

Change IC (2013) The physical science basis. In: Working group 1 contribution to the fifth assessment report of the intergovernmental panel on climate change.

Costello A, Abbas M, Allen A, Ball S, Bell S, Bellamy R, Patterson C (2009) Managing the health effects of climate change: lancet and University College London Institute for Global Health Commission. The Lancet 373(9676):1693–1733

Cutter SL, Barnes L, Berry M, Burton C, Evans E, Tate E, Webb J (2008) A place-based model for understanding community resilience to natural disasters. Glob Environ Chang 18(4):598–606

Duindam D (2022) Transitioning to sustainable healthcare: decarbonising healthcare clinics, a literature review. Challenges 13(2):68

Ebi K (2011) Climate change and health risks: assessing and responding to them through 'adaptive management.' Health Aff 30(5):924–930

Field CB, Barros VR, Dokken DJ, Mach KJ, Mastrandrea MD, Bilir TE, White LL (2014) Inter-governmental panel on climate change. In: Climate change 2014: Impacts, adaptation, and vulnerability. Part A: global and sectoral aspects

Finlay SE, Moffat A, Gazzard R, Baker D, Murray V (2012) Health impacts of wildfires. PLoS Curr 4

Guenther R, Vittori G (2008) Sustainable healthcare architecture. John Wiley & Sons

Hajat S, Vardoulakis S, Heaviside C, Eggen B (2014) Climate change effects on human health: projections of temperature-related mortality for the UK during the 2020s, 2050s and 2080s. J Epidemiol Community Health

Hoban, E., Haddock, R., Woolcock, K. (2021). Transforming the health system for sustainability: environmental leadership through a value-based health care strategy

Kotcher J, Maibach E, Montoro M, Hassol SJ (2018) How Americans respond to information about global warming's health impacts: evidence from a national survey experiment. GeoHealth 2(9):262–275

Kraemer MU, Sinka ME, Duda KA, Mylne AQ, Shearer FM, Barker CM, Hay SI (2015) The global distribution of the arbovirus vectors Aedes aegypti and Ae. albopictus. elife 4:e08347

Lee SM, Lee D (2022) Developing green healthcare activities in the total quality management framework. Int J Environ Res Public Health 19(11):6504

Lousley C (2015) Narrating a global future: Our common future and the public hearings of the World Commission on Environment and Development. In Global ecologies and the environmental humanities. Routledge, pp 263–285

Masson-Delmotte V (2018) Global warming of 1.5 °C: An IPCC special report on impacts of global warming of 1.5 °C above pre-industrial levels and related global greenhouse gas emission pathways, in the contex of strengthening the global response to the there at of climate change, sustainable development, and efforts to eradicate poverty. (No Title)

Molero A, Calabrò M, Vignes M, Gouget B, Gruson D (2021) Sustainability in healthcare: perspectives and reflections regarding laboratory medicine. Ann Lab Med 41(2):139–144

Osorio-González CS, Hegde K, Brar SK, Avalos-Ramírez A, Surampalli RY (2020) Sustainable healthcare systems. Sustain Fundament Appl 375–396

Patz JA, Campbell-Lendrum D, Holloway T, Foley JA (2005) Impact of regional climate change on human health. Nature 438(7066):310–317

Robine JM, Cheung SLK, Le Roy S, Van Oyen H, Griffiths C, Michel JP, Herrmann FR (2008) Death toll exceeded 70,000 in Europe during the summer of 2003. CR Biol 331(2):171–178

Senay E, Liu I, Mickel G, Bialowitz J, Cort T, Sherman JD (2023) Proposed SEC climate rules: implications for the US health care sector. NEJM Catal Innov Care Deliv 4(6)

Smith KR, Chafe Z, Woodward A, Campbell-Lendrum D, Chadee DD, Honda Y, Haines A (2015) Human health: impacts, adaptation, and co-benefits. In Climate change 2014 impacts, adaptation and vulnerability: part a: global and sectoral aspects. pp 709–754

van Heerden N, Grobbelaar SSS, Meyer I (2023) One health studies that adopt a formal modelling approach: a scoping review. Environ Sci Policy 145:93–103

Watts N, Amann M, Ayeb-Karlsson S, Belesova K, Bouley T, Boykoff M, Costello A (2018) The Lancet Countdown on health and climate change: from 25 years of inaction to a global transformation for public health. The Lancet 391(10120):581–630

Watts N, Amann M, Arnell N, Ayeb-Karlsson S, Belesova K, Boykoff M, Montgomery H (2019) The 2019 report of The Lancet Countdown on health and climate change: ensuring that the health of a child born today is not defined by a changing climate. The Lancet 394(10211):1836–1878

World Health Organization. (2015). Operational framework for building climate resilient health systems. World Health Organization

World Health Organization. (2018). Climate change and health fact sheet.

World Health Organization. (2021a). Young people's fight against climate change: actions we can take for health and well-being

World Health Organization. (2021b). 2021 WHO health and climate change global survey report

Mental Health and Well-Being in Sustainable Practices

Vivek Barik, M. D. Samsuddin, Piyali Khamkat, and Bhakti Bhusan Barik

Abstract In adherence to the principles of sustainable healthcare, mental health services have an emphasis on prevention and design acute services around the triple bottom line of high quality, low carbon footprint, and low cost, using natural solutions when applicable. People who are in good mental health are better equipped to handle stress, make wise decisions, and give back to their communities. Additionally, they are less likely to develop chronic illnesses, which might place a pressure on the resources available for healthcare. In this study, baseline experiences, facilitators and challenges to service user-caregiver participation in India's developing mental health system, as well as stakeholders' ideas on how more involvement may be accomplished, were explored. The goal of India's national mental health programme was to expand community-based mental health treatment within the country's existing primary health care system. India was one of the first low- and medium-income nations (LAMICs) to do so, although there has been progress in moving away from institution-based treatment and towards a system of primary mental health care, implementation issues have led to inconsistent service delivery locally. As a result, there is a significant treatment gap, with just 10% of people with mental health issues able to get quality care. In India, the prevalence of mental health issues is also rising. The substantial costs of this unmet demand for mental health treatment include pain, impairment, stigma and discrimination, unemployment, early mortality, and the possibility of human rights abuses.

Keywords Sustainable healthcare · Mental health service · Primary mental health · Community based mental health · Low- and medium-income nations

V. Barik · M. D. Samsuddin · P. Khamkat (✉) · B. B. Barik
Department of Pharmaceutical Technology, Brainware University, 398, Ramkrishnapur Road, Barasat, Kolkata 125, India
e-mail: pk.pt@brainwareuniversity.ac.in

P. K. Prabhakar and W. Leal Filho (eds.), *Preserving Health, Preserving Earth*, World Sustainability Series, https://doi.org/10.1007/978-3-031-60545-1_10

1 Introduction

People nowadays are very busy and also stressed due to several reasons be it work pressure, family pressure, health reasons and many more. This is why having a good mental health is very important for an individual's well-being. If any individual is suffering from mental health problems it can affect their ability to work studying and maintain relationships (Feijt et al. 2020). Therefore, mental health has always been a rising topic in the healthcare industry along with that a recent trend which is seen in this industry is sustainable health care. This healthcare is an approach in which the healthcare organization provides high-quality care to their customers but also makes sure to minimize the environmental impact, promote social equity and also ensure long-term financial stability. Further, sustainable health care can also be linked with mental health services by providing treatments that have been successful in improving mental health conditions. This can also help them to reduce the need for medication and other interventions that can have negative environmental impacts.

2 Sustainable Mental Health Practices

2.1 Triple Bottom Line: A Sustainable Approach to Mental Health

A person needs to maintain their mental well-being in order to make sure that they are working properly and their relationships are going strong. Therefore, in order to place more effort on mental health services, Authorities have committed to a triple bottom line approach, which involves a holistic approach (Rashidi et al. 2020). The approach structure is simple which consists of three major factors that is high quality, low carbon footprint and low cost which helps Industries to shape their mental health practices in a long running market.

- **High-Quality**. The high quality makes sure that the healthcare industries in the market provides reporting for each treatment they provide to the public. As well as interventions that ensure the healthcare services are effective for patients. In addition, this principle also focuses on the different cultural needs of people from diverse cultural backgrounds (Ranjbari et al. 2021). Further, they also emphasize on continuous improvement feedback system and outcome evaluation to make sure that the overall quality of the care is continuously enhanced.
- **Low Carbon Footprint**. This principle of sustainable healthcare services encourages eco-friendly and energy-efficient practices within the healthcare organisation. In order to do so they use technology for remote consultation and telepsychiatry to reduce the need for physical travel of the patients (Astell-Burt and Feng 2019). It also focuses on using sustainable infrastructure and resource management strategies to minimise the environmental impact of healthcare services

Implementation of sustainable infrastructure and resource management strategies to minimise environmental impact.

- **Low Cost**. This principle helps the organization to develop cost-effective mental health interventions to increase their accessibility. They also focus on implementing preventive measures to reduce the economic burden which comes with longtime mental health issues on the customers. This principle also encourages the organization to collaborate with community resources to improve the existing infrastructure to enhance their financial efficiency.

Therefore, by providing mental health services that connect with the triple bottom line makes sure that the organization not only ensures the well-being of their patients but also contributes to improving ecological and economic sustainable goals. Therefore, it can be said that this framework promotes a balance between the delivery of high-quality mental health care, environmental responsibility and economic efficiency.

2.2 Integration of Natural Solutions in Mental Health Care

People usually run to the hospital when they have any problem and they usually trust the medical processes to solve their issues be they physical or mental problems. However recently people are trusting natural solutions to treat their mental health problems and care for their mental well-being. One of the main reasons for this is that natural solutions are often less expensive than traditional therapies like medication and personal psychotherapy. Due to this, there are several natural-based activities are available which are free or low cost and thus I am making them accessible to a broader range of individuals.

- **Nature Therapy**. Recently healthcare organizations have encouraged outdoor activities which expose individuals with healthcare problems to the nature environment which helps them to calm themselves (Kotera et al. 2020). This therapy usually includes eco therapy which involves activities such as gardening hiking or natural works which helps to reduce stress and enhance the overall well-being of that individual. Further, this therapy is particularly sustainable as it implements natural elements to create a calming and supportive environment.
- **Green Spaces in Healthcare Facilities**. Creating green open spaces like gardens is especially helpful in creating a calming environment for a patient suffering from mental health problems however several healthcare organizations do not have access to such large areas (Slater et al. 2020).Therefore, as an alternative they can use biophilic design principles to implement in nature-inspired elements within the clinical settings this would not only give the organization a sustainable outlook but would also help the patients to get a stress-free and calming environment within the hospital setting.

- **Animal-Assisted Therapy**. Another most famous or used natural solution for mental health services is animal-assisted therapy. Therapy usually includes interaction with trained animals as it would help the patients to reduce their anxiety and also improve their mood (Jones et al. 2019). This is particularly helpful and most effective as it focuses on creating a human-animal bond which improves the emotional well-being of the patient and makes a social connection.

2.3 Link Between Mental Well-Being and Reduced Risk of Chronic Illnesses

A person's mental health is not only related to their daily activities or their overall well-being but it can also have connections with the prevention of chronic illness. Therefore, it is very important to understand the connection between mental health and physical health to provide effective holistic health care to the patients (Long et al. 2020). Among the several connections between mental health and chronic illness, one of the most common is stress. This is because chronic stress has always been linked to dysregulation of the immunity system which also results in the increase of various diseases. Therefore, it is very important to prioritize mental well-being through stress management techniques which would help to form a healthier immune response and also result in reducing the risk of chronic illness. Further positive healthcare can be maintained very easily if a person is constantly associated with healthier lifestyle choices like performing regular exercises having balanced nutrition and getting adequate sleep every day. These lifestyle choices are very important to prevent serious illnesses like cardiovascular diseases, diabetes, and also obesity.

Therefore, it can be said that people who are receiving mental care services both preventive and treatment have a better mental well-being. As individuals with good mental health are more likely to engage in preventive health care measures which lead to early detection and management of potential chronic conditions (Simon et al. 2021). Further recognizing and addressing this connection between mental health and physical well-being is very important to encourage a preventive approach to health care which is effective.

2.4 Global Disparities in Mental Health Data

Patients who are suffering from mental health problems need access to mental health services within sustainable health care. However, this current landscape suffers from a significant bias towards high-income nations due to which there is a large gap in understanding and addressing the different needs of populations across the globe (Wang et al. 2021). Therefore, it is very important to understand the gap as Several low and middle-income countries have limited access to trained mental health professional's medication and infrastructure and immediately need alternative interventions that are both culturally appropriate and also resource efficient. Further, there are several factors which play a significant role in mental health like poverty violence and social injustice which require a holistic approach which helps them to solve the basic cause of suffering. Therefore, if the needs of such countries are ignored it can lead to a cycle of unequal global progress in mental health. So, to make sure to provide a global approach it is very important to understand the different cultural factors which influence mental health and then develop effective, accessible, and culturally responsible interventions for patients.

Also, the organizations need to invest in training and getting more resources for mental health professionals in low and middle-income countries to make sure there is long-term sustainability (GBD 2019 Mental Disorders Collaborators 2022). This would help them to empower local communities and also make sure to address the specific needs and challenges of the patients. Further, there is a need to encourage collaboration and knowledge exchange between middle and low-income countries to make sure there is progress in mental health research treatment and prevention strategies. Also, for this research to take place and to also address the mental health gap needs global attention and also resource allocation.

3 India's Mental Health System

3.1 Overview of India's National Mental Health Program and Its Goals

Mental health issues are a problem which is increasing in number day by day in multiple nations including India. Due to the increasing number of problems related to the mental health is witnessed in India. After a thorough discussion between authorities, the government have established India's National Mental Health Program (NMPH) which aim is to address the increasing number of issues in the public related to mental health (Gautham et al. 2020). The establishment was done in 1982 and from then it has constantly worked to achieve its one and only goal to decrease the mental health issues and make everyone aware of the precautions and things to live a mentally healthy life. The NMHP has taken many steps after stepping in the sector as India is a low and medium-income nation (LAMICs) the aim was to make the mental health services available and accessible to everyone without discrimination or difficulty. NMHP has recognised the importance of community participation and created many marketing strategies involving them into it to spread awareness.

In the presence and lead of a national mental health program India has to transform the institution-based mental treatment into community-based mental health treatment represents a shift in the approach to providing mental health care (Roy et al. 2021). For that the authorities in charge have taken many steps like spreading awareness among individuals to make the mental health treatment a normalisation. Increased the outpatient care services in which counselling, therapy, medication management and other supporting services are being provided to a community setting with care. Also, enhanced those centres and equipped them with the emergency crisis handling measures to treat any person at any time in an emergency, in which a mobile crisis team is being gathered with appropriate qualification and skills to treat and stabilise the situation as soon as possible.

3.2 Challenges in Implementation Leading to Inconsistent Service Delivery

After all the hard work and other creation of program like NMHP, policies and regulation to assist the mental health care sector and the people who are suffering from it. There still remains many challenges which need the attention of the nation to solve soon to achieve the set goal and objectives. The resource constraints, Infrastructure and accessibility, Information and awareness, Policy dissemination and Policy implementation are the major challenges faced by the Indian authorities in developing the mental health sector (Masefield et al. 2020). The resource constraints involve inadequate funding which makes the financial allocation unstable among the sector and hamper the development and sustainability of mental health services. Infrastructure and accessibility involve the cultural sensitivity (Kapologwe et al. 2020). India have a diverse range of culture inside the states and each culture and tradition have their own briefs about mental health contribution which make the collaboration lack the support and many people does not know about the services and overall decrease the awareness among them. Other two key challenges are related to the policies created by the authorities in which they face problems in implementation it due to a lack of workforce and unable to monitor it which is risky.

3.3 Service User-Caregiver Participation

Mental health structures in different countries have different settings and arrangements in which they face many challenges plus benefits from it. For example, a developing nation like India's mental health system has many benefits and for those benefits, the nation has faced numerous challenges as it develops the system from the fundamental landscape. If those experience are been analysed in this part India's mental health structure often carry a strong social brand (Abayneh et al. 2020a) which raised the challenge for the authorities to spread awareness and decrease the stigma. Also, there is always a significant shortage of mental health professionals for which authorities have to develop plans for training, resourcing and construction of mental health facilities. However, India is a diverse and big nation reaching out to everyone equally after trying their best is not possible. Urban areas have access but in small rural areas which are situated regions becomes hard for them to reach out which is a major challenge mental health authority face and for that, they try their best to create camps and emergency crisis services.

In the mental health sector of India, there are many perspectives which are important to take for understanding the correct situation. Not only in India in each country there are multiple stakeholders in the mental health sector whose perspectives are very important due to some reason. The top stakeholder who is the core present at the core of any mental health sector is the national government which decides the rules regulations and resource allocation in the sector. The perspective of the government is always to prioritize the people's health for which they create policies in which many beneficial things are being given to the public from the government end. Also, the government provide funding to the sector to maintain its services on the ground and treat each and every person who has or does not have a stable financial status (Abayneh et al. 2020b). The second stakeholder in line will be the professionals who have their own perspective in which they are responsible for treating people who are suffering from mental disorders. The professionals in the mental health sector have studied the subject and trained themselves to support people and treat them for their bad health. They are the people who put all their collaborative efforts with the other stakeholders to produce not only treatment but awareness for those who are not aware of the services.

4 Treatment Gap in India

India as a low-middle-income country has made an impressive progress in shifting towards community-based primary mental health care through their National Mental Health Programme (NMHP). However, there are still several challenges that may restrict the effectiveness of the program and leave a large treatment gap. The program mainly faces challenges because it is a three-tier system which usually has issues like staff shortage, particularly in rural areas. Therefore, due to the presence of limited mental health facilities and uneven distribution of resources, it contributes to an inconsistent service delivery across the country (Mahajan et al. 2019). Also, the program trains the primary healthcare programs however the training duration and the depth of knowledge provided to the workers might not be sufficient enough for them to handle complex mental health cases. This limited knowledge and experience can lead to misdiagnosis and can also result in ineffective treatment of the patients.

Also, the NMHP makes a lot of effort to make the patients and also the local public aware of the mental health problems. However, despite there are giving strong efforts there is still misinformation about mental health in certain areas which prevents individuals from seeking help and also hinders community acceptance of mental health services. Lastly among the several obstacles, there is one major challenge which is there is limited funding which restricts the program from expanding its infrastructure development and also to recruit professional mental health workers. Due to all these reasons, there is still a gap of 10% of individuals who require adequate mental health care. This gap shows that there is a disconnect between service availability and actual utilization.

Due to this treatment gap, India is presently suffering from a rising issue of mental health. This is because when people do not treat their mental illness it can cause an immense pain and suffering which not only impacts individuals' emotional health but it also affects their social and physical well-being. It can further lead to anxiety depression substance abuse and even suicidal thoughts which significantly reduce the quality of life and productivity of that individual. Also, in several parts of the country when an individual suffers from mental illness they are often subjected to discrimination which leads to social isolation, and difficulty in accessing health care and education. This also increases the existing problems of the patient and also limits the recovery chances of that individual. Also, when an individual doesn't treat their mental illness it can affect their ability to work and eventually lead to unemployment and loss of income. Due to this, there can be a significant impact on the family's economic condition and also place extra pressure on that individual. It is also important to note that individuals were suffering from mental illness are often at a risk of human rights violation which includes neglect abuse and discrimination within their institution or even communities. This can often lead to serious ethical concerns and requires immediate attention to this issue.

5 Opportunities for Improvement

The program launched by India which is known as the National Mental Health Program (NMHP) has made several progresses but there are still some areas that need extra attention.

- **Strengthening Infrastructure and Resources-** The program provides effective care in rural areas but there is still a need to increase the number of healthcare facilities over there to make sure that the people there have wider access to the services. Further, the program also needs to address the shortage of professionals who are experts in mental health care (Maulik et al. 2020). In order to do so they need to recruit professionals and also apply retention strategies which include scholarship and comparative salaries so that they take part in the program effectively. The facilities also need to include important diagnostic tools and technology to enhance the service quality and efficiency provided to the patients.
- **Enhancing Training and Skill Development-** There is a need to improve the quality of training and knowledge provided to primary healthcare workers in order to make them ready to offer better skills for diagnosis, intervention, and referral to patients. They should also focus on developing and implementing specialised training programs for mental health care professionals that would help them to provide care to diverse patients with specific needs and also to handle complex care (Mukhalalati and Taylor 2019). Further, they can also encourage network or

mentorship programs for the workers to help them exchange their knowledge and increase their continuous learning opportunities.
- **Addressing Stigma and Promoting Awareness-** The people in several rural areas still have some miscommunications about mental health so there is a need to organize public awareness campaigns through various media platforms and communities to make the people aware of mental health and address their miscommunication. They can also implement mental health education into schools and workplaces which would help them to promote early awareness of mental health problems and also result in positive attitudes. Also, they can collaborate with several community leaders, religious organizations, and even NGOs to raise awareness and encourage people to seek help from health care systems.
- **Improving Service Delivery and Ensure Sustainability-** The program can also improve itself by exploring several telemedicine and digital platforms which will help them to provide remote consultation and self-management tools to understand and study the unreserved areas. Further in order to improve themselves they need to allocate sufficient funding for mental health services within the national budget which is important considering the rising burden of mental illness. They can also look for alternative funding sources like public–private partnerships, proper social responsibility initiatives, and even crowdfunding platforms (Long et al. 2020; Kapologwe et al. 2020).

6 Conclusion

The report emphasizes that the intersection of mental health and sustainable healthcare practices is very important to address the growing challenges faced by individuals globally. This is because there is an increase in mental health issues that not only affects the individual's lifestyle but also their relationship as a whole. Therefore, it is very important to establish a comprehensive and sustainable mental health solution. This is important because sustainable mental health practices are often based on the triple bottom line principle which emphasizes high-quality care, focuses on reducing carbon footprint, and also provides low cost. Further, there is a growing trend of using natural solutions such as nature therapy or animal-assisted therapy to care for mental illness as they are not only cost-effective but have better recovery chances.

References

Abayneh S, Lempp H, Alem A, Kohrt BA, Fekadu A, Hanlon C (2020a) Developing a theory of change model of service user and caregiver involvement in mental health system strengthening in primary health care in rural Ethiopia. Int J Ment Health Syst 14(1):1–17. https://doi.org/10.1186/s13033-020-00383-6

Abayneh S, Lempp H, Hanlon C (2020b) Participatory action research to pilot a model of mental health service user involvement in an Ethiopian rural primary healthcare setting: study protocol. Res Involv Engagem 6:1–14. https://doi.org/10.1186/s40900-019-0175-x

Astell-Burt T, Feng X (2019) Association of urban green space with mental health and general health among adults in Australia. JAMA Netw Open 2(7):e198209–e198209. https://doi.org/10.1001/jamanetworkopen.2019.8209

Feijt M, De Kort Y, Bongers I, Bierbooms J, Westerink J, IJsselsteijn W (2020) Mental health care goes online: practitioners' experiences of providing mental health care during the COVID-19 pandemic. Cyberpsychol Behav Soc Netw 23(12):860–864. https://doi.org/10.1089/CYBER.2020.0370

Gautham MS, Gururaj G, Varghese M, Benegal V, Rao GN, Kokane A, Chavan BS, Dalal PK, Ram D, Pathak K, Lenin Singh RK (2020) The National Mental Health Survey of India (2016): prevalence, socio-demographic correlates and treatment gap of mental morbidity. Int J Soc Psychiatry 66(4):361–372. https://doi.org/10.1177/0020764020907941

GBD 2019 Mental Disorders Collaborators (2022) Global, regional, and national burden of 12 mental disorders in 204 countries and territories, 1990–2019: a systematic analysis for the Global Burden of Disease Study 2019. Lancet Psychiatry 9(2):137–150. https://doi.org/10.1016/S2215-0366(21)00395-3

Jones MG, Rice SM, Cotton SM (2019) Incorporating animal-assisted therapy in mental health treatments for adolescents: a systematic review of canine assisted psychotherapy. PloS one 14(1):e0210761. https://doi.org/10.1371/journal.pone.0210761&fbclid=IwAR

Kapologwe NA, Meara JG, Kengia JT, Sonda Y, Gwajima D, Alidina S, Kalolo A (2020) Development and upgrading of public primary healthcare facilities with essential surgical services infrastructure: a strategy towards achieving universal health coverage in Tanzania. BMC Health Serv Res 20(1):1–14. https://doi.org/10.1186/s12913-020-5057-2

Kotera Y, Richardson M, Sheffield D (2020) Effects of shinrin-yoku (forest bathing) and nature therapy on mental health: a systematic review and meta-analysis. Int J Ment Health Addict 1–25. https://doi.org/10.1007/s11469-020-00363-4

Long KN, Kim ES, Chen Y, Wilson MF, Worthington EL Jr, VanderWeele TJ (2020) The role of hope in subsequent health and well-being for older adults: an outcome-wide longitudinal approach. Glob Epidemiol 2:100018. https://doi.org/10.1016/j.gloepi.2020.100018

Mahajan PB, Rajendran PK, Sunderamurthy B, Keshavan S, Bazroy J (2019) Analyzing Indian mental health systems: reflecting, learning, and working towards a better future. J Curr Res Sci Med 5(1):4–12. https://doi.org/10.4103/jcrsm.jcrsm_21_19-

Masefield SC, Msosa A, Grugel J (2020) Challenges to effective governance in a low income healthcare system: a qualitative study of stakeholder perceptions in Malawi. BMC Health Serv Res 20(1):1–16. https://doi.org/10.1186/s12913-020-06002-x

Maulik PK, Thornicroft G, Saxena S (2020) Roadmap to strengthen global mental health systems to tackle the impact of the COVID-19 pandemic Int J Ment Health Syst 14:1–13. https://doi.org/10.1186/s13033-020-00393-4

Mukhalalati BA, Taylor A (2019) Adult learning theories in context: a quick guide for healthcare professional educators. J Med Educ Curric Dev 6:2382120519840332. https://doi.org/10.1177/2382120519840332

Ranjbari M, Esfandabadi ZS, Zanetti MC, Scagnelli SD, Siebers PO, Aghbashlo M, Peng W, Quatraro F, Tabatabaei M (2021) Three pillars of sustainability in the wake of COVID-19: a systematic review and future research agenda for sustainable development. J Clean Prod 297:126660. https://doi.org/10.1016/j.jclepro.2021.126660

Rashidi K, Noorizadeh A, Kannan D, Cullinane K (2020) Applying the triple bottom line in sustainable supplier selection: a meta-review of the state-of-the-art. J Clean Prod 269:122001. https://doi.org/10.1016/j.jclepro.2020.122001

Roy A, Singh AK, Mishra S, Chinnadurai A, Mitra A, Bakshi O (2021) Mental health implications of COVID-19 pandemic and its response in India. Int J Soc Psychiatry 67(5):587–600. https://doi.org/10.1177/0020764020950769

Simon J, Helter TM, White RG, van der Boor C, Łaszewska A (2021) Impacts of the Covid-19 lockdown and relevant vulnerabilities on capability well-being, mental health and social support: an Austrian survey study. BMC public health 21(1):1–12. https://doi.org/10.1186/s12889-021-10351-5

Slater SJ, Christiana RW, Gustat J (2020) Peer reviewed: recommendations for keeping parks and green space accessible for mental and physical health during COVID-19 and other pandemics. Prev Chronic Dis 17:200204. https://doi.org/10.5888/pcd17.200204

Wang Q, Xu R, Volkow ND (2021) Increased risk of COVID-19 infection and mortality in people with mental disorders: analysis from electronic health records in the United States. World Psychiatry 20(1):124–130. https://doi.org/10.1002/wps.20806

Unani Medicine's Integration in Healthcare: Harmonizing Sustainable Food Wisdom for Holistic Well-Being

M. S. F. Sapra

Abstract In a time of increasing health issues and environmental concerns, the combination of sustainability and healthcare becomes a critical path to overall wellbeing. chapter examining the mutually beneficial interaction between healthcare systems and sustainable food practices, which are based on the ideas of traditional Unani medicine. The chapter explores the fundamental connection between a healthy diet and overall well-being, highlighting the important lessons that traditional Unani medicine can impart. It focuses on utilizing traditional knowledge to address current health concerns by implementing *Ijai bil Ghiza* (Unani dietary principles) into contemporary healthcare. The concept of Mizaj, or temperament, and its impact on personalized nutrition are central to this investigation. It shows how dietary choices that are specifically tailored can enhance health outcomes while also supporting ecological conservation. As the world's population grows and puts more demand on the planet's resources, the chapter highlights how urgently we need an ecologically conscious healthcare paradigm. It describes tactics that are based on Unani customs, such as promoting plant-centric diets and farm-to-fork programs. These tactics encourage sustainable procurement and waste reduction, which not only helps patients heal but also restores the world. The chapter also emphasizes how Unani teachings can be used to raise awareness and educate healthcare professionals about the importance of sustainable food systems. The concept envisions a future in which Unani medicine is integrated into healthcare institutions. These institutions are viewed as bulwarks of sustainable living, encouraging preventative wellness through diet and cultivating a healthier population and planet.

Keywords Unani medicine · Sustainable healthcare · Ijai bil Ghiza · Mizaj

M. S. F. Sapra (✉)
Department of Unani Pharmacology, Faculty of Indigenous Medicine, University of Colombo, Colombo, Sri Lanka
e-mail: fathimasapra@fim.cmb.ac.lk

P. K. Prabhakar and W. Leal Filho (eds.), *Preserving Health, Preserving Earth*, World Sustainability Series, https://doi.org/10.1007/978-3-031-60545-1_11

1 Introduction

Unani medicine, also known as Greco-Arab medicine, traces its origins to ancient Greece but is predominantly practiced in the Indian Subcontinent. This traditional system boasts a rich history and fundamental principles governing medicinal approaches. Often viewed as a holistic framework, it relies on natural remedies derived from various sources. However, when not meticulously prepared, identified, and administered, these remedies may present potential risks.

Greek medicine, spanning from 500 BC to 500 AD, was attributed to Apollo, the healer who purportedly banished all ailments. Throughout its evolution, healers like Hippocrates, Dioscorides, and Galen contributed to its practices. Hippocrates, notably, introduced an extensive materia medica. His theory of four humours—Sanguin (Dam), Phlegm (Balgam), Yellow Choler (Safrā), and Melancholia (Saudā)—parallels Ayurvedic medicine's concept of the three doshas: Phlegm, Choler, and Wind. Dioscoridius, in 70 AD, authored the first comprehensive illustrated book cataloging major medicinal plants (Rahman 2006).

Our world is in constant flux—a dynamic interplay of evolving environments, growing populations, and advancing scientific knowledge. Amidst these shifts, our bodies and personalities transform, shaping our unique needs and aspirations. Yet, within this mosaic of change, a delicate equilibrium is essential for fostering a life of vitality and well-being (Bashir and Akhtar 2018). It's within this context that our exploration of nutrition and healthcare emerges, emphasizing the promotion of health as a realistic and vital goal (Roth 2011).

Central to this pursuit is the understanding that a well-rounded diet, abundant in essential nutrients, serves as the cornerstone for both nourishment and disease prevention. Throughout history, the awareness of the profound impact of diet on our health has been deeply ingrained (Anonymous 2013). In today's context, poor dietary choices manifest in an array of lifestyle disorders, from hypertension to obesity, underscoring the critical link between what we consume and our overall well-being (Glynn et al. 2013).

Ijai bil Ghiza, (Dietotherapy) stands as a beacon, aiming to restore the body's balance disrupted by unhealthy lifestyles (Taher et al. 2017). In the realm of Unani medicine, the foundation rests upon *Asbab-e-Sitta Zarooriya*—the six essential prerequisites for life. These factors encompass atmospheric air, foods and drinks, physical and mental activity, sleep patterns, and elimination processes—inescapable necessities woven into the fabric of human existence (Zaman et al. 2013, Jurjani 2010, Hamdani 1980).

The Unani approach to disease management is holistic, addressing not just the body but also the mind and soul. Among the core modes of treatment, Ilaj bil Ghiza—dietotherapy—stands out. It diverges from medicinal therapy, employing dietary modulation such as fasting or altering the quantity and quality of food intake (Hamdani 1980, Taher et al. 2017). Its primary objective lies in catering to the body's nutritional demands, thus minimizing the reliance on pharmacotherapy, which, being external to the body, may trigger adverse reactions.

Ilaj bil Ghiza embodies the philosophy that nourishing the body through dietary means is a harmonious and effective way to maintain health, aligning with the natural needs of our bodies without resorting solely to pharmaceutical intervention.

2 Historical Background

Dietotherapy, an ancient practice, finds its origins in the Zhou Dynasty circa 1000 BC, notably recognized by Zhang Ji, a distinguished Han dynasty physician, for its therapeutic applications in managing various illnesses (Jurjani 2010). Additionally, Ayurvedic texts such as the Charaka Samhita and Sushrutha Samhita outlined dietetics as a distinct branch. In the Unani tradition, luminaries like Hippocrates, Aristotle, Celus, and Galen advocated for the vital role of dietary elements in maintaining health (Zaman et al. 2013).

2.1 Quotes from Unani Physicians on Diet

Throughout history, eminent Unani physicians have extolled the virtues of food and beverages for both health preservation and disease treatment. Hippocrates, renowned for advocating "let food be your medicine, and medicine be your food," emphasized the profound healing potential residing within diet (Glynn et al. 2013). Avicenna, echoing similar sentiments, regarded the stomach as a focal point for ailments and emphasized the healing power of a balanced diet (Jurjani 2010). Aristotle's regimen incorporating grape juice, bread, and mutton reinforced the importance of diet coupled with physical exercise for overall well-being (Ibn Abi Usaiba 1990). Notably, scholars like Haris bin Kalda and Razi contributed invaluable insights into the therapeutic qualities of specific foods and the pivotal role of nutrition in managing various ailments (Tabri 2010; Ibn Sina AAHIA 1993). **Four Edible Categories According to Unani Physicians.

In Unani medicine, edible substances are categorized into four distinct groups: Dawa (drug), Ghiza (diet), Ghiza-e-Dawai (diet cum drug), and Dawa-e-Ghizai (drug cum diet), each carrying unique properties and functions (Anonymous 2013; Taher et al. 2017; Jalinus 2008; Baig et al. 2015).

2.2 Unani Principles of Dietotherapy

In the Unani system, food not only fortifies the body but also influences pathological elements. In certain scenarios, complete abstinence or a regulated reduction in diet is advised (Ibn Hubal Baghdadi 2005).

Tark-e-Ghiza (Abstinence from Diet): Recommended for patients with sufficient stamina, complete abstinence assists in expelling morbid matter from the body (Taher et al. 2017; Tabri 2010; Ibn Sina 2010).

Taqleel-ul-Ghiza (Reduced Diet): Quantitative or qualitative reduction of food aims to preserve physical strength and aid the body's self-healing mechanisms (Anonymous 2013; Glynn et al. 2013; Jalinus 2008).

- Reduction in kammiyat (quantity) of the diet: Involves decreasing the diet's quantity while maintaining its nutritional value, beneficial for individuals with weak digestion (Ibn Sina 2010).
- Reduction in kaifiyat (quality) of the diet: Diminishes the nutritional value of the diet, suitable for scenarios where the body's vascular system is affected by morbid matter (Ibn Sina 2010).
- Reduction in kammiyat and kaifiyat (quantity and quality) of the diet: Applied when bodily faculties are not significantly disturbed (Hamdani 1980, Ibn Sina 2010).

This elaboration aims to provide a more detailed insight into the historical roots, philosophical underpinnings, and categorizations within dieto-therapy in the Unani system

2.3 The Concept of "Sitt-e Zarooriya" in Unani Medicine

The concept of "Sitt-e Zarooriya" in Unani medicine refers to the six essential factors vital for human existence and well-being. These factors encompass various aspects crucial for maintaining life and health. Among these six essentials, "Ghiza" (food) holds a paramount position due to its pivotal role in sustaining life and preserving health (Ibn Sina AAHIA 1993).

Sitt-E Zarooriya:

1. Hawa-e-Muheet (Atmospheric Air): Signifies the importance of clean and pure air for breathing, fundamental for life.
2. Makoolat wa Mashroobat (Foods and Drinks): Encompasses the significance of food and beverages, essential for nourishment and sustenance.
3. Harkat wa Sukoon--Badni (Physical Activity and Repose): Balancing movement and rest to maintain physical well-being.
4. Harkat wa Sukoon wa Nafsani (Mental Activity and Repose): Emphasizes the harmony between mental activity and relaxation for psychological health.
5. Naum wa Yaqza (Sleep and Wakefulness): Acknowledges the essential role of adequate sleep and wakefulness in preserving health.
6. Ehtibas wa Istifragh (Retention and Elimination): Highlights the importance of proper bodily functions, including retention and elimination, for maintaining balance (Ibn Sina AAHIA 1993).

Importance of Food in Sitt-e Zarooriya:

Food, being one of the Sitt-e Zarooriya, holds a pivotal place in this holistic system of maintaining health. It is considered not just a source of nourishment but a fundamental element in the regulation of overall well-being.

1. **Nourishment and Vitality:** Food provides essential nutrients necessary for the body's growth, repair, and sustenance. It supplies vital energy and nutrients required for bodily functions, ensuring vitality and strength.
2. **Balance and Harmony**: In Unani medicine, the type and quality of food are seen as influential factors in maintaining the body's balance. A well-balanced diet helps in preserving the equilibrium of bodily humors (Akhlāt), thereby preventing diseases.
3. **Disease Prevention and Treatment:** Specific foods are believed to possess therapeutic properties in the Unani system. Certain diets and dietary habits are recommended not only to prevent diseases but also to aid in their treatment.
4. **Temperament and Mizaj:** Unani medicine deeply emphasizes the concept of "Mizaj" (temperament) of foods, considering that each food item possesses its own inherent nature. Understanding the Mizaj of food helps in prescribing diets that complement an individual's body temperament, thereby maintaining health and preventing diseases.
5. **Holistic Healing:** Incorporating food as a therapeutic tool, the Unani system adopts a holistic approach to healing. Dietotherapy (Ilaj bil ghiza) is used to restore balance and treat ailments without solely relying on medicinal interventions.

In essence, within the framework of Sitt-e Zarooriya, the significance of food goes beyond mere sustenance. It embodies a holistic approach to health, emphasizing the quality, quantity, and balance of food intake to preserve health, prevent diseases, and aid in the healing process according to the principles of Unani medicine.will be converted by our typesetters (Table 1).

2.4 Distinctive Dietary Regimens and Their Therapeutic Applications

- **Barley Water (Ma-ul-shaeer)**: This concoction, recommended by ancient scholar Buqrat (Hippocrates), involves soaking barley in water and boiling it until the water turns reddish. It's highly regarded for its cleansing properties, ease of absorption, and moderate nature, making it a suitable dietary choice for ailments associated with a hot temperament. Its gentle yet effective cleansing action contributes to its therapeutic value in certain health conditions (Ibn Sina AAHIA 1993; Razi 1994; Wadud 2004).
- **Milk (Jubn)**: Acknowledged for its moderate and slightly cool nature, milk is esteemed for its nourishing properties. Its characteristics of maintaining body

Table 1 Dietary terms used in Unani medicine (Bashir and Akhtar 2018; Anonymous 2013; Taher et al. 2017; Wylie-Rosett et al. 2002; Baig et al. 2015; Ibn Hubal Baghdadi 2005; Razi 1994; Wadud 2004)

Terms	Description
Ghiza-e-Lateef (Light and soft diet)	The foods which are digested easily and reduce the viscosity of *khilt* (humour) inside the body. For e.g. *Aab-e-anar* (pomegranate juice), *Aab-e-naranji* (orange juice), *Ma-ul-shaeer* (barley water), tea, coffee, wine, etc
Ghiza-e-Lateef Kaseer-ut-Taghzia (Attenuated highly nutritious diet)	The foods which are digested easily and rich in calories such as *Ma-ul-lahem* (meat distillate) and *Zardi baiz-e-murgh neem birasht* (half boiled egg yolk)
Ghiza-e-Lateef Qaleel-ul-Taghzia (Attenuated less nutritious diet)	The foods which are digested easily and have low calories, e.g. Fruits
Ghiza-e-Lateef Kaseer-ut-Taghzia jayyad al-Kaimus (Attenuated, highly nutritious and good chyme forming diet)	The foods which are easily digested, rich in calories and produce good chyme like meat distillate, half boiled egg yolk etc
Ghiza-e-Lateef Kaseer-ut-Taghzia raddi al-Kaimus (Attenuated, highly nutritious and bad chyme forming diet)	The foods which are digested easily, having high calories and produce bad chyme. For e.g.: liver and lung
Ghiza-e-Lateef Qaleel-ut-Taghzia jayyad al-Kaimus (Attenuated, less nutritious and good chyme forming diet)	The foods which are digested easily, having low calories and produce good chyme. For e.g: fruit juices and fruits
Ghiza-e-Lateef Qaleel-ut-Taghzia raddi al-Kaimus (Attenuated, less nutritious and bad chyme forming diet)	The foods which are digested easily, provides less energy and produces bad chyme. For e.g. radish, mustard, lettuce, etc
The foods which are digested easily, provides less energy and produces bad chyme. For e.g. radish, mustard, lettuce, etc	The foods which are not easily digested, form *ghaleez khilt* (viscous humour) and produces *sudda* (obstruction) in organs
Ghiza-e-Kaseef Kaseer-ut-Taghzia (Heavy and highly nutritious diet)	The foods which are not digested easily and having high calories such as beef
Ghiza-e-Kaseef Qaleel-ul-Taghzia (Heavy and less nutritious diet)	The foods which are not digested easily and having high calories such as beef
Ghiza-e-Kaseef Qaleel-ul-Taghzia (Heavy and less nutritious diet)	The foods which are not digested easily and having high calories such as dry beef
Ghiza-e-Kaseef Kaseer-ut-Taghzia jayyad al-Kaimus (Heavy, highly nutritious and good chyme forming diet)	Heavy and highly nutritious foods which produce good chyme, e.g., fully boiled egg, meat of young sheep
Ghiza-e-Kaseef Kaseer-ut-Taghzia raddi al-Kaimus (Heavy, highly nutritious and bad chyme forming diet)	The foods which are not easily digested and always produce bad chyme, e.g., duck meat and horse meat
Ghiza-e-Kaseef Qaleel-ul-Taghzia jayyad al-Kaimus (Heavy, less nutritious and good chyme forming diet)	Less nutritious foods which produce good chyme, e.g., meat of young beef

(continued)

Table 1 (continued)

Terms	Description
Ghiza-e-Kaseef Qaleel-ul-Taghzia raddi al-Kaimus (Heavy, less nutritious and bad chyme forming diet)	Less nutritious and heavy foods, which are not easily digested and produce bad chyme, e.g., dry beef meat
Ghiza-e-Motadil (Moderate diet)	Foods that produce *khilt* (humour) of moderate viscosity which is not harmful to the human body, e.g., *Khichdi*, cow"s milk, etc
Aghzia Barida/Aghziya Mubarrida	The food of cold temperament having ability to produce those humours which induce coldness in the body, e.g., *Khas* (*Andropogan muricatus* Retz.), *Kasni* (*Cichorium intybus* Linn.), etc
Aghziya Hamiza	Sour foods e.g. lemon, sour pomegranate, sour apple, orange, vinegar, curd, etc. They change the thick waste products of stomach into smaller particles
Aghziya Harra	The foods of hot temperament having ability to produce the humours which induce heat in the body, e.g., onion, garlic, etc
Aghziya Hirrifa	Spicy (pungent) foods that may produce heat in the body e.g. pepper
Aghziya Lazija	The foods which increase the viscosity of humours, e.g., *Khurfa*, Till, etc
Aghziya Mahmuda	Foods that produce good humours
Aghziya Maliha	Salty foods
Aghziya Mubakhkhira	The food liable to produce flatus and vapours, e.g., cabbage, pea, brinjal, cauliflower, beans, turnip, etc
Aghziya Mughalliz-e-Dam	Foods producing Dam (Sanguine) of thick consistency
Aghziya Mugharriya	Glutinous foods
Aghziya Mujaffifa	Foods causing desiccation of body fluid, e.g., roasted gram
Aghziya Muraqqiq-e-Dam	Foods that cause thinning of *Dam* (Sanguine)
Aghziya Murattiba/Aghziya Rataba	Foods of moist temperament having ability to produce the humours which give rise to moisture in the body, e.g., milk, cucumber, watermelon, clarified butter, almond oil, etc
Aghziya Musakhkhina	Foods producing heat in the body, e.g., spices
Aghziya Musammina	The foods which help to put on body weight, e.g., dry fruits, milk products, etc

(continued)

Table 1 (continued)

Terms	Description
Aghziya Mutaaffina	Putrefied food stuffs
Aghziya Muwallid-e-Balgham	The foods which produce excessive quantity of phlegm in the body, e.g., sheep milk, beet root, cucumber, etc

heat and addressing dryness within the body make it a beneficial dietary element in various health contexts (Ibn Sina AAHIA 1993; Razi 1994).

- **Milk Water (Ma-al-jubn)**: This preparation, achieved by boiling milk with water and a sour agent, is recommended in cases of bilious conditions like jaundice. Its light and easily digestible nature, coupled with its purgative effects, make it a favorable choice for cleansing and providing relief in specific health issues (Ibn Sina AAHIA 1993; Razi 1994).
- **Meat Distillate (Ma-ul-lahm)**: This distilled form of meat, known for its high caloric content and easy digestibility, is often prescribed for individuals dealing with malnutrition, tuberculosis, cerebral, or cardiac ailments. Its richness in nutrients and ease of absorption make it an important dietary inclusion in such cases (Wadud 2004).
- **Buttermilk (Ma-ul-raib)**: With its cooling and hydrating properties, buttermilk is frequently suggested to quench thirst and provide gentle hydration, particularly in hot weather conditions (Wadud 2004).
- **Honey Water (Ma-ul-asl)**: The boiled mixture of honey and water finds application in various health contexts, particularly for its soothing effects in fever, cough, and addressing dryness in the body (Looqa 2007).
- **Fruit and Vegetable Juices (Ma-al-Fawakeh, Ma-al-buqool)**: These juices, derived from fruits and vegetables like apple, pomegranate, orange, and others, serve as sources of essential nutrients and are often recommended to address deficiencies and weaknesses in the body (Hamdani 1980).
- **Other Preparations**: The Unani system features a wide array of specialized dietary offerings such as fermented mustard (Aabkama), spice-free mutton soup (Asfeedaj), and various soups, jams, and diets tailored to address specific health conditions. These unique dietary formulations, detailed in Unani pharmacopoeias, play pivotal roles as adjuncts to pharmacotherapy, emphasizing a comprehensive approach to health management (Zaman et al. 2013; Ibn Sina AAHIA 1993; Razi 1994; Anonymous 2006; Kabeeruddin.

These dietary practices, deeply rooted in historical knowledge and affirmed by Unani scholars, form an integral part of the holistic healing approach advocated by the Unani system of medicine, catering to diverse health needs through tailored dietary interventions.

2.5 The Crucial Role of Digestive Correction in Unani Medicine

- **Essential Nutrient Assimilation**: Digestion isn't solely about breaking down food; it's the gateway through which the body absorbs vital nutrients. Unani philosophy emphasizes that effective digestion ensures the transformation of ingested food into essential elements that nourish tissues, cells, and organs. This process directly influences overall health and vitality.
- **Humoral Balance**: The equilibrium of bodily humors lies at the core of Unani medicine. Proper digestion plays a pivotal role in maintaining this balance. When digestion falters, it can disrupt these humors, leading to an imbalance associated with various health issues.
- **Preventing Imbalances and Toxin Accumulation**: Incomplete or improper digestion can lead to the accumulation of toxins (ama) within the body. According to Unani principles, this toxic buildup disturbs the body's equilibrium, serving as a breeding ground for diseases and imbalances.
- **Treatment Modalities**: Unani practitioners focus on correcting digestion to restore harmony within the body. This involves a multifaceted approach, including dietary modifications tailored to individual constitutions, herbal remedies, lifestyle adjustments, and specific regimens designed to optimize digestion and eliminate toxins.
- **Herbal Remedies and Therapeutic Regimens**: Unani medicine harnesses the power of herbs and natural remedies rich in digestive aids and carminatives. These formulations aim to enhance digestion, alleviate digestive issues, and restore the body's balance. Additionally, personalized dietary recommendations and regimens are prescribed to support optimal digestive function.
- **Holistic Perspective**: Unani medicine perceives the body as an interconnected system, wherein digestion influences various bodily functions. Addressing digestive issues isn't isolated but is part of a broader approach aimed at restoring overall harmony and balance within the body.

In essence, digestive correction in Unani medicine isn't limited to treating stomach ailments; it's a cornerstone principle essential for maintaining holistic well-being. By prioritizing optimal digestion and humoral balance, Unani medicine aims to prevent diseases and promote overall health and vitality.

3 Conclusion

Unani Medicine weaves a complex web of knowledge—dietotherapy—that dates back to the time of great medical figures such as Hippocrates, Galen, Razi, and Avicenna. It is drawn from a vast body of Unani literature and demonstrates the acute understanding that traditional healers had of the critical importance of food and drink. Not only is diet therapy a cure, but it is also a fundamental component in the management and avoidance of many physical ailments.

Its importance goes beyond simply curing a disease; it also facilitates a quicker recovery, improves medication tolerance with little dietary adjustments, and is a readily available, reasonably priced, and astonishingly non-reactive therapeutic choice. The careful choice of diets, either on their own or in conjunction with medication, becomes a powerful ally in the modern world when lifestyle diseases such as diabetes, hypertension, and dyslipidemia are prevalent.

Beyond the domain of lifestyle disorders, dietotherapy provides a natural and efficient treatment for conditions like anemia, malnutrition, and vitamin deficiencies. Nevertheless, a clarion call is necessary given the absence of standardized clinical application, despite its enormous value. It calls for extensive clinical research to confirm the efficacy of this traditional treatment, boosting its scientific standing and making it more widely available to the public for its amazing advantages. Essentially, Unani Medicine works nonstop to steer people toward a life of health and little illness. Based on a comprehensive view of health, it promotes the use of immunomodulators to strengthen immunity, the delicate balance of humors, the control of life's six vital components, and the implementation of healthy routines and diets.

By adopting the paradigm of Unani Medicine, we are adopting not only a healthcare system but also an ideology that promotes the development of robust health and harmony both inside oneself and across the universe. It's a call to create a route that leads not only to healing but also to long-term wellbeing, continuing the enduring harmony of health and

References

Anonymous (2013) Diet, nutrition & inflammatory bowel disease. Crohn"s and Colitis Foundation of America, New York, p 8

Anonymous (2006) National formulary of unani medicine part I. Department of AYUSH, Ministry of H & FW, Government of India, New Delhi, p 220

Baig MG, Quamri MA, Alam MA, et al (2015) Ilaj bil Ghiza (dietotherapy): a core principle of Unani treatment- an appraisal. J Biol Sci Opin 1(3):52–56

Bashir F, Akhtar J (2018) Ilaj Bil Ghiza (Dietotherapy) in Unani system of medicine–an appraisal. Eur J Pharm Med Res 5(12):582–588

Glynn J, Bhikha-Vallee N, Bhikha R (2013) Dietotherapy: background and theory. Ibn Sina Institute of Tibb

Hamdani MK (1980) Usool-e-Tibb. Syed Mohammed Kamaluddin Hamdani, Muslim University, Aligarh, pp 139, 168, 169, 271, 299, 300, 425, 426–428

Ibn Abi Usaiba (1990) Uoyun al-Amba -Atibba vol. 1 (Urdu translation by CCRUM). Dept. of AYUSH, Ministry of H & FW, New Delhi, p 74, 75, 97, 138, 220

Ibn Hubal Baghdadi (2005) Kitab al-Mukhtarat Fit-Tibb, vol. 2 (Urdu translation by CCRUM). Dept. of AYUSH, Ministry of H & FW, Govt. of India, New Delhi, pp 3–21

Ibn Sina (2010) Al-Qanoon fit-Tibb vol. 2 (Urdu translation by Kantoori GH). Ejaz Publishing House, New Delhi, p 203

Ibn Sina AAHIA (1993) Al-Qanun Fil Tibb (English Translation of the Critical Arabic Text). Jamia Hamdard, New Delhi, (1)154–156:166–167.

Jalinus (2008) Kitab fil-Mizaj (Urdu translation by Rehman SZ). Ibn Sina Academy, Aligarh, pp 161–162

Jurjani AH (2010) Zakheera Khawarizm Shahi (Urdu translation by Khan HH) vol. 3. Idarah Kitab-us-Shifa, New Delhi, pp 1–3

Kabeeruddin Sharah Asbab vol. 2. Hikmat Book Depot; YNM, Hyderabad, p 163, 172, 200

Looqa Q (2007) Risal-e-Nabeez (Urdu translation by Rehman Z). Ibn Sina Academy, Aligarh, pp 57–58

Rahman HSZ (2006) Historical perspective of traditional medicine with special reference to ADRs. In: National symposium on relevance of pharmacovigilance for indian system of medicine. Department of AYUSH, Ministry of Health and Family Welfare, Government of India and Society of Pharmacovigilance, India, pp. 53–61. [Google Scholar]

Razi (1994) Kitab al-Murshid (Urdu translation by Nadvi MRI). Director Taraqqi Urdu Beuro, New Delhi, 93–95

Roth RA (2011) Nutrition & diet therapy, 10th edn. Delmar Cengage learning, USA, pp 4–5

Tabri R (2010) Firdaus al-Hikmat (Urdu translation by Shah MA). Idarah Kitab-us-Shifa, New Delhi, p 345, 347, 349

Taher MA, Nyeem MAB, Ahmed MM et al (2017) Ilaj bil Tadbeer (Regimental Therapy): a review. Int J Med Health Res 3(10):54–56

Wadud A (2004) Ashrah al-Advia (Kulliyat-e-Advia). Printed by Mumtaz Screen Printer, Burhanpur, p 80

Wylie-Rosett J, Segal-Isaacson CJ, Deen D et al (2002) Nutritional education: understanding the past to prepare for future medical practice. Einstein Quart J Biol Med 19:59–67

Zaman R, Basar SN, Farah SA (2013) Dietotherapy in Unani system of medicine. Int J Pharmaceut Chem Biol Sci 3(4):1035–1039

Advances in Cosmetic Products Towards a New Future

Parnika Neytal, Nilima Maji, and Suman Maji

Abstract Cosmetics are defined by the Federal Food, Drug & Cosmetic Act (FD&C) as "articles intended to be rubbed, poured, sprinkled, or sprayed on, introduced into, or otherwise applied to the human body or any part thereof for cleansing, beautifying, promoting attractiveness, or altering the appearance". We use various cosmetic products every day as a part of healthy lifestyle for different purposes. They not only enhance our beauty but also protect our skin and hair from environmental damage and stress thus maintaining human health and wellbeing. As consumers become more aware of the harmful effects of pollution and aging, they demand more effective and safe cosmetics. This leads to new research and innovation in the cosmetic industry, using natural extracts, nature-identical ingredients. Development of cosmeceuticals or cosmetic pharmaceuticals, and integration of artificial intelligence suggesting new formulations are going to drive this industry. The trends in cosmetics are driven by consumer preferences and popularized by social media, but they also need to be sustainable and ethical.

Keywords Cosmetics · Make-up · Beautifying products · Skin care · Cosmeceuticals

1 Introduction

Humans are a visual being and makes an impression based on the appearance of anything that comes to our sight. This also makes us conscious regarding our own appearance and hence we have always focused on our looks. To maintain it or give a different perspective we use cosmetics (Wallen-Russell 2018). Cosmetics can highlight or subdue certain features of our look giving a different appearance. Cosmetic products have become an indispensable feature of our life. There is a wide range of

P. Neytal · N. Maji · S. Maji (✉)
School of Chemical Engineering and Physical Sciences, Lovely Professional University, Phagwara, Punjab, India
e-mail: suman.18432@lpu.co.in

P. K. Prabhakar and W. Leal Filho (eds.), *Preserving Health, Preserving Earth*, World Sustainability Series, https://doi.org/10.1007/978-3-031-60545-1_12

Fig. 1 Cosmetic box of the royal butler

cosmetic products available such as skincare, haircare, body care, perfumes, beautifying (make-up products) etc. It is believed that cosmetics were first used in Egypt nearly 6000 years ago as a symbol of wealth and to please God (El-Kilany and Raoof 2017; Lucas 1930) (Fig. 1).

A substance name unguent was excessively used by both men and women so that their skin didn't get dry and wrinkled (Jones 1993). Some studies suggest that Jews adopted make up from Egyptians. Plautus wrote, "a woman without paint is like food without salt"(Duarte 2014). Antimony was used to make kohl and eyeshadow, chalk powder was used for whitening complexion, pearl powder was also used for skin lightening and whitening, and rouge was used as cheek tint. Henna was a common dying agent. In some cultures, it was believed that dying face and hair with henna allows people to summon the majesty of earth (Fig. 2).

But contrary to this in some country and cultures, make up was either considered hazardous for health or a sin. The hazardous elements used in cosmetics were lead and mercury (Raza-Naqvi et al. 2022). Lead exposure can lead to anemia, weakness, kidney, and brain damage. Exposure to mercury can cause neuromuscular disorders, headaches, insomnia, and emotional changes. One major ailment caused due to mercury is the nephritic syndrome (Li et al. 2010). Nephritic syndrome is a clinical syndrome that presents elevated blood pressure, decreased urine output and edema (Lamba et al. 2020). In current world use of lead, and mercury is restricted in some countries due to their toxic nature. In some area's cosmetics or make-up was associated to prostitution and actresses and people didn't differentiate between the two. The Book of Jeremiah argued against the use of cosmetics (Quick 2019). It was considered as a symbol of wickedness, selfishness and vulgarity. People preferred using home remedies and natural substances to enhance their beauty. Flour was used as face powder, beetroot as lip and cheek tint. Some women even pinched their cheeks to give them some redness. Cosmetic products especially make-up had a different significance with different cultures and the love-hate relationship

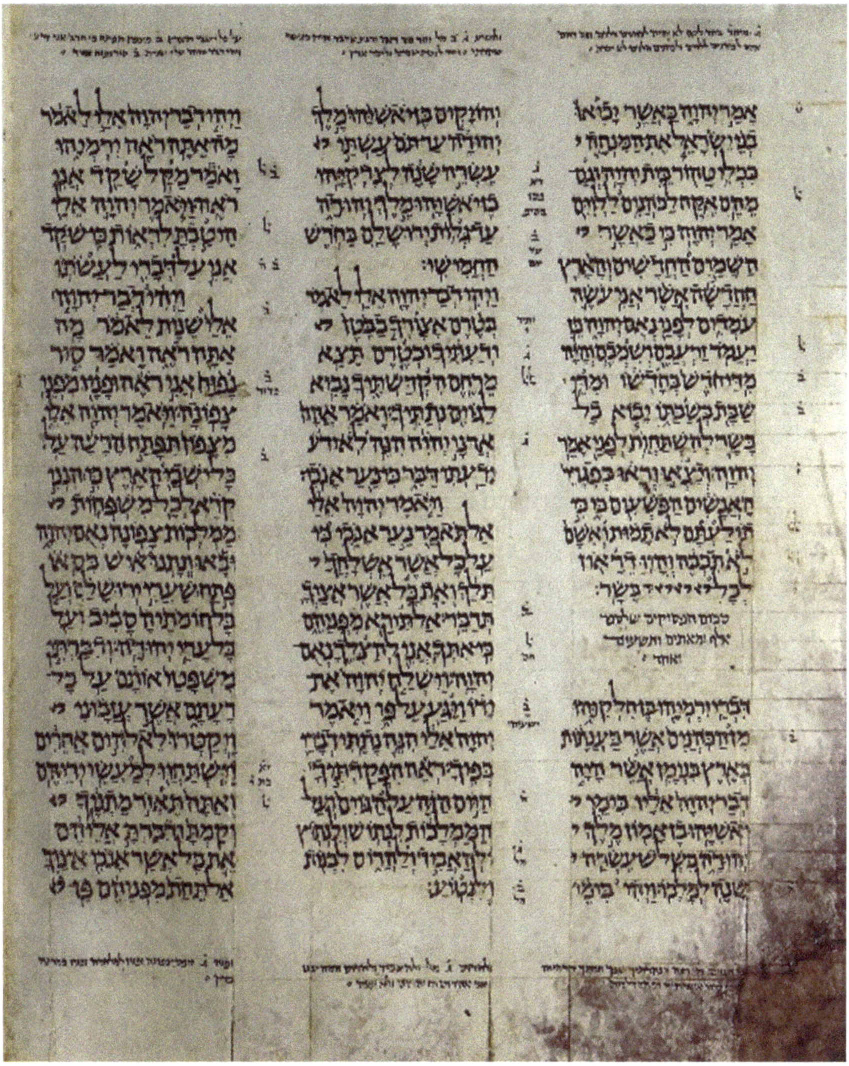

Fig. 2 The book of Jeremiah

of cosmetics with people went on for a long time. But it wasn't until 1920s when the use of cosmetic products was normalized. Some people started their own brand/company using well known home-remedies and make-up hacks and become established as reputed cosmetic brands. In late nineteenth century and twentieth century cosmetic industry had emerged and created a lot of opportunities for women. This was the era when companies like Maybelline etc. were started. The cosmetic companies back then were more focused on beauty products, i.e., the sole purpose of these products was to beautify the appearance and very less attention was paid on protecting and caring about our body and skin's health. People had very little knowledge about the chemicals or techniques of preparation and were apathetic against their harmful effects.

But later, products started emerging in market with proper scientific study and technical knowhow. A special line emerged known as 'skincare' and 'haircare'. Products containing more of natural ingredients aloe vera, curcumin, salicylic acid, azelaic acid etc. were bought (Mahesh et al. 2019). People had started taking care of their skin due to widespread availability of diverse skincare products and increased access to information as well as a practice of broader self-care. Our skin has its own defense mechanism which works against pathogens and bacteria, protecting our body (Esteban 2012). But with prolonged used of products containing harsh chemicals, heavy metals etc. our natural skin barrier is broken. It has been established that skin is an important part of innate immune system and is crucial in bodily functions such as preventing fluid loss, maintaining body temperature, transferring sensory inputs etc. It provides physical as well as biochemical protection from pathogens, microbial infections, bacteria etc (Nguyen and Soulika 2019).

2 Basic Anatomy of Skin:

Skin is made up of three layers- (i) Epidermis, (ii) Dermis and (iii) Subcutaneous fat tissue (Gilaberte et al. 2016). Each layer plays an important role in protecting our skin and internal organs from bacteria, pathogens etc. it does so by forming physical barriers (Fig. 3).

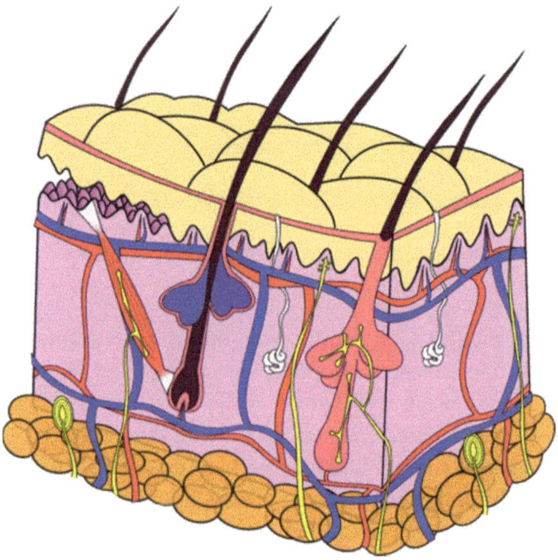

Fig. 3 Basic anatomy of skin

2.1 Epidermis

It is the outermost layer of the skin. It has five sub layers- stratum basale (the deepest part), stratum spinosum, stratum granulosum, stratum lucidum and stratum corneum (the surface layer) (Koster 2009). Stratum basale contains basale keratinocytes, immune cells and melanocytes that provide pigmentation to the skin (Menon 2015). Keratinocytes in simple words are the cells that produce keratin (Hirobe 2014). They also play a critical role in repairing wounds and deploying immune response against pathogens. Stratum spinosum is composed of polyhedral keratinocytes and the function of this layer is to allow keratinocytes to mature. Stratum corneum contains corneocytes (keratinocytes). Corneocytes are differentiated and enucleated keratinocytes (Évora et al. 2021). They protect underlying mitotically active cells from UV- radiations, maintain hydration etc. Stratum lucidum contains dead keratinocytes which contains eleidin that provides layer with transparent appearance (Matsui 2023). Stratum granulosum contains cystine and histidine-rich granules that bind keratin filaments together (Holbrook 1989). The thickness of epidermis depends on the area of the body part for example- it is thin near the eyelids but is thick around our stomach and sole of feet.

2.2 Dermis

This is the second layer of skin i.e., inner layer of the two other layer (Brown and Krishnamurthy 2022) It is formed of two layers- papillary dermis (the thin layer at top) and reticular dermis (the thicker layer at bottom). Papillary dermis forms extensions containing capillaries that reach up to epidermis to facilitate nutrient transportation (Ciarletta and Ben 2012). In reticular dermis certain parts of skin such as sebaceous glands, hair follicles and sweat glands are located. Both these layers have fibroblasts, myofibroblasts and immune cells such as macrophages, lymphocytes, and mast cells (Woodley 2017) These tissues further contain three layers containing varied amount of collagen, elastic tissues, and reticular fibers.

2.3 Subcutaneous Fat

This is the bottommost layer of skin made up of fat and connective tissues. The major components are fibrocytes and adipocytes, rich in proteoglycans and glycosaminoglycans (Justine 2016). It also has larger blood vessels and nerves. This layer helps in regulating the body temperature as fat acts as good insulator of heat. It also has energy reserves and multiple immune cells that forbade entry of pathogens in our body.

These layers collectively form the natural defense barrier of skin and provide skin with nutrition, protection, and immunity from foreign materials. Our skin is a major target of oxidative stress, and all the skin ageing problems and diseases result from reactive oxygen species. To avoid all these problems our skin provides us with physical as well as biochemical protection. For e.g.- melanocyte, found in papillary dermis helps in protection from UV rays by forming a barrier and reticular dermis helps the skin to repair itself (Abdel-Malek et al. 2010). Skin inflammations like acne, bumps, papules, pustules are the results of reaction of langerhans and mast cells found in dermis with allergens and bacteria (Bischoff and Krämer 2007).

But it is also very important on our part to take care of our skin and not only depend on itself to repair and heal on its own, otherwise the natural defense barrier of skin breaks and causes skin problems. If the skin barrier breaks the skin becomes dehydrated, dry, and irritated. It also causes rashes, acne, redness, flakiness, fine lines, itchiness, and increased breakouts. To avoid these a person must take care of their skin by avoiding using harsh chemicals on their skin, avoiding excessive sun exposure, alkaline soaps, and detergents, too much dry or humid environment and avoid taking stress (Fig. 4).

If a person feels his skin barrier is starting to get damaged, he should immediately stop using excessive products and simplify his skincare regime, using products with a suitable pH, face wash containing salicylic acid, moisturizers containing hyaluronic acid may help (Mawazi et al. 2022). Not only this but using certain products containing natural ingredients such as aloe vera, neem and curcumin will also be beneficial for the skin as these have antibacterial, antifungal, and antiviral properties. They also help in recovering skins natural dense barrier and are good for healing purposes.

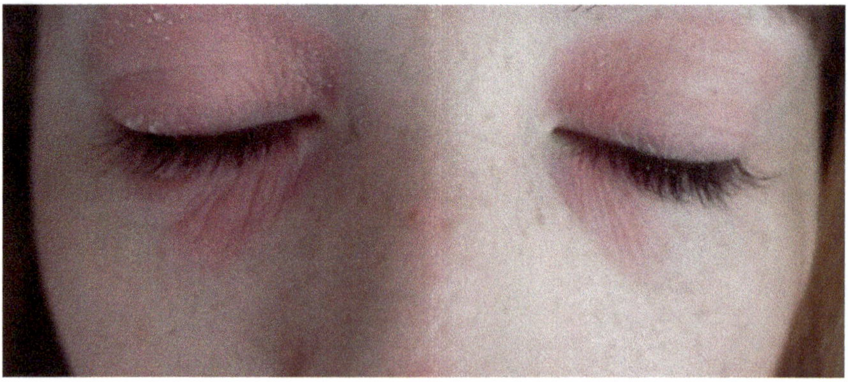

Fig. 4 Skin problems

3 Cosmetics

Cosmetic products can be separated into two parts (i) natural and (ii) synthetic and they can be further categorized into two parts (i) skin and (ii) hair. We will be mainly focusing on skin products. There are different kind of skin products available such as skincare, beautifying products, soaps, moisturizers etc. Let us broadly categorize the ingredients used in making these products (Jones and Selinger 2020).

Broad categorization of ingredients used in cosmetics:

- Water
- Surfactants
- Thickeners
- Emollients
- Antioxidants
- Pigments
- Fragrances
- Preservatives

3.1 Water

Almost on every product you pick (aqueous phase emulsions) water is present in the highest concentration and generally is written as "AQUA". It is generally used to dissolve or suspend other ingredients in the mix. It also helps in conditioning, nourishing, and cleansing the skin.

3.2 Surfactants

Surfactants are used in cosmetics as emulsifiers and for cleansing purposes (Corazza et al. 2010). Earlier surfactants were only used in soaps and detergents for cleaning purposes but gradually were introduces in cosmetic products (Rieger 2017). Surfactant can be of two type natural and man-made (chemical) which are synthesized in laboratories. Natural surfactant or biosurfactant tend to cause less damage to skin and environment as compared to chemical surfactants (Banat et al. 2000; Lourith and Kanlayavattanakul 2009). Some of the commonly used chemical surfactants are-

- Sodium lauryl sulfate—SLS is a foaming agent found in many cleaning and personal care products like shampoos, soaps, and toothpaste. It is derived from coconut and is effective in producing high foam. It can be irritating to some skin types and replaced with milder forms like sodium laureth sulfate (SLES).

- Ammonium laureth sulfate—ALES is similar to sodium lauryl sulfate (SLS) but gentler on skin. It's a foaming agent in shampoos, cleansers, and other products. While generally safe, it may cause dryness or irritation in some people.
- Disodium lauryl sulfosuccinate—DLS is a gentle, coconut-derived, cleansing agent and foamer used in shampoos, soaps, etc. It's milder than SLS but less than SLES, creating rich, creamy lather. Generally safe, but potential irritation exists.
- Cocoamphocarboxyglycinate—it is a mild amphoteric surfactant derived from coconuts.
- Decyl polyglucoside—DPG is a mild, non-ionic surfactant widely used in various personal care products like shampoos, body washes, and facial cleansers. It is derived from sustainable sources like glucose (from starch) and fatty alcohols (often coconut or corn) and may not be as effective for heavy cleansing in specific applications.
- Cocamidopropyl betaine—CAPB is a common ingredient in personal care products like shampoos, soaps, and body washes. It helps to lift dirt and oil from skin and hair, aiding in cleaning. It is generally safe for most people, but some may experience skin irritation due to impurities formed during manufacturing. It can be derived from coconut oil and propylene glycol (synthetic). However, its use is restricted now due to potential harm to aquatic life.
- Decyl glucoside—gentle, plant-based cleansing and foaming agent found in personal care products. It is made from renewable sources like glucose and coconut-derived alcohols. It is suitable for sensitive skin due to its non-irritating nature while being easily biodegradable. Forming a rich, stable lather it is preferably used in hair and body cleansing.
- Similarly other vegetable-based products like glyceryl cocoate, almond glycerides, sodium methyl cocoyl taurate, sucrose cocoate, polysorbate 20 & polysorbate 80 –are also commonly used.
- Cetearyl alcohol and stearyl alcohol—both the fatty alcohols are derived from plants (like coconut oil) or animals. They are thick, waxy solid, and white in color used in moisturizing lotions, creams, and conditioners. They are non-irritating, and helps products like lipsticks spread smoothly.

Biocosmetics are changing the beauty game by offering a natural touch to our daily routine. Unlike conventional cosmetics filled with potentially harmful chemicals, biocosmetics use 100% natural ingredients sourced from plants, animals, microbes, and even insects. These ingredients are grown organically, free from pesticides and chemical fertilizers, ensuring purity and sustainability (Novak et al. 2014). Abandoning the synthetic oils and mineral-based formulas often found in mainstream skincare, biocosmetics instead, embrace the power of nature's elements, offering gentle yet effective care for our skin, hair, face, and even oral health (Goyal and Jerold 2021).

Many cosmetic manufacturers throughout the world are investigating the methods for making natural and eco-friendly cosmetics as a result of the worldwide hunt for high-performing, secure, and sustainable components. There are many concerns about the safety of using chemicals in cosmetic formulas. Several news articles and social media posts discuss potentially harmful ingredients found in cosmetic formulations, such as lead, aluminum, mercury, parabens, etc.

In addition, certain biosurfactants or natural surfactants are also becoming very popular. These are surface active agents which are produced naturally by bacteria, fungi or yeast as secondary metabolites. Employing different separation methods like extraction, precipitation, and distillation they are obtained for further applications. The naturally derived surfactants can have wide range of molecular weights. Compounds having low molecular weights effectively reduce the surface tension and interfacial tension while on the other hand surfactants with high molecular weight are better emulsifiers. Some examples of biosurfactants are –

- Sophorolipids- Sophorolipids are a diverse group of glycolipid biosurfactants produced by various yeasts and filamentous fungi. They possess a unique structure consisting of a sugar moiety (usually sophorose) linked to a fatty acid chain through a glycosidic bond. This structure grants them amphiphilic properties, having both hydrophilic (water-loving) and hydrophobic (water-hating) properties. This unique characteristic allows them to act as emulsifiers, dispersants, and wetting agents, making them valuable ingredients in a variety of applications, including cosmetics (Fig. 5).

  Sophorolipids are used as emulsifiers, detergents, solubilizers, foaming agents and have good compatibility with the skin. Sophorolipids are also used in treatment of skin problems like acne, dandruff, and body odors.
- Rhamnolipids- made up of one or two rhamnose sugar units linked to a beta-hydroxy fatty acid chain. There can be mono- or di-rhamnolipids, depending on the number of sugar units. The fatty acid chains vary in length and saturation, influencing properties like surface activity and biodegradability. These are produced by various bacteria strains like Pseudomonas aeruginosa cultivated on glucose, glycerol or triglycerides. They are the leading commercial biosurfactant which are suitable from a combined application/cost perspective and is produced on an industrial scale(29,24) (Fig. 6).

Fig. 5 Sophorolipids

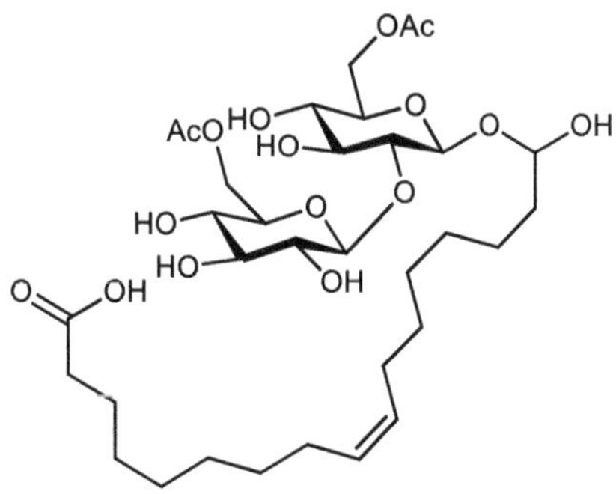

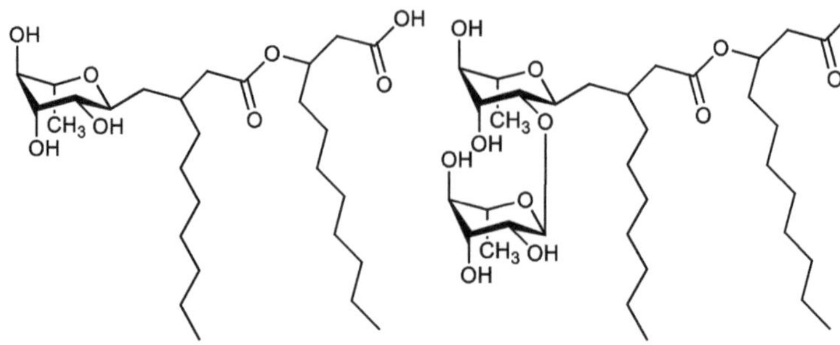

Fig. 6 Structure of monorhamnolipid (left) and dirhamnolipid (right)

Fig. 7 Mannosylerythritol lipid

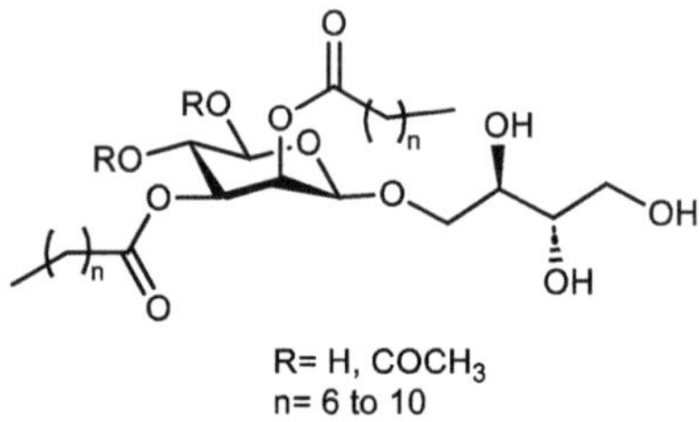

Rhamnolipids also have exceptional antimicrobial activities including several fungi and bacteria. They are used in products like insect repellents, antacids, acne pad, contact lenses solution, deodorants, nailcare and toothpastes.

- Mannosylerythritol lipid- Mannosylerythritol lipids (MELs), produced by yeasts and fungi mainly belonging to the genera Pseudozyma and Ustilago. The Core structure consists of 4-O-β-D-mannopyranosyl-meso-erythritol, a sugar-alcohol unit. The fatty acid chain attached to the hydroxyl group of the erythritol unit, vary in length (C10-C18) and saturation. The presence of acetyl groups on the mannose moiety creates different MEL subtypes with varying properties. Mannosylerythritol lipids containing skin care formulations are used as anti-wrinkle cosmetics and prevent skin roughness (Fig. 7).

3.3 *Thickener*

Thickeners help in regulating the viscosity of the product and provides it consistency. Each product requires different viscosity, some are liquids, some are thick and there are gel product too (Moravkova and Filip 2014). Viscosity can be altered by changing the concentration of the solute by adding more solids, or by addition of thickening agents. Thickeners are of two types- aqueous thickeners and non-aqueous thickeners.

- Aqueous thickeners—aqueous thickeners work by dissolving in water, increasing the viscosity of the water-based phase in creams, lotions, and gels. Common examples include cellulose and its derivatives, plant-derived gums like guar gum and xanthan gum, as well as synthetic polymers like carbomer (Mudgil et al. 2014). These thickeners offer a range of textures, from light and flowy to rich and luxurious.

- Nonaqueous thickeners—these on the other hand, dissolve in oils and other non-aqueous ingredients. They're crucial for thickening oil-based products like makeup removers, sunscreens, and balms. Popular choices include silicones, organoclays, fatty alcohols, and beeswax. These thickeners impart different sensory experiences, from silky smooth to rich and buttery.

3.4 Emollients

Emollients are used to provide skin with a softening effect and moisture by reducing the water loss from skin. They're ideal for restoring skin barrier function. they are sometimes refereed as moisturizers (Lodén 2003). Emollients work by forming a thin protective film to cover the skin and trap in the moisture. This helps in reduction of water loss through perspiration. Emollients are found in creams, ointments, lotions, gels and balms (Chao et al. 2018). They mainly contain lipids like paraffin wax, urea etc. as they prevent water loss through skin (Stojiljković et al. 2013). Products containing high concentrations of emollient are best suited for people with dry skin. Emollients are further of four types given in table below.

Types of Emollients:

Dry emollients	Fatty emollient	Astringent emollients	Protective emollients
Petroleum jelly, mineral oil, lanolin, beeswax, decyl oleate, isostearyl alcohol, isopropyl palmitate	Jojoba oil, olive oil, avocado oil, coconut oil, glyceryl stearate, propylene glycol, octyl stearate,	Dimethicone, cyclomethicone, isopropyl myristate, octyl octanoate	Zinc oxide, titanium dioxide, octinoxate, oxybenzone, diisoprpyl dilinoeate, isopropyl isostearate

3.5 Antioxidants

We know that our skin is a major target of oxidative stress and that results in degradation of our skin quality. It can cause skin wrinkling, synthesis of melanin, cause cancer and several allergy and skin diseases. Antioxidants are responsible for chain breaking and for inhibition the oxidation reaction (Kusumawati and Indrayanto 2013). They intervene the free radical production responsible for skin damage. These free radicals like reactive oxygen species (ROS) have one or more unpaired electron and are unstable and highly reactive. They react easily with other biomolecules, through a chain process which can kill or damage healthy cells. Antioxidants can prevent oxidative damage to the skin. Regular application of antioxidants may lessen the lentigines (sunspots) and help in slow skin aging (Lupo 2001). Best antioxidants for skin are-

- Vitamin C—it boosts collagen production which helps in making skin firm and it also helps fading dark spots. It prevents sun damage and helps with hyperpigmentation.
- Niacinamide—Niacinamide, a form of vitamin B3 is a powerful antioxidant, shielding skin from free radical damage and environmental aggressors. It acts as a texture refiner, minimizing pores and smoothing unevenness, and evens skin tone by tackling hyperpigmentation, resulting in a brighter, more radiant complexion. It also helps in reduction of fine line and wrinkles. This vitamin is often used to treat skin conditions like rosacea and acne and is not known to cause any adverse reactions, making it ideal for sensitive types.
- Resveratrol—it is also known as longevity molecule because of its excellent anti-aging properties. It is found in grapes and berries. It also has anti-inflammatory effect and anti-fungal property. It should be used in evening or at night as UV exposure hampers its effectiveness.
- Vitamin E—it I known for its healing action. It is found in almost every product such as creams and lotions formulated for the treatment of dry skin. It also helps in stabilizing other antioxidants such as vitamin c and increase their effectiveness.
- Vitamin A (Retinol)—Vitamin A, in the form of retinol, is found in most anti-aging cosmetics. These small molecules penetrate deep into the skin and acts as a cell communicator, stimulating collagen production and skin cell turnover. This results in a firmer, smoother appearance with reduced fine lines and wrinkles. Retinol also targets sun damage, minimizing hyperpigmentation and uneven skin tone. It even helps control breakouts by unclogging pores and regulating sebum production.

3.6 Pigments

Pigments play a vital role in cosmetics, serving two main purposes: coloration and visual effects. They come in two main types: organic and inorganic. Organic pigments are primarily derived from plants, minerals, or insects, these offer vibrant colors like reds, oranges, and yellows. Examples include carmine (red), iron oxides (reds, yellows, browns), and ultramarines (blues, violets). Inorganic pigment on the other hand often are derived from minerals or metals, and these provide opacity and durability. Examples include titanium dioxide (white), iron oxide (black), and mica (metallic sheen) (Mohana Priya et al. 2020).

Based on the visual effects they can be classified as pearlescent pigments like mica-based pigments. They create a shimmery, luminous effect, highlighting features like cheekbones and eyelids. Glitters on the other hand adds sparkle and shine, creating playful makeup looks. Interference pigments play with light, creating iridescent or duochrome effects depending on the viewing angle. Examples include bismuth oxychloride and chameleon pigments (Faulkner and Hollenberg 2021). Some of the makeup products which use pigments majorly are:

- Lipstick: A combination of organic and inorganic pigments creates the desired color and opacity.
- Foundation: Titanium dioxide provides opacity and evens skin tone, while iron oxides adjust the shade.
- Eyeshadow: Mica and interference pigments offer shimmer and dimension, while organic pigments deliver bold colors.

Some pigments offer additional benefits. Titanium dioxide, for example, also protects against UV rays. Iron oxides can soothe sensitive skin.

Natural pigments are ecofriendly, non-toxic, non-carcinogenic etc. Natural versus synthetic pigments is a personal choice, each with its advantages and disadvantages.

Overall, pigments are essential for creating the colors, textures, and finishes that define the world of cosmetics, allowing for self-expression and individual style (Becker 2016).

3.7 *Fragrances*

Fragrances are essential components of cosmetic products that not only mask unpleasant odors, but also create sensory and emotional experiences for the users. Fragrances can affect the mood, self-esteem, and attractiveness of the users, as well as their preferences and choices for different products (Sharmeen et al. 2021). Fragrances can be derived from natural or synthetic sources, and can have different characteristics and effects depending on their composition and structure. Fragrances also require careful formulation and testing to ensure their safety, quality, and efficacy, as well as their sustainability and environmental impact. Furthermore, fragrances can offer personalized and innovative solutions for the future of cosmetics, using biotechnology and advanced delivery systems. Fragrances are more than just scents; they are powerful tools that shape our perception, emotions, and choices (Milotic 2003).

3.8 *Preservatives*

The role of preservatives is to provide durability to the product. Our surroundings contain various kinds of microorganisms that may react with the product and make it go bad (Hrádková et al. 2021) varied environment either too hot or cold can also deteriorate the quality of product. Oxidation, reduction all these processes tend to decrease the shelf life of a product that is why preservatives are added into products. They give the product protection against bacteria, fungi and micro- organism and also have disinfectant properties (Uzdrowska and Górska-Ponikowska 2023). Preservatives are of three kinds- antimicrobial, antioxidant, and UV light absorbers.

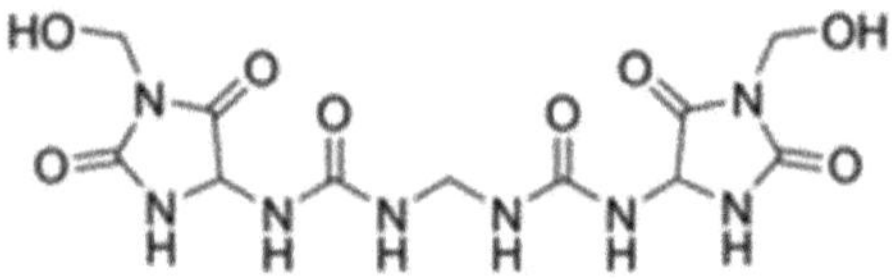

Fig. 8 Imidazolidinyl urea

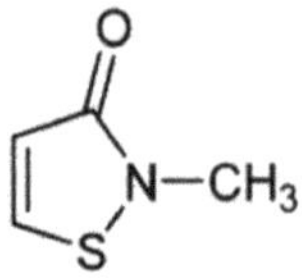

Fig. 9 Methylisothiazolinone (MIT)

Antimicrobial preservative is of two types-

(i) Formaldehyde releasing- these preservatives work by slowly releasing formaldehyde, a potent biocide that kills bacteria, fungi, and other microbes. Some common FRPs include: DMDM hydantoin, quaternium-15, diazolidinyl urea, imidazolidinyl urea, bronopol (Fig. 8).
(ii) Non-formaldehyde releasing- non-formaldehyde releasing preservatives (NFRPs): These preservatives use mechanisms other than formaldehyde release to kill microbes. Some common NFRPs include: parabens, phenoxyethanol, benzoic acid, sorbic acid, methylisothiazolinone (MIT) (Fig. 9)

4 Emerging Trends

Cosmetic pharmaceuticals, often known as cosmeceuticals, are combination of products from pharmacy and cosmetics, thus having benefits both in cosmetics field and therapeutic field (Rangaraj et al. 2023). Although this phrase is used in marketing and has no official legal definition. At low quantities, some substances are acceptable in cosmetics, but at greater concentrations, they are considered medications. There are always demands for products with multiple benefits (Samaranayaka and Li-Chan 2011; Anunciato and Rocha Filho 2012). Like in the case of a lipstick or lip balm, the primary role will be adding great color to the lips, but at the same time if it rejuvenates or regenerates the cracked lips, that will be advantageous and the need of today's cosmetic industry (Ramasubramania Raja 2016). Some popular cosmeceuticals categorized by their key benefits:

- Retinol: Acts as an antiaging agent, stimulates collagen production, reduces wrinkles, and improves skin texture (Birru et al. 2023).
- Hyaluronic ccid: Acts as a hydrating agent, plumps and hydrates the skin, reducing the appearance of fine lines and wrinkles. It also plays a crucial role in wound healing by controlling inflammation and redirecting blood flow to damaged tissue (Fallacara et al. 2018; Nam et al. 2023).

- Niacinamide: Reduces hyperpigmentation, brightens the skin, reduces redness, and improves skin texture (Kawada et al. 2008).
- Vitamin C: An antioxidant that protects the skin from free radical damage and brightens the complexion. It also acts as an antiaging and sun protecting agent (Enrique Lorente Prieto 2023).
- Salicylic acid: Used for Acne treatment, exfoliates the skin, unclogs pores, and reduces inflammation (Wiśniewska et al. 2023).
- Benzoyl peroxide: Acts as a bactericidal, it works by reducing the acne-causing bacteria and by causing the skin to dry and peel.
- Azelaic acid: Reduces inflammation, unclogs pores, and brightens the skin (Pérez-Bernal et al. 2000).
- Hydroquinone: A skin lightening agent, it bleaches the skin and help reducing hyperpigmentation of different forms (Draelos 2007).
- Kojic acid: A natural skin lightening agent derived from fermented rice. It may also be used to treat skin conditions, such as sun damage, scars, and age spots, by reducing melanin production.
- Licorice root extract: Brightens the skin and reduces hyperpigmentation. It neutralizes free radicals to prevent damage to the skin and keep looking youthful and radiant. It can help reduce acne-related inflammation and prevent the growth with its antibacterial activity (Cerulli et al. 2022).

AI powered skincare is a technology that uses artificial intelligence to analyze the skin condition and needs of each individual, and provide personalized recommendations for skincare products and routines (Sureshkumar et al. 2023; Wei et al. 2023). AI can measure various skin parameters, such as hydration, pigmentation, texture, and pore size, and compare them with a large database of skin images and data. Based on the analysis, AI can suggest the best formulation of ingredients and active agents for each skin type and concern (Vatiwutipong et al. 2023). Some examples of AI powered skincare brands are Sephora, Nivea, and L'Oreal, which offer virtual skin analysis tools and customized product suggestions.

Another emerging technology is 3D bioprinting that uses 3D printers to create biological structures, such as tissues and organs, by depositing layers of living cells and biomaterials. 3D bioprinting can mimic the natural structure and function of the body's tissues, and can be used for various applications, such as tissue engineering, drug testing, and cosmetic treatments (Vijayavenkataraman et al. 2016). 3D bioprinting uses bioinks, which are solutions or gels that contain cells and biocompatible substances, such as polysaccharides, proteins, and synthetic polymers. Depending on the application, different types of bioinks and bioprinting techniques can be used to create customized and functional tissue constructs (Zhang et al. 2021). For example, 3D bioprinting can be used to create skin substitutes for wound healing, burn treatment, or skin grafting. It can also be used to create depigmented patches for vitiligo patients, which can be treated with growth factors, stem cells, and cytokines to stimulate melanin production and skin regeneration.

So far, we have discussed about a few components of ingredients used in cosmetics. And what we can observe is although there are many chemicals which are still widely used in formulations of cosmetic products, people are now shifting towards natural products like natural surfactants, natural emollients and cosmeceuticals. This change is necessary as many chemicals not only damage the skin but also the environment. The process of synthesizing these chemical compounds is very critical and hazardous so a shift to biomaterials or natural products is beneficial for all the aspects of the society. Presence of heavy metals in high amounts is also dangerous for our own health. None of the skincare products that we use are 100% safe or guarantee 100% results because chemicals react on varied skin textures in different ways. One product may work wonders for one person but the same can deteriorate someone else's skin condition. People are now more conscious and some prefer minimalist approach with occasional use of natural home-grown aloe vera, neem, turmeric, chickpea flour, clarified butter, eggs and other products (Arora et al. 2019). Aloe vera is known for its antiviral, antifungal, and antimicrobial property. It also helps in healing the skin and is rich in antioxidants. it is most natural remedy for sunburns. Turmeric has curcumin which is a powerful antioxidant and also has anti-inflammatory properties. It helps in reducing wrinkles and helps in lightening dark spots and skin brightening. It improves collagen deposition and increased fibroblast and vascular density in wounds thereby enhancing both normal and impaired wound-healing. Neem is used for many purposes, like antiallergen, ant-inflammatory, and antifungal agents. Neem helps with acne; it fights with acne causing bacteria. It is extremely useful for oil control. Neem has so many antioxidants, moisturizing triglycerides and vitamins that help with skin ageing. It reduces wrinkles and fine lines.

As the demands are changing companies are also moving in the same direction with recycling natural by-products and waste from food and agriculture to create active ingredients for cosmetics. For example, some startups use coffee grounds, fruit peels, or olive pits to extract antioxidants, oils, or exfoliants for skincare products. Cosmetics which take care of skin microbiome, which is the community of microorganisms that live on the skin, to enhance or balance the skin flora through use of probiotics, prebiotics, or postbiotics to boost skin health, hydration, or immunity (Paetzold et al. 2019). Bio coated materials with controlled delivery of active ingredients are also being actively used.

5 Conclusion

The cosmetics industry is undergoing a major transformation, as consumers are becoming more aware and demanding of the environmental and health impacts of the products they use. They are looking for cosmetics that offer not only beauty, but also wellness, sustainability, and social responsibility. This creates new opportunities and challenges for startups and innovators who want to cater to these needs and preferences, by developing products that are customized, natural, vegan, clean, and high-performance. To achieve this, they need to leverage the latest advances in technology and science, and explore new sources and methods of extracting and delivering natural ingredients.

One of the key challenges in the cosmetics industry is to find solvents that are compatible with both the natural ingredients and the final formulations, and that do not compromise the stability, safety, or efficacy of the products. Traditional solvents, such as water, ethanol, or glycerol, have limited polarity ranges and solubility capacities, and may also have negative environmental or health effects. Therefore, there is a need for alternative solvents that are bio-based, biodegradable, and biocompatible, and that can dissolve a wide range of natural compounds with different polarities and functionalities. One of the promising solutions is natural deep eutectic solvents (NADES), which are mixtures of natural metabolites, such as sugars, amino acids, or organic acids, that form liquid phases at room temperature (Vanda et al. 2019). NADES have several advantages over conventional solvents, such as low toxicity, high biodegradability, high thermal stability, low volatility, and high solvation power. Moreover, NADES can be tailored to specific applications by adjusting the composition and ratio of the components, and can also act as carriers or enhancers of the bioactive properties of the solutes. NADES have been used to extract and formulate various natural ingredients for cosmetics, such as antioxidants, pigments, fragrances, and antimicrobials, from plants, microalgae, fungi, and bacteria (Rente et al. 2022).

Another challenge in the cosmetics industry is to find natural ingredients that have multifunctional and synergistic effects on the skin, such as anti-aging, anti-inflammatory, anti-microbial, moisturizing, whitening, and sun protection. These ingredients should also have high bioavailability, stability, and penetration, and low side effects and allergenicity. To overcome these challenges, researchers and innovators have been exploring various sources and strategies of obtaining and delivering natural ingredients, such as:

- Short antimicrobial peptides (SAMPs), which are small proteins that have potent and selective antimicrobial activity against various pathogens, and can also modulate the immune system, promote wound healing, and enhance skin barrier function (Kardani and Bolhassani 2021). SAMPs can be derived from natural sources, such as plants, animals, or microorganisms, or synthesized artificially. SAMPs have been used as preservatives, anti-acne agents, anti-dandruff agents, and anti-inflammatory agents in cosmetics (Miazga-Karska et al. 2020).
- Plant-cell culture, which is a biotechnological technique that allows the production of plant cells, tissues, or organs in vitro, under controlled conditions. Plant-cell culture can produce high-quality and consistent natural ingredients, such as phenolic compounds, terpenoids, alkaloids, and polysaccharides, that have various cosmetic benefits, such as antioxidant, anti-wrinkle, skin-firming, and skin-lightening effects (Georgiev et al. 2018; Barbulova et al. 2014). Plant-cell culture can also overcome the limitations of conventional plant extraction, such as low yield, seasonal variation, environmental impact, and ethical issues. Plant-cell culture has been used to produce natural ingredients from various plants, such as ginseng, grape, rose, orchid, and saffron (Krasteva et al. 2021).

- Marine algae, which are aquatic organisms that can produce a variety of bioactive compounds, such as polysaccharides, polyphenols, carotenoids, phycobiliproteins, and fatty acids, that have diverse cosmetic applications, such as UV protection, skin hydration, pigment production, and anti-aging effects (Wang et al. 2015). Marine algae can also adapt to different environmental conditions, such as salinity, temperature, and light, and can be cultivated in large scale, with low environmental impact and high sustainability. Marine algae have been used to produce natural ingredients from various species, such as brown algae, red algae, green algae, and cyanobacteria (Fonseca et al. 2023).

One of the common challenges in delivering natural ingredients to the skin is to overcome the skin barrier, which is composed of the stratum corneum, the outermost layer of the epidermis, and prevents the penetration of foreign substances into the skin. Many natural ingredients have poor bioavailability, low solubility, and uncontrolled release, which limit their cosmetic efficacy and safety. Therefore, there is a need for novel delivery systems that can enhance the skin penetration, stability, and targeting of natural ingredients, and also provide controlled and sustained release, and improved sensory and aesthetic properties. Some of the novel delivery systems that have been used for natural ingredients in cosmetics are:

- Hydrotalcites, which are layered minerals that can absorb and release molecules, such as drugs, vitamins, or sunscreens, depending on the pH or temperature. Hydrotalcites can act as carriers or reservoirs of natural ingredients, and can also protect them from degradation, oxidation, or photolysis (Trujillano 2022). Hydrotalcites can also improve the stability, texture, and appearance of cosmetic formulations, and can provide additional benefits, such as anti-inflammatory, antimicrobial, or anti-acne effects. Hydrotalcites have been used to encapsulate natural ingredients, such as fragrances, vitamins, or sunscreens, and to deliver them to the skin (Kumari et al. 2023; Perioli et al. 2008).
- Organoclays, which are organically modified bentonite clays that have enhanced rheological and thixotropic properties, and can act as thickeners and flow-control additives in cosmetic formulations (Calabi Floody et al. 2009). Organoclays can also form complexes or intercalates with natural ingredients, such as oils, pigments, or antioxidants, and can improve their dispersion, stability, and delivery to the skin. Organoclays can also impart film-forming, water-resistant, and color-retention properties to cosmetic products, and can improve their sensory and aesthetic attributes. Organoclays have been used to formulate natural ingredients, such as nail paints, eye shadows, and lipsticks (Patel et al. 2006).

These are some of the examples of how technology and innovation are transforming the cosmetics industry, and creating new possibilities for wellbeing and attractiveness with natural substances. However, there are still many challenges and opportunities for further research and development, such as ensuring the safety, quality, and efficacy of the natural ingredients and products, complying with the regulatory and ethical standards, and meeting the consumer expectations and demands.

References

Abdel-Malek ZA, Kadekaro AL, Swope VB (2010) Stepping up melanocytes to the challenge of UV exposure. Pigment Cell Melanoma Res 23:171–186. https://doi.org/10.1111/j.1755-148X.2010.00679.x

Anunciato TP, da Rocha Filho PA (2012) Carotenoids and polyphenols in nutricosmetics, nutraceuticals, and cosmeceuticals. J Cosmet Dermatol 11. https://doi.org/10.1111/j.1473-2165.2011.00600.x

Arora R, Aggarwal G, Dhingra GA, Nagpal M (2019) Herbal active ingredients used in skin cosmetics. Asian J Pharmaceut Clin Res https://doi.org/10.22159/ajpcr.2019.v12i9.33620

Banat IM, Makkar RS, Cameotra SS (2000) Potential commercial applications of microbial surfactants. Appl Microbiol Biotechnol 53

Barbulova A, Apone F, Colucci G (2014) Plant cell cultures as source of cosmetic active ingredients. Cosmetics 1

Becker D (2016) Color trends and selection for product design: every color sells a story

Birru PW, Hilmi IL, Salman S (2023) Article review : retinol in cosmetics. J Pharmaceut Sci 6

Bischoff SC, Krämer S (2007) Human mast cells, bacteria, and intestinal immunity. Immunol Rev 217:329–337. https://doi.org/10.1111/j.1600-065X.2007.00523.x

Brown TM, Krishnamurthy K (2022) Histology, dermis. StatPearls [Internet] StatPearls Publishing

Calabi Floody M, Theng BKG, Reyes P, Mora ML (2009) Natural nanoclays: applications and future trends—a Chilean perspective. Clay Miner 44. https://doi.org/10.1180/claymin.2009.044.2.161

Cerulli A, Masullo M, Montoro P, Piacente S (2022) Licorice (Glycyrrhiza glabra, G. uralensis, and G. inflata) and their constituents as active cosmeceutical ingredients. Cosmetics 9

Chao C, Génot C, Rodriguez C, et al (2018) Emollients for cosmetic formulations: towards relationships between physico-chemical properties and sensory perceptions. Colloids Surf A Physicochem Eng Asp 536. https://doi.org/10.1016/j.colsurfa.2017.07.025

Ciarletta P, Ben AM (2012) Papillary networks in the dermal–epidermal junction of skin: a biomechanical model. Mech Res Commun 42:68–76. https://doi.org/10.1016/j.mechrescom.2011.12.001

Corazza M, Lauriola MM, Zappaterra M et al (2010) Surfactants, skin cleansing protagonists. J Eur Acad Dermatol Venereol 24:1–6. https://doi.org/10.1111/j.1468-3083.2009.03349.x

Draelos ZD (2007) Skin lightening preparations and the hydroquinone controversy. Dermatol Ther 20

Duarte CM (2014) Red ochre and shells: Clues to human evolution. Trends Ecol Evol 29

El-Kilany E, Raoof E (2017) Facial cosmetics in ancient Egypt. Egypt J Tour Stud 16

Enrique Lorente Prieto MD (2023) Cosmetic Topical Use of Vitamin C. In: Ascorbic acid–biochemistry and functions [Working Title]

Esteban MÁ (2012) An overview of the immunological defenses in fish skin. ISRN Immunology 2012:1–29. https://doi.org/10.5402/2012/853470

Évora AS, Adams MJ, Johnson SA, Zhang Z (2021) Corneocytes: relationship between structural and biomechanical properties. Skin Pharmacol Physiol 34:146–161. https://doi.org/10.1159/000513054

Fallacara A, Baldini E, Manfredini S, Vertuani S (2018) Hyaluronic acid in the third millennium. Polymers (Basel) 10

Faulkner EB, Hollenberg JC (2021) Coloring the cosmetic world: using pigments in decorative cosmetic formulations

Fenner J, and Clark RAF (2016) Anatomy, physiology, histology, and immunohistochemistry of human skin. In: Skin tissue engineering and regenerative medicine, vol. 1

Fonseca S, Amaral MN, Reis CP, Custódio L (2023) Marine natural products as innovative cosmetic ingredients. Mar Drugs 21

Georgiev V, Slavov A, Vasileva I, Pavlov A (2018) Plant cell culture as emerging technology for production of active cosmetic ingredients. Eng Life Sci 18

Gilaberte Y, Prieto-Torres L, Pastushenko I, Juarranz Á (2016) Anatomy and function of the skin. In: Nanoscience in dermatology. Elsevier, pp 1–14
Goyal N, Jerold F (2021) Biocosmetics: technological advances and future outlook. Environ Sci Pollut Res 30:25148–25169. https://doi.org/10.1007/s11356-021-17567-3
Hirobe T (2014) Keratinocytes regulate the function of melanocytes. Dermatol Sin 32:200–204. https://doi.org/10.1016/j.dsi.2014.05.002
Holbrook KA (1989) Biologic structure and function: perspectives on morphologic approaches to the study of the granular layer keratinocyte. J Investig Dermatol 92:S84–S104. https://doi.org/10.1038/jid.1989.36
Hrádková I, Kumherová M, Šmidrkal J (2021) Preservatives in cosmetics. Chem Listy 115. https://doi.org/10.1016/b978-0-444-63508-2.00009-6
Jones DW (1993) Phoenician unguent factories in dark age Greece: social approaches to evaluating the archaeological evidence. Oxf J Archaeol 12:293–303. https://doi.org/10.1111/j.1468-0092.1993.tb00334.x
Jones O, Selinger B (2020) The chemistry of cosmetics. Australian Acad Sci 6
Kardani K, Bolhassani A (2021) Antimicrobial/anticancer peptides: bioactive molecules and therapeutic agents. Immunotherapy 13
Kawada A, Konishi N, Oiso N, et al (2008) Evaluation of anti-wrinkle effects of a novel cosmetic containing niacinamide. J Dermatol 35. https://doi.org/10.1111/j.1346-8138.2008.00537.x
Koster MI (2009) Making an epidermis. Ann N Y Acad Sci 1170:7–10. https://doi.org/10.1111/j.1749-6632.2009.04363.x
Krasteva G, Georgiev V, Pavlov A (2021) Recent applications of plant cell culture technology in cosmetics and foods. Eng Life Sci 21
Kumari S, Sharma V, Soni S, et al (2023) Layered double hydroxides and their tailored hybrids/composites: Progressive trends for delivery of natural/synthetic-drug/cosmetic biomolecules. Environ Res 238
Kusumawati I, Indrayanto G (2013) Natural antioxidants in cosmetics. In: Studies in natural products chemistry
Lamba P, Nam KH, Contractor J, Kim A (2020) Nephritic syndrome. Primary Care Clin Office Pract 47:615–629. https://doi.org/10.1016/j.pop.2020.08.003
Li S-J, Zhang S-H, Chen H-P et al (2010) Mercury-induced membranous nephropathy. Clin J Am Soc Nephrol 5:439–444. https://doi.org/10.2215/CJN.07571009
Lodén M (2003) Role of topical emollients and moisturizers in the treatment of dry skin barrier disorders. Am J Clin Dermatol 4
Lourith N, Kanlayavattanakul M (2009) Natural surfactants used in cosmetics: glycolipids. Int J Cosmet Sci 31
Lucas A (1930) Cosmetics, perfumes and incense in ancient Egypt. J Egypt Archaeol 16:41. https://doi.org/10.2307/3854332
Lupo MP (2001) Antioxidants and vitamins in cosmetics. Clin Dermatol 19. https://doi.org/10.1016/S0738-081X(01)00188-2
Mahesh SK, Fathima J, Veena VG (2019) Cosmetic potential of natural products: industrial applications. In: Natural bio-active compounds. Springer, Singapore, pp 215–250
Matsui T (2023) Epidermal barrier development via corneoptosis: a unique form of cell death in stratum granulosum cells. J Dev Biol 11:43. https://doi.org/10.3390/jdb11040043
Mawazi SM, Ann J, Othman N et al (2022) A review of moisturizers; history, preparation, characterization and applications. Cosmetics 9:61. https://doi.org/10.3390/cosmetics9030061
Menon GK (2015) Skin basics; structure and function. In: Lipids and skin health. Springer International Publishing, pp 9–23
Miazga-Karska M, Michalak K, Ginalska G (2020) Anti-acne action of peptides isolated from burdock root—preliminary studies and pilot testing. Molecules 25. https://doi.org/10.3390/molecules25092027
Milotic D (2003) The impact of fragrance on consumer choice. J Consum Behav 3. https://doi.org/10.1002/cb.131

Mohana Priya M, Chidambara Rajan P, Lavanya M (2020) Use of natural pigments as colorants in cosmetics—a review. J Emerg Technol Innov Res 7
Moravkova T, Filip P (2014) The influence of thickeners on the rheological and sensory properties of cosmetic lotions. Acta Polytech Hungarica 11. https://doi.org/10.12700/aph.11.06.2014.06.11
Mudgil D, Barak S, Khatkar BS (2014) Guar gum: processing, properties and food applications–a review. J Food Sci Technol 51
Nam G, Lee HW, Jang JS, et al (2023) Novel conformation of hyaluronic acid with improved cosmetic efficacy. J Cosmet Dermatol 22. https://doi.org/10.1111/jocd.15555
Nguyen AV, Soulika AM (2019) The dynamics of the skin's immune system. Int J Mol Sci 20:1811. https://doi.org/10.3390/ijms20081811
Novak AC, Sydney EB, Soccol CR (2014) Biocosmetics. Biotransformation of waste biomass into high value biochemicals. Springer, New York, New York, NY, pp 389–411
Paetzold B, Willis JR, Pereira De Lima J, et al (2019) Skin microbiome modulation induced by probiotic solutions. Microbiome 7. https://doi.org/10.1186/s40168-019-0709-3
Patel HA, Somani RS, Bajaj HC, Jasra RV (2006) Nanoclays for polymer nanocomposites, paints, inks, greases and cosmetics formulations, drug delivery vehicle and waste water treatment. Bull Mater Sci 29. https://doi.org/10.1007/BF02704606
A Pérez-Bernal MA Muñoz-Pérez F Camacho 2000 Management of facial hyperpigmentation Am J Clin Dermatol 1 https://doi.org/10.2165/00128071-200001050-00001
Perioli L, Nocchetti M, Ambrogi V, et al (2008) Sunscreen immobilization on ZnAl-hydrotalcite for new cosmetic formulations. Microporous Mesoporous Mater 107. https://doi.org/10.1016/j.micromeso.2007.02.021
Quick L (2019) Decorated women: a sociological approach to the function of cosmetics in the books of Esther and Ruth. Biblic Interpret 27
Ramasubramania Raja R (2016) Nutraceuticals and cosmeceuticals for human beings-an overview. Am J Food Sci Health 2
Rangaraj S, Sasikanth V, Ammashi S, Rathinavel T (2023) Nutraceuticals and cosmeceuticals: an overview. In: Nutraceuticals: sources, processing methods, properties, and applications
Raza-Naqvi SA, Idrees F, Sherazi TA et al (2022) Toxicology of heavy metals used in cosmetics. J Chil Chem Soc 67:5615–5622. https://doi.org/10.4067/S0717-97072022000305615
Rente D, Cvjetko Bubalo M, Panić M, et al (2022) Review of deep eutectic systems from laboratory to industry, taking the application in the cosmetics industry as an example. J Clean Prod 380
Rieger M (2017) Surfactants in cosmetics, 2nd edn. Routledge
Samaranayaka AGP, Li-Chan ECY (2011) Food-derived peptidic antioxidants: a review of their production, assessment, and potential applications. J Funct Foods 3
Sharmeen JB, Mahomoodally FM, Zengin G, Maggi F (2021) Essential oils as natural sources of fragrance compounds for cosmetics and cosmeceuticals. Molecules 26
Stojiljković D, Arsić I, Kostov MT, et al (2013) Investigation of the effects of different emollients on the structure and skin moisturizing potential of the cosmetic creams. Acta Facultatis Medicae Naissensis 30. https://doi.org/10.2478/afmnai-2013-0011
Sureshkumar P, Rishika S, Vyshakh VS, Udhaya M (2023) Advanced AI based cosmetic suggestion system. Int J Prog Res Eng Manag Sci https://doi.org/10.58257/ijprems30949
Trujillano R (2022) 30 years of vicente rives' contribution to hydrotalcites, synthesis, characterization, applications, and innovation. ChemEngineering 6. https://doi.org/10.3390/chemengineering6040060
Uzdrowska K, Górska-Ponikowska M (2023) Preservatives in cosmetics technology. Aesthet Cosmetol Med 12
Vanda H, Verpoorte R, Klinkhamer PGL, Choi YH (2019) Natural deep eutectic solvents: from their discovery to their applications. In: Deep eutectic solvents: synthesis, properties, and applications
Vatiwutipong P, Vachmanus S, Noraset T, Tuarob S (2023) Artificial intelligence in cosmetic dermatology: a systematic literature review. IEEE Access 11
Vijayavenkataraman S, Lu WF, Fuh JYH (2016) 3D bioprinting of skin: a state-of-the-art review on modelling, materials, and processes. Biofabrication 8

Wallen-Russell C (2018) The role of every-day cosmetics in altering the skin microbiome: A study using biodiversity. Cosmetics 6:2. https://doi.org/10.3390/cosmetics6010002

Wang HMD, Chen CC, Huynh P, Chang JS (2015) Exploring the potential of using algae in cosmetics. Bioresour Technol 184

Wei Y, Simay AE, Agárdi I, et al (2023) Using artificial intelligence to promote branded color cosmetics: evidence from Indonesia. J Promot Manag 29. https://doi.org/10.1080/10496491.2022.2163036

Wiśniewska J, Klasik-Ciszewska S, Duda-Grychtoł K (2023) Salicylic acid and its use in cosmetology. Aesthet Cosmetol Med 12. https://doi.org/10.52336/acm.2023.011

Woodley DT (2017) Distinct fibroblasts in the papillary and reticular dermis. Dermatol Clin 35:95–100. https://doi.org/10.1016/j.det.2016.07.004

Zhang J, Wehrle E, Rubert M, Müller R (2021) 3d bioprinting of human tissues: biofabrication, bioinks and bioreactors. Int J Mol Sci 22

The Role of Analytics in Improving Health Outcomes: Are Accredited Social Health Activists (ASHA) at the Forefront in Social Health Protection?

Vijayetta Sharma

Abstract Rural health analytics is perceived to be a perennial breakthrough to detect and adapt the required changes in response to the service bottlenecks in the rural health sector. The paper walks through the prospective uses of analytics in the healthcare sector to improve health outcomes in rural areas. In the midst of covid pandemic, ASHA has been entrusted the responsibility of social health protection in rural areas. The role of ASHA in covid pandemic preparedness, providing necessary support to the rural populace, maintaining records, and facilitation of treatment has been discussed. The primary data comprised 100 ASHA from the state of Punjab and assessed the avenues for the usage of analytics and its relation to social health protection during covid-19. The paper signifies the importance of health data analytics and mining for covid pandemic which serves the broader aim of social health protection in the testing times. Various analytic strategies in relation to healthcare such as text analytics, speech analytics, patient treatment journey analytics have been discussed to make improvements in rural healthcare sector. The paper impinges to bring forth a revolution for the use of analytics and data mining using the novel concepts of business intelligence to benefit the rural health sector.

Keywords Social health protection · Rural healthcare · Healthcare analytics · Covid-19 · ASHA

1 Introduction

Social health protection in age of crisis demands an accelerated attention to save the humanity at the time of pandemics and disaster-prone circumstances. Analytics is the indispensable approach to enhance the social outreach and effectiveness of social health protection initiatives. The rights-based approach to social health protection

V. Sharma (✉)
Associate Professor, Public Policy, School of Behavioural and Social Sciences, Manav Rachna International Institute of Research and Studies, Faridabad, Haryana, India
e-mail: vijayetta@gmail.com

P. K. Prabhakar and W. Leal Filho (eds.), *Preserving Health, Preserving Earth*, World Sustainability Series, https://doi.org/10.1007/978-3-031-60545-1_13

guarantees social security and universal access to healthcare providing to its citizen freedom to live their lives free from any disease and illness. The social health protection fosters the basic premise of the affordability of health to the people and provision of healthcare services to raise their state of wellbeing. "Universal social health protection ensures that all people in need have effective access to at least adequate care and is thus a key mechanism for achieving social and economic growth" (Social Health Protection 2008). The social health protection also helps in building human capital and utilization of their services for social health benefits and realizing economic development of the nation. It has been reinforced by various international bodies that implementing universal social health protection might turn out to be a milestone to implementing MDGs and SDGs.

In 2015, the United Nations General Assembly adopted the 2030 Sustainable Development Goals (SDG). Goal 1 specified to end poverty in all its forms everywhere, calls for implementation of "nationally appropriate social protection systems and measures for all, and by 2030 achieve substantial coverage of the poor and the vulnerable". Goal 3 promised to deliver good health and well-being, calls for the achievement of "universal health coverage, including financial risk protection, access to quality essential health-care services and access to safe, effective, quality and affordable essential medicines and vaccines for all". Further, the core agenda of Universal Health Coverage (UHC) implies that there should be equitable access and use of quality services by all those who need them (SEARO, 2019). The International Labour Organisation (Social Health Protection 2008) provides for social health protection and its promotion via strengthening financial systems of the countries that provide:

(a) Universal and equitable access to health services.
(b) Financial protection in case of sickness; and
(c) Overall efficient and effective delivery of health services.

The prevalence of social health protection can be gauged by calibrating the financing situation globally, nationally and locally. The countries which provide effective access to health services through social health protection are Germany, Luxembourg, Belgium, France and the United Kingdom, through tax based national health services and witnessed least OOP expenditure. The situation in the high-income countries provide a positive link between the increasing national income, GDP and spending on social health protection. It may be noted here that social health protection is a worthy and viable option for low-income countries to protect the lives of the population from various diseases, although, the extent of population coverage has been independent of income levels of the countries. India foresees 20% coverage of population in social health protection via different insurance schemes. The Human Development Report (1997) estimated that the majority of the poor lacking access to health services live in developing countries, out of which 34% live in South Asia, 27% in sub-Saharan Africa and 19% dwell in South-East Asia and the Pacific regions (UNDP 1997). The deficiency or inequitable access to health services calls for a paradigm shift in social health protection strategies by adopting propitious means involving data analytics and business intelligence. Table 1 provides the health expenditure related social health protection indices in India and Punjab.

The data clearly shows the abysmal situation with respect to the health expenditure and out of pocket payments in India and Punjab. The situation of Punjab is better than India, the statistics of India falls behind many developing and developed nations, which outlines the need for identifying the ground realities and provide policy directions for carrying out improvements in social health protection measures of the population.

The indicators which define the outputs of health policies with respect to maternal and child health stipulates the first approach to measure effective access to health services which form part of social health protection (ILO, 2008). In India, the maternal and child health services are provided through a vast network of Accredited Social Health Activists (ASHA) which operate at the village level. These are the grassroot level workers being trained in the basic maternal and childcare through a series of training and induction programmes by the government. The ASHA model has achieved success in increasing the maternal and child health outcomes by improvement in the rate of institutional deliveries, decreasing mortality and morbidity, enhancing the immunization coverage for children below 5 years of age etc. After the spread of the pandemic—corona virus in 2019, ASHA has been entrusted responsibility of safeguarding the health of people residing in rural areas by making checks for active cases in the vicinity and for further guiding the populace for schedules testing at the health centres. The detailed guidelines were prepared by Ministry of Health and Family Welfare for instituting ASHA in gaining access to the rural populace, knowing about the status of rural people on the spread of the communicable disease that ventured in India from the carriers travelling from the west and east, and providing effective solution for its containment or further spread etc. The success achieved by ASHA in facilitating other services that has been entrusted to her has been documented from time to time such as containment of malaria, kala azhar, Japanese encephalitis, hypertension, maternal and newborn care (Acharya et al. 2015; Das et al. 2014; Abell-All et al. 2018; Mondal and Murhekar 2018; Nyamathi et al. 2018; Padda et al. 2013; Rajvanshi et al. 2021; Tripathy et al. 2011). The success in the provision of health services is associated to the acquaintance of ASHA with the locale or nearby health centres. This adds to the advantage of having ASHA as an important person for community decision on the corona virus social stocktaking. It further marks as an immediate requirement implying social health protection of rural masses. Some studies have highlighted that the use of information

Table 1 Social health protection indices (India and Punjab)

S. No	Social health protection	India	Punjab
1	Total health expenditure (percentage of GDP)	1.35	2.9
2	Out-of-pocket payments (percentage of total health expenditure)	47.1	65.5
3	Out-of-pocket payments (percentage of private health expenditure)	97	79

Source NHA and SHA, Punjab, 2023

technology, and technology led programmes have resulted in improved performance of ASHA (Patel et al. 2019; Shah et al. 2016; Sharma et al. 2021). Although, the ASHA has been the immediate respite to save the people for the further spread of corona virus but the extent of availability of tools and techniques with her determine the success in achieving a detailed understanding of the situation to further decide on the containment measures. Apart from the inventory in the form of drug and equipment kit provided to her, training on the maintenance of data, modes and type of interaction, and communication skills etc., an advanced use of analytics and data mining strategies would usher the smooth workflow process. It would further lead to early detection of problems in the operational procedure, easy identification of delays, service insufficiencies i.e. recurrent lags in same service or at different intervals, spillover effect of one service drawback on the other services or health outcomes, and other bottlenecks in the process of work facilitation. The forecasting and decision making using the data would yield effective achievement of intended outcome with least drawbacks. Studies highlight that there is significant scope for improving the way healthcare is delivered by resorting to the power of big data (Malik et al. 2018). Hence, rural health analytics is perceived to be a perennial breakthrough to detect and adapt the required changes in response to the service bottlenecks in the rural health sector.

2 Research Methodology

The cluster random sampling technique has been utilized and data has been collected from 3 districts of Punjab i.e. Hoshiarpur, Ropar and Mohali. A sample of 100 ASHA has been secured who have been regularly visiting the PHCs for the discharge of their duties, giving equal weightage to all the districts. While selecting the sample, report from ANM for the active ASHA in the village who have been entrusted responsibility of curtailment of pandemic and having better rapport with the village populace and medical staff has been considered. The rapport or congeniality of relations with the beneficiaries in the villages, medical and paramedical staff define their work efficiency and is an indicator of achieving proposed and planned effectiveness in result. The sample consisted of ASHA who were educated, working since last two years and were having a mobile phone. These parameters signify ASHA amenableness to the use of analytics—an advanced technique in decision making. Further, the mature ASHA equipped with the required inventory and communication skills were a resource in providing quick and intelligible feedback for corona pandemic containment. For this purpose, data from random 6 to 7 PHCs from each district has been collected to have a sample of 33–34 ASHA from each district for the study.

3 Big Data Analytics and Healthcare

Many studies have demonstrated that the knowledge discovery from the big data leads to improvement of the delivery of healthcare services (Malik et al. 2018; Anderson and Chang 2015; Bonacina et al. 2005; Ceglowski et al. 2007; Koh and Tan 2011). It has also been established that data mining and visual support has augmented the decision making and optimum utilization of health care resources (Lavrač et al. 2007). The data modelling has achieved huge success in maintaining electronic health records (Menon et al. 2014), thereby augmenting the decision support modelling and making rational, quick and verified choices at the times of crisis. Further, performance measurement and organizational success are intricately linked to the meticulous use of data mining and predictive analytics (Moullin 2007). Artificial intelligence has embarked its foot in healthcare quality improvement and ushered a paradigm of online healthcare access and delivery (Ng et al. 2006). A few scholars talked about value-based competition and business process analysis to improve the delivery of healthcare to patients (Porter and Teisberg 2006; Rebuge and Ferreira 2012). Further, there is a huge success in drawing statistical inference from text and transcripts (Shapiro and Markoff 1997). The healthcare business intelligence is amenable to provide long term care with superior quality protocols (Spruit et al. 2014). A large number of longitudinal, systematic and review studies have demonstrated how big data can make big impact. The gaining significance of business intelligence in arriving at improved and quick decisions has made data mining and predictive analytics an important tool for healthcare decision making. Hence, it can be stated that data science, predictive analytics, and big data can usher a revolution that has the potential to achieve the complete metamorphosis of health care service delivery and patient satisfaction.

Data mining is defined as the 'nontrivial extraction of implicit, previously unknown and potentially useful information from data' which provide a strong basis for informed decision making (Glowacka et al. 2009). In the last decade there has been a paradigm shift in strategies and models which aim at exploring possibilities of utilizing data mining tools and techniques for designing and managing healthcare operations (Wamba et al. 2015). Likewise other industries, the healthcare sector has access to large volumes of data in the current age such as the medical information systems, decision support models, inter and intra organization linkages etc. These provide opportunities for more informed and resounding decision making relating to clinical and administrative setups. Data mining has been used to early prediction of patients requiring longer hospital care, which aid in short- or medium-term hospital resource planning (Ng et al. 2006; Lavrač et al. 2007). This helps in management of inpatient and outpatient hospital care with the careful utilization of resources. Many times, the historic data has been used to predict the patient attendance behaviour for their appointments, thereby helping in optimized capacity allocation to outpatient clinics (Samorani and Laganga 2015). Furthermore, data mining patterns have been used to identify patterns for daily and weekly arrival rate of blood donors which facilitated work schedules for the health facility (Testik et al. 2012). Data mining principles have been used to discrete event simulation model for scheduling of work force in an emergency ward (Ceglowski et al. 2007). Further, business intelligence and data mining has been used to predict various financial performance measures, staffing level, operations per facility which help in planning revenue generation and controlling expenditure (Spruit et al. 2014).

In rural areas there is an intense need to revisit the use of analytics and data mining techniques for making plausible decisions where large proportion of population dwells and the need of healthcare access is prominent. There is a significant scope for improving the way healthcare is delivered in rural areas with the intervention of community health workers by resorting to the power of big data and artificial intelligence. The community health workers can act as catalysts in the provision of healthcare access and aiding in the information requirement for the preparing data models, analysis and preparing credible future forecasts. The community health workers also play the role of storehouse of all demographic, social and epidemiological information. They are conversant with pattern of flow of patients, norms prevalent in the society impacting the absorption of any policy promulgation, and instant as well as long term changes happening in the villages in the structure of socio-cultural and political mechanisms intertwined with the effects of policy implementation and its administration at every step of the health interventions and their social assimilation. The community health workers further aid in the process of business intelligence by having a vast repository of beneficiaries' base which serves the purpose of classification, clustering, positioning and segmentation. These can serve the purpose of training and validation data for having future forecasts. At the time of pandemic, when quick and informed decisions are required for effective containment of the communicable disease such as corona virus, the ingenuine support by community health workers can go a long way for having fine-tuned results. The use of business intelligence by these community health workers can further improve the effectiveness of desired outcomes by offering legible means to achieve informed outcomes. The various forms of health analytics which can be used to achieve intended results in rural areas at the time of pandemic are: text analytics, speech analytics, and patient treatment journey analytics.

4 Findings and Discussion

This section provides an analysis of the healthcare situation in the field during the covid pandemic. The important discussion areas are covid pandemic preparedness after its sudden eruption in 2020, services delivered by ASHA, digital engagement during the lockdown in the rural areas, use of digital data sources by ANMs and doctors, linking corona patients to the health centres, data analytics and sharing of records.

1. **Covid pandemic preparedness**

The ASHA workers were at the forefront of securing the healthcare of the masses in the rural areas and were entrusted with the huge responsibility to tackle the wrath caused by mammoth virus i.e. covid-19 which was unpredictably causing the mortality and increasing the morbidity among the already ailing population. The healthcare services went haphazard during the initial phase of pandemic eruption creating many patterns of un-identified, un-reported, mis-reported, mis-guided, mis-allocated, mis-referred category of patients who were either already taking treatment or fall under the category of beneficiaries under various schemes to be covered then or in future. The preparation for the pandemic involving routinization of services in the proper sequence, maintaining the proper record of patients, linking the corona patients to the health centres, identifying the blank spots in the service delivery, and integration of analytics to achieve service effectiveness was the challenge faced by the community health workers. The task of pre-planning before going to the field became more daunting with the eruption of the pandemic. While the larger stress was on securing the lives of the patients afflicted with corona virus, the provision of the general service delivery came to the halt.

Table 2 provides the descriptive statistics for covid pandemic preparedness in Punjab.

The data in Table 2 shows that there is a greater standard deviation from the mean for routinization of services (standard deviation—26.56) and linking the corona patients to the health centres (standard deviation—26.56) which shows variation in these two services after the eruption of the corona pandemic. The following qualitative insights can be noted in this regard.

The major challenges faced by the ASHA were in routinization of services as many patients who were already under some medications were affected by the sudden eruption of the pandemic. ASHA faced a lot of problems in maintaining the routine routinization of services as well as integration of the covid pandemic protocols. The major struggle was in making the people's lives safe and at the same time providing services in line with the covid guidelines. The threshold of convincing people and making them conversant with the recent covid safety protocols were stressing for ASHA keeping in view their own safety at stake. A lot of hassles were faced by ASHA including the retarding behaviour of the village people in welcoming their suggestions and complying with the necessary covid safety routine measures including hygiene, distancing etc. Linking of corona patients to the health centres amid the ongoing crisis were overcome by ASHA with difficulties posed by socio-cultural and pro-fear field situations which impact people's mindset.

Table 2 Covid pandemic preparedness

S.No	Covid pandemic preparedness	Mean	Stdev	Mode
1	Routinization of services	4.11	26.56	4
2	Maintaining the proper record of patients	3.65	16.54	4
3	Linking the corona patients to the health centres	4.11	26.56	4
4	Identifying the blank spots in the service delivery	2.52	16.02	3
5	Integration of analytics to achieve service effectiveness	3.46	11.58	4

2. **Services delivered by ASHA and bottlenecks faced**

Digital data ingestion, cross team alignment, along with patient data analytics tool is perennial to understand community behavoiur changes during covid pandemic. It can be noted that the role of analytics is more critical during the pandemic situation as the need to detect and adapt to changes is more pronounced. The *Aarogyasetu* app helps in location-based analytics as few unwell people may be travelling. Hence, knowing their location can add to the adherence of precautionary measures for the people in 500 m or 1 km vicinity. ASHA played a crucial role in streamlining the healthcare services, providing a valid database of infected people in the villages, interlinking them to the health centes, spreading awareness about the precautionary measures, keeping the ailing but non- covid infected population at a bay, team alignment with paramedical and medical staff, and preventing the situation of socio-political breakdown.

Table 3 provides the services delivered by ASHA to the rural people during the pandemic.

The data in Table 3 shows that interlinking the patients to the health centes (standard deviation- 17.25) and keeping the ailing but non- covid infected population at a bay (Standard deviation- 16.76) have higher dispersion in the services during the corona pandemic. The following qualitative insights can be noted in this regard.

The ASHA faced a lot of challenges which foster them to become more resilient to face the travesties posed by nature and provide strength to healthcare system to regain its functioning. The challenges ranged from interlinking of beneficiaries to the health centres due to ostracization by the beneficiaries to enter their households, blockage of services in terms of non-availing or limited usurpation of services due to fear of getting infected, isolated heath centres, shortage of items in ASHA drug kit, absence of medical and paramedical staff to provide services to the patients registered under various schemes, and transportation of patients to nearby health centres. The significant concern faced by ASHA was in keeping the ailing and non-covid infected population at bay, attaching the different cords emulating provision of healthcare in a unified whole. Further, tightening the service delivery by bringing different stakeholders and healthcare inventories together to prevent the situation of socio-political breakdown was another challenge faced by the ASHA.

Table 3 Services delivered by ASHA

S. No	Services delivered by ASHA	Mean	St.dev	Mode
1	Providing a valid database of infected people in the villages	3.23	9.51	4
2	Interlinking them to the health centes	3.89	17.25	4
3	Spreading awareness about the precautionary measures	3.8	15.68	4
4	Keeping the ailing but non- covid infected population at a bay	3.81	16.76	4
5	Team alignment with paramedical and medical staff	3.82	13.49	4
6	Preventing the situation of socio-political breakdown	2.69	12.35	4

3. **Digital engagement during the lockdown in the rural areas-Use of mobile phones**

Digital engagement during the lockdown emerged as a new phenomenon to the pandemic adversity posed by nature and human carriers of virus. The novice concepts of digital engagement got introduced to the rural areas with ease, though at the primitive level due to the necessity of the times to stimulate the physical structure to the working mode. The ASHA workers were faced with the urgent necessity to mould the services in digital mode to curtail the spread of virus and ease the achievement of desired health outcomes. Table 4 provides the digital engagement during the lockdown in the rural areas, specifically using the mobile phones.

The data in Table 4 shows that the use of text messages to beneficiaries (standard deviation—20.77) and sending service-related messages to the group of beneficiaries (standard deviation—20.77) have higher dispersion for the digital engagement during the lockdown in the rural areas. The following qualitative insights can be noted in this regard.

The network of beneficiaries providing the necessary information to each other via text messages as discussed with ASHA while sitting at their homes eased the process of healthcare inclusion. The ASHA were sending the text messages to beneficiaries and created the group of beneficiaries as per ANC and PNC status and immunization coverage. They also created a separate group of the village ladies to enrol the prospective beneficiary to the relevant scheme and give counselling on general protocols to be followed during the covid pandemic. At the initial stage, the messages of all sort were floated to all the beneficiaries in a single group which later on took the shape of item wise listing as per the components of different services. These groups gradually took the momentum and were used as the primal mode of spreading the required information to be used for healthcare advantage and safeguard against the infection of covid 19 virus.

4. **Use of digital data sources by ANMs and doctors**

The digital data repository and digital communication method was utilized by the paramedical and medical officials to keep the medical service delivery in momentum. The use of whatsapp messages, con-calls, creation of digital message service delivery repository, digital communication with the ASHA regarding the disbursal of medicines, trainings, and sharing of communication content with beneficiaries

Table 4 Digital engagement during the lockdown in the rural areas- Use of mobile phones

S.No	Digital engagement during the lockdown in the rural areas	Mean	Stdev	Mode
1	Use of text messages to beneficiaries	4.2	20.77	5
2	Sending service related messages to the group of beneficiaries	4.2	20.77	5
3	Separate group of the village ladies to disburse information on schemes and counselling on covid preparedness	3.48	14.40	4
4	Floating and item-wise listing of messages	3.16	13.58	4

was utilized in the online platform. Table 5 provides the use of digital data source by ANMs and doctors.

The data in Table 5 shows that whatsapp messages, con-calls, and LAN services have greater dispersion (standard deviation- 19.89) for the use of digital data sources by ANMs and doctors. The following qualitative insights can be noted in this regard.

It has been witnessed that the use of whatsapp and LAN services were utilized by the healthcare officials to deliver services in the villages with the aid of ASHA in the field. It became easy for the ANMs and doctors to converse with the wide set of community health workers and give directions regarding the to-dos in a quick, clear and detailed manner with lesser possibilities of errors.

5. **Linking corona patients to the health centres**

ASHA being the frontline health workers were entrusted with the task of linking of corona patients to the health centres to safeguard the lives of the population inhabiting the village. For this purpose, ASHA visited the households with the corona virus testing kits, screened the households for the spread of virus, isolated the corona patients from their family members, conducted the virus screen test after the recovery of corona patients along with the screening of family members for possible infection, conducted the mass awareness drives via digital medium, propagated the sanitary and other pro health measures to be adopted during the pandemic. Table 6 provides the provisions related to linking of corona patients to the health centres.

The data in Table 6 shows that the household visits regarding the screening of corona patients has higher dispersion (standard deviation- 22.19) for linking of corona patients to health centres. The following qualitative insights can be noted in this regard.

Table 5 Use of digital data sources by ANMs and doctors

S.No	Use of digital data sources by ANMs and doctors	Mean	Stdev	Mode
1	Whatsapp messages, con-calls, and LAN services	4.13	19.89	4
2	Creation of digital message service delivery repository	2.71	9.49	3
3	Digital communication pertaining to services and sharing of communication content	3.64	12.86	4

Table 6 Linking corona patients to the health centres

S. No	Linking corona patients to the health centres	Mean	Stdev	Mode
1	Household visits regarding the screening of corona patients	4.17	22.19	5
2	Isolation of corona patients from their family members	4.16	18.75	5
3	Virus screening test for corona patients	4.13	18.26	5
4	Screening of family members for possible infection	4.16	18.75	5
5	Conducting the mass screening drive for corona patients	4.13	18.26	5
6	Screening of family members for possible infection	4.16	18.75	5

The linking of patients for treatment was done by ASHA and proper record was maintained till the treatment formalities are completed along with the isolation of patients from their family members. The linking of corona patients to the health centres was a challenge initially, however it was eased with the direction and recurrent trainings by the government on providing necessary support to ASHA to screen the village for the spread of corona virus. The ASHA was also a hope to corona patients by maintaining their optimism in getting treated and linking them to the health centres for the oxygen requirements and lifesaving interventions.

6. **Data analytics and sharing of records**

The sharing of records by the ASHA with the ANM and medical doctors was resorted through an online mode. The record making process which was hitherto undertaken by ASHA in registers, forms and taking stock of different cards provided by the health centres took a backseat in the corona hit country and online information was requested from the beneficiaries. The data in Table 7 shows the use of data analytics and sharing of records during the corona pandemic.

The data in Table 7 shows that there is higher dispersion for comparison of records received from different sources (standard deviation—12.41), leveraging data to provide special field visits to critical patients (standard deviation—12.86), deciding on further forecasts (standard deviation—12.35) for the use of data analytics and sharing of records during the corona pandemic. The following qualitative insights can be noted in this regard.

The enrolment of the new beneficiaries was done by sending the pamphlet to the prospective clients asking them to save the form with them and provide the details requested in the form to send the information further to the health centre for enrolment. The online information gathered from the beneficiaries helped the ANMs to have better record keeping and compare the records received from the different sources, places, and rural healthcare seekers. These records were used by the ANM to perform a comparative assay of health needs of different patients, identify cases needing special attention, provide review of the situation in the field to the senior officials, identify the next level of care in an impeccable manner, present the finding in lucid and clear way. The data analytics techniques also help in providing special field visits to the serious patients, have better track record of the performance of ASHA in the field, and do future forecasts of the beneficiaries' requirements, inventories and disease prevalence.

Table 7 Data analytics and sharing of records

F	Data analytics and sharing of records	Mean	Stdev	Mode
1	Sharing of records by ASHA, ANM, and doctors	3.02	9.59	3
2	Enrolment of the new beneficiaries	3.24	8.15	4
3	Comparison of records received from different sources	2.6	12.41	4
4	Leveraging data to provide special field visits to critical patients	3.62	12.86	4
5	Deciding on further forecasts	2.41	12.35	2

5 Use of Analytics in the Form of Text, Speech, and Patient Treatment Journey to Aid Community Services

Various analytical strategies such as text, speech and patient treatment journey can make viable improvements in the maternal health. Integration of these techniques to the structural processes can be made feasible at the local level to grassroot functionaries along with the necessary outreach methods to achieve a grand success. The following discussion provides the description, benefits and steps of analytical techniques to be integrated into the rural health analytics.

1. ***Text and speech analytics***: Text and speech analytics are more accurate and insightful modes of data analytics and business intelligence. It helps in meticulously responding to changing emotions, attitudes, and values of communities during the crisis situations. It would aid in generating internal patient feedback and external data such as community affinity, administrative response to situation and social media utility. It helps in identifying emerging issues and diligently balancing changing community sensitivities.
2. ***Patient treatment journey analytics:*** It is used in identifying which patient treatment journeys are most prominent. It helps in the application of analytics to the identified route of patient treatment journeys and execution of the prepared plan of action. The health administration can fix the frictions in fetching treatment and prioritize on the rigid and most common bottlenecks based on the treatment journey route followed.

Steps for predictive analytics for rural health coverage

Data visualization is one of the farsighted approach lending hands in data cleaning, exploratory analysis and diagnosis. The decision making via predictive analytics makes use of machine learning, data mining and statistics to solve unusual problems which require deep thinking. The following steps of predictive analytics can be utilized for rural health coverage:

1. **Data pre-processing:** It entails understanding data and preparing it for analysis and involves data cleaning, coding and transformation. Linking community health workers to the field, providing them with aids and sensation of the situational demographics, collection of data from them and cleaning for the required analysis.
2. **Feature engineering:** It encompasses feature generation and feature selection techniques and uses parallel coordinates, scatter plots, and matrix and tree representation. Selection of variables, probable equities and inequities in the prescribed paths of variables as per the desk analysis of situation, featuring some emerging new branches to the variable tree fitting the situation, further expansion of subbranches of the variable tree which point towards newly emerging features. Feature engineering can aid in the providing a clear image of different demographic, disease indices, and patients utilization details of the community health services.

3. **Model building:** The predictive analytics involve applying explicit modelling to the data to fetch desired outcomes such as decision trees, hierarchical clustering, linear regression etc. Model building proves to be a boon in providing a valid data base description and forecast of the future trends.
4. **Model selection and exploring results:** Selection of appropriate model basis the context, internal and external environment and its utility along with comparison among the model inputs or agents constitute an important step of model selection. For the purpose of displaying the results, scatterplots, line charts, and other statistical graphics are used as the primary means of visualization. The detailed reports providing the trends in the healthcare usurpation by the patients during the corona pandemic, general service uptake, and further exploration of results depends on the selection of the model and outcomes required for the desired variables.
5. **Validation:** The performance of the model is evaluated using the validation data and depends on the value of mean square error. The residual plots, receiver operating characteristic (ROC) curve, autocorrection function (ACF) plot provides a discrete view of the validation process. The validation of data in case of detecting the healthcare access provides a clear understanding of the quality and accuracy of the data and provides for the consistency of results.

6 Conclusion and Recommendations

Social health protection is a significant concern in the rural areas and needs urgent attention to improve the prevailing healthcare indices. Its need was felt more pronouncedly during the covid pandemic. There is a need to strengthen the pandemic preparedness measures to keep hold of the situation in rural areas in case of outburst of crisis. A diligent strategy adopted for routinization of services, linking corona patients to the health centres, identifying black spots in the service delivery, maintaining the proper record of patients and integration of analytics to achieve service effectiveness would yield augmented outcomes. The services delivered by ASHA further needs to go through a transformative change and urgent attention to meet the changing needs of the people. The foremost challenge needs to be addressed in spreading awareness about the precautionary measures as it consumes a greater time and requires field integration. A safer and pro-care approach of linking the patients to the health centres, keeping the ailing but non- covid infected population at a bay, providing a valid database of infected people in the villages, team alignment with paramedical and medical staff, and preventing the situation of socio-political breakdown would help in bringing social health protection with the aid of community health workers.

Mobile phones played a greater role in digital engagement in the rural areas. A planned and itemized strategy is monumental in bringing gains in augmenting digital communication pertaining to services and sharing of communication content. The creation of digital message service delivery repository would aid in decision making by identifying similar patterns and drawing evidence driven insights to be integrated into policy provisions. The use of analytics in linking corona patients to the health centres and maintaining records pertaining to household visits regarding the screening of corona patients, isolation of corona patients from their family members, virus screening test for corona patients, screening of family members for possible infection, conducting the mass screening drive for corona patients, screening of family members for possible infection would help in achieving speedy interlinking of the steps for better and fast action. Data analytics have further advantages in making a fast and easy sharing of records. The pathways to propitious social health protection action by adopting a laid out analytical plan for sharing of records by health workers and professionals, enrolment of the new beneficiaries, comparison of records received from different sources, leveraging data to provide special field visits to critical patients, deciding on further forecasts would fructify if the governments lay out a valid strategy and implementation plan for effective execution, monitoring, and fostering quick amendments based on the feedback. Thus, it can be rightly said that the integration of novel concepts of business intelligence to health sector has a farsight potential to benefit the rural health sector.

References

Abell-All M, Thrift AG, Riddell M et al (2018) Evaluation of training program of hypertension for accredited social health activists (ASHA) in rural India. BMC Health Serv Res 18(320)

Acharya A, Lalwani T, Dutta R et al (2015) Evaluating a large-scale community-based intervention to improve pregnancy and newborn health among the rural poor in India. Am J Public Health 105(1):144–152

Anderson JE, Chang DC (2015) Using electronic health records for surgical quality improvement in the era of big data. JAMA Surg 150(24):29. https://doi.org/10.1001/jamasurg.2014.947

Bonacina S, Masseroli M, Pinciroli F (2005) Foreseeing promising bio-medical findings for effective applications of data mining. Biological and Medical Data Analysis, Springer

Ceglowski R, Churilov L, Wasserthiel J (2007) Combining data mining and discrete event simulation for a value-added view of a hospital emergency department. J Oper Res Soc 58(246):254. https://doi.org/10.1057/palgrave.jors.2602270

Das VNR, Pandey RN, Pandey K, Singh V, Kumar V et al (2014) Impact of ASHA training on active case detection of Visceral Leishmaniasis in Bihar, India. PLOS Negl Trop Dis 8(5)

Glowacka KJ, Henry RM, May JH (2009) A hybrid data mining/simulation approach for modelling outpatient no-shows in clinic scheduling. J Oper Res Soc 60(1056):1068. https://doi.org/10.1057/jors.2008.177

Koh HC, Tan G (2011) Data mining applications in healthcare. J Healthc Inf Manag 19:65

Lavrač N, Bohanec M, Pur A, Cestnik B, Debeljak M, Kobler A (2007) Data mining and visualization for decision support and modeling of public health-care resources. J Biomed Inform 40(438):447. https://doi.org/10.1016/j.jbi.2006.10.003

Malik MM, Abdallah S, Ala'raj M (2018) Data mining and predictive analytics applications for the delivery of healthcare services: a systematic literature review. Ann Oper Res 270(1/2):287–312

Menon AK, Jiang X, Kim J, Vaidya J, Ohno-Machado L (2014) Detecting inappropriate access to electronic health records using collaborative filtering. Mach Learn 95(87):101. https://doi.org/10.1007/s10994-013-5376-1

Mondal N, Murhekar MV (2018) Factors associated with low performance of accredited social health activist (ASHA) regarding maternal care in Howrah district, West Bengal 2015–16, an unmatched case control study. Clin Epidemiol Global Health 6(1):21–28.
Moullin M (2007) Performance measurement definitions: Linking performance measurement and organisational excellence. Int J Health Care Qual Assur 20(181):183. https://doi.org/10.1108/09526860710743327
Ng S-K, McLachlan GJ, Lee AH (2006) An incremental EM-based learning approach for on-line prediction of hospital resource utilization. Artif Intell Med 36(257):267. https://doi.org/10.1016/j.artmed.2005.07.003
Nyamathi AM, Carpenter CL, Ekstrand ML et al (2018) Randomised controlled trial of a community based intervention on HIV and nutritional outcomes at six months among women living with HIV/AIDS in rural India. AIDS 32(18):2727–2737
Padda P, Devgun S, Gupta V et al (2013) Role of ASHA in improvement of maternal health status in northern India: an urban rural comparison. Indian J Community Health 25(4):465–471
Patel AB, Kuhite PN, Alam A et al (2019) M-SAKHI–Mobile health solution to help community providers promote maternal and infant nutrition and health using a community based cluster randomized controlled trial in rural India: a study protocol. Matern Child Nutr
Porter ME, Teisberg EO (2006) Redefining health care: creating value-based competition on results. Harvard Business Press, Brighton
Rajvanshi H, Saha KB, Shukla MM et al (2021) Assessment of ASHA for knowledge, diagnosis and treatment on malaria in Mandla district of Madhya Pradesh as part of the malaria elimination demonstration project. Malaria J 2(78)
Rebuge Á, Ferreira DR (2012) Business process analysis in healthcare environments: a methodology based on process mining. Inf Syst 37(99):116. https://doi.org/10.1016/j.is.2011.01.003
Samorani M, Laganga LR (2015) Outpatient appointment scheduling given individual day dependent no-show predictions. Eur J Oper Res 240(245):257. https://doi.org/10.1016/j.ejor.2014.06.034
Shah MP, Kamble PA, Agnihotri SB (2016) Tackling child malnutrition: an innovative approach for training health workers using ICT a pilot study. In: IEEE conference publication
Shapiro G, Markoff G (1997) Methods for drawing statistical inferences from text and transcripts. Text Analysis for the Social Sciences. Lawrence Erlbaum Associates, Mahwah, NJ, pp 3–31
Sharma S, Arora K, Chandrashekhar et al (2021) Evaluation of a training program for life skill education and financial literacy to community health workers in India: a quasi-experimental study. BMC Health Serv Res 21(46)
Social Health Protection (2008) An ILO strategy towards universal access to health care, Social security policy briefings. Paper 1/International Labour Office, Social Security Department, ILO, Geneva
Spruit M, Vroon R, Batenburg R (2014) Towards healthcare business intelligence in long-term care: an explorative case study in the Netherlands. Comput Hum Behav 30(698):707. https://doi.org/10.1016/j.chb.2013.07.038
Testik MC, Ozkaya BY, Aksu S, Ozcebe OI (2012) Discovering blood donor arrival patterns using data mining: a method to investigate service quality at blood centers. J Med Syst 36(579):594. https://doi.org/10.1007/s10916-010-9519-7
Tripathy P, Nair N, Mahapatra R et al (2011) Community mobilization with women's groups facilitated by Accredited Social Health Activists (ASHAs) to improve maternal and newborn health in underserved areas of Jharkhand and Orissa: study protocol for cluster-randomised control trial. Trials 12(182)
UNDP (United Nations Development Programme) (1997) Human development to eradicate poverty. Human Development Report, New York
Wamba SF, Akter S, Edwards A, Chopin G, Gnanzou D (2015) How 'big data'can make big impact: findings from a systematic review and a longitudinal case study. Int J Prod Econ 165(234):246. https://doi.org/10.1016/j.ijpe.2014.12.031

Mindful Approaches to Sustainable Healthcare: Nurturing Mental Health and Well-Being

Rahul Saxena, Suyash Saxena, and Ajit Pal Singh

Abstract Being in a state of health is defined as being free from all diseases and physical defects, as well as having strength and stability in one's body, mind, and social environment. Infectious disease pandemics have been shown in the previous several centuries to considerably increase the death rate by negatively affecting the population's physical and mental health. Even in the form of infectious diseases, this tragic occurrence is still occurring in the current era. Globally, there has been a dramatic loss in human health as a result of inadequate measures taken to combat infectious agents. In the post-COVID era, it has been discovered that social deprivation, fear of unexpected health complications in unaffected people, and a number of treatment side effects in COVID 19 affected patients are all linked to compromised health conditions, such as anxiety, depression, and financial loss. We have not had much success in providing a comprehensive approach to the physical and mental health, despite the use of all the most cutting-edge medical therapy approaches in reducing the symptoms of infectious diseases. Spiritual care works well in reducing the physical and mental health sustainability, in addition to following standard hygiene practices, WHO recommendations, and avoiding disease related risk factors. Numerous research have recently focused on the importance of spiritual care practices in treating and avoiding physical and mental sickness brought on by infectious disease pandemics. Additionally, it has been suggested that incorporating the 4S formula for spirituality into sustainable healthcare is an excellent way to foster social and physical well-being. It is feasible that receiving spiritual care can help a person cope with the effects of infectious diseases by giving them inner strength and social support. Therefore, the current chapter offers a thorough analysis of mental health and wellbeing in sustainable healthcare, as well as the contribution of spiritual care to the treatment of infectious diseases.

R. Saxena (✉) · S. Saxena
Department of Biochemistry, School of Allied Health Sciences, Sharda University, Greater Noida 201310, UP, India
e-mail: rahul.saxena@sharda.ac.in

A. P. Singh
Department of Medical Laboratory Technology, School of Allied Health Sciences, Sharda University, Greater Noida 201310, UP, India

P. K. Prabhakar and W. Leal Filho (eds.), *Preserving Health, Preserving Earth*, World Sustainability Series, https://doi.org/10.1007/978-3-031-60545-1_14

Keywords Mental health · Well-being · Spiritual care · Sustainable healthcare · Anxiety · Depression

1 Introduction

Improved environmental conditions and a healthy lifestyle may ultimately result in greater noticeable increases in life expectancy than additional advancements in clinical therapy can provide. It's interesting to note that these improvements will be responsible for ensuring longer lifespans. Interest in health and wellbeing has existed since antiquity, and many scientific endeavors have been made to achieve a prosperous and healthy existence. Nevertheless, the mystery of immortality remains unsolved to this day, and the steady decline in both physical and mental well-being, accompanied by a host of health-related issues and eventual death, is acknowledged as a normal part of life (Saxena and Batra 2020). It is interesting to note that everyone wants to have a healthy life, which includes not only avoiding illness and remaining physically strong but also having a high quality of life that includes preserving mental and physical well-being as well as cognitive abilities. Several holistic strategies have been linked to leading a prosperous and healthy life by implementing strict measures for early disease detection and prevention. Despite this, the flimsy defense against early detection, efficient treatment, and prevention of infectious diseases has demonstrated the validity of Nobel laureate Joshua Lederberg's foretelling.

Effect of Pandemics on human health: Infectious disease pandemics have been shown throughout history to have a substantial impact on both the physical and mental health of the populace, which has increased the death rate. This terrible trend has persisted into the current millennium as well, manifesting itself as COVID-19, which is brought on by the novel zoonotic SARS-COV-2. Clinically, COVID-19 manifested as a complex illness since the virus impacts numerous essential organs and raises the risk of morbidity and death. In the post-COVID era, social deprivation, fear of disease symptoms in non-affected individuals, and various treatment side effects in virally-affected patients have all been linked to compromised health conditions, including anxiety, depression, and financial loss (Contini et al. 2020).

It is imperative that we all recognize that the epidemic affects people's mental health in addition to their physical health and wellness. There are numerous negative impacts of this disease on people's mental health. In hospitalized patients, delirium is frequently reported during the acute stage of infection. Owing to the disease's high contagiousness, sufferers must be kept in socially isolated environments, under quarantine, and away from other people. Following the quarantine period, the patient may experience emotional difficulties, depression, insomnia, irritability, and post-traumatic stress disorder due to these circumstances as well as complications and ambiguity over the disease. The period of confinement during an infection can usually last anywhere from 10 to 14 days; this prolonged period of social isolation can increase feelings of loneliness and social disengagement, which can be made worse by the economic downturn and society's tendency to avoid the affected person out of fear of infection.

While the long-term effects of acute infections are not well-documented, there is potential for a wide range of complications, including anxiety, anger, depression, post-traumatic stress disorders, and alcohol abuse, if we compare the current incidence of acute infections with past pandemics and chronic infections. We have seen the sense of dishonor and shame that survivors of contagious diseases have experienced in a number of ways, including persecution, social avoidance, typecasting, and, in certain cases, physical assault. These victims can also be experiencing suicide thoughts, depression, and emotional instability. Thus, there is a close relationship exist between physical and mental health (Ohrnberger et al. 2017).

2 Safety Measures for Physical Health

This would remain a hot topic of conversation because it represents a significant issue for society or the global community. There are a few post-infectious disease preventative measures that we must follow because as the globe recovers from the horrific pandemic, several regulations are put in place to help the world recover.

First, continuing and adhering to the general precautionary guidelines recommended by the World Health Organization (WHO), which include frequent hand washing and sanitization with sanitizers, avoiding crowded areas, wearing masks and disposing of them after use for a while, cleaning and sanitizing places of work, and gatherings hosted by the local government. Vaccination on a periodic basis is also crucial. The vaccination would benefit them and restore their world to its previous state if it were made available to everyone and not restricted to any certain set of individuals. The world, and Indians in particular, first opposed the vaccine, but after widespread publicity about the campaign to vaccinate, they have now come around to it.

The media has emerged as a tool for creating awareness and sensitization, which can be utilized to raise awareness and cultivate a new mindset for long-term health, as many individuals are finding it challenging to adopt the new normal. Rehabilitation of survivors and warnings are also necessary because the virus can recur in people who have already experienced it. Thus, it is recommended that these individuals have regular examinations and tests, for which the government might help pay the associated costs. Additionally, testing those who exhibit signs of this virus and directing them to seek medical attention should be promoted.

Any pandemic disease has posed a huge challenge to global health systems, resulting in millions of deaths and affecting people globally at the same ratio. Post-infectious disease difficulties remain a big worry, despite hopeful statistics showing that the recovery rate is on the rise as experts worldwide find new ways to treat the condition. Many patients have complained of experiencing conditions like depression, fatigue, fever, brain fog, loss of smell or taste, chest pain, cough, shortness of breath, joint muscle pain, and other symptoms that only worsen with physical activity, even though they have recovered slowly or asymptomatically. Numerous safety precautions have been recommended for these circumstances, some of which include routinely cleaning hands for at least twenty seconds every twenty minutes, avoiding facial contact, particularly with the sensory organs, using masks while entering public areas, and sanitizing commonly touched objects (Cipriani et al. 2020).

Other safety precautions include avoiding crowds as much as possible, calling a doctor for physical issues, and maintaining mental well-being. Many health-related difficulties can be handled as long as safety precautions are followed. Following the effects of an infectious disease, health needs to be given the utmost attention because there are a number of secondary issues that must be resolved. Extreme caution must be used when it comes to health since infectious pathogens can cause lung fibrosis, kidney, heart, and neurological issues. One of the most frequent side effects seen whenever steroids are taken is diabetes. Mucormycosis, a fungal infection, has also proven lethal to some who had recovered from the illness. Even yet, there are still several actions we can do to resume our normal lives.

3 Safety Measures for the Mental Health

A sizable portion of the global population has been impacted by infectious diseases since their inception. Scenario of COVID-19 is one of the suitable examples. Since it was a new disease, lockdown, social distancing, isolation, and quarantine have been the main strategies employed by most governments to combat it. People have been shut up in their houses for a very long time as a result, feeling depressed, anxious, and distrustful of one another (Galea et al. 2020). Previous SARS and MERS health-related endemics have made it clear that there are acute and post-illness consequences of these viral infections on the psyche and spine. The effects of viral infection in the brain are always under dispute when a disease caused by a virus penetrates human cells through any receptor and that receptor is not very present in brain cells. There are theories that a virus can enter and multiply anywhere, including in neural cells, but the results of these infections have not been well-defined and require sufficient data to be supported. So the area of concern affecting mental health should be the social and environmental factors. There are 2 partners in maintaining the mental health i.e. the society and the individual himself.

4 Safety Measures Required from the Society

1. **Short and adequate duration of quarantine:** Quarantining is unquestionably a crucial strategy for halting the spread of viral infections. However, there is no denying that patients who remain in the quarantine for longer will experience higher symptoms of stress. Therefore, rather than taking an overtly precautionary

stance, it is always advisable to base the length of the quarantine on scientific rationales such as the virus's incubation period (Kato et al. 2020).
2. **Proper and adequate supplies:** The authorities, neighbors, friends, and society as a whole should ensure that the individual placed under quarantine and his family receive timely and sufficient supplies of necessities in order to calm and reassure the patient and his family.
3. **Educate the patient with adequate and accurate information:** The patient experiences a range of anxieties and worries regarding the disease's symptoms and prognosis while under quarantine. The inadequate and false information gathered from multiple sources exacerbates these anxieties several times over. Calming restless minds can be achieved through removing uncertainties and offering genuine facts.
4. **Guidance to the patient and his family to tackle the loneliness:** Loneliness and isolation will make you feel distressed. It is important to provide useful tips to individuals under quarantine so they can keep themselves from being bored and lonely. It's essential to have the tendency to talk about feelings with loved ones and pals. When it comes to keeping in touch with loved ones who live far away and cannot be physically present, social media is invaluable.
5. **Tele psychiatry and psycho-social support:** Health professionals and psychiatrists who can provide them support and guidance if they exhibit any symptoms should also be regularly reached by phone or video chat for those under quarantine. The patient will be reassured by this that they will receive care when needed. These comforting words might help lessen feelings of worry, agitation, and terror (Saxena et al. 2021).
6. **Support groups:** Numerous data points to the ease with which major problems can be handled by support groups. On a volunteer basis, those who have recovered from their viral infection and are no longer at risk should serve as mentors for newly admitted patients. Talking to others who have experienced similar things can be reassuring and consoling. It can also help them discover the support they feel they are not getting from other people.

5 Safety Measures Required from the Individual

1. **Create routines:** Creating and adhering to a busy routine is the greatest way to minimize boredom, which is an inevitable problem during quarantine.
2. **Eat well:** Since there is no known etiopathophysiology for any sad disease, a person's immune system serves as their most important weapon in the struggle, so they should be disciplined to eat a healthy diet (https://www.niddk.nih.gov/health-information/weight-management/keep-active-eat-healthy-feel-great).
3. **Exercise regularly:** A daily regimen should include some mild exercise because it helps to tone up the muscles and releases endorphins, which in turn lessen pain perception and promote happy feelings in the head. In addition, regular exercise

has been found to be associated with longevity as it counteracts the aging effect (Carapeto 2021). However, the workout should not be very taxing.

4. **Work enough from home:** once the virus has finished its incubation period. If the patient's body and mind permit, they should engage in sufficiently non-exhausting work. This can assist in managing feelings of loneliness and boredom.
5. **Reduce the stress through breathing practice:** In addition to providing the body and mind with an additional oxygen supply, breathing exercises are an excellent way to strengthen the respiratory muscles. The body's stress signals can be lowered by a brain that receives enough oxygen.
6. **Sustain good sleep:** Nothing soothes a restless mind and relaxes a weary body more than a good night's sleep. It lowers stress hormones, aids in growth hormone secretion, strengthens the immune system, enhances breathing, lowers blood pressure, and promotes cardiovascular health.
7. **Communicate with your loved ones:** Among the most crucial issues, communications with family members have received much attention. It helps to feel safer and less anxious to share your feelings with those who are close to you (https://youaremom.com/parenting/family-communication/). Thankfully, there are many ways for us to contact with people who live far away these days. Thus, we ought to communicate with people and propagate positivism through all available channels.

6 Conclusion

All humans, therefore, have the desire to live a healthy life. A pathologic state is characterized by declining mental and physical health and affects the body's overall homeostasis. Its impact on the state of the economy cannot be disregarded either. As a result, it is evident that routine health monitoring and retesting of routine tests at prearranged times are necessary. A regular exercise should begin with a routine that is adjusted based on each person's suitability. Most likely, you should begin the activity gently and with less effort. Everyone should focus on improving the safety of the environment, following the rules set forth by health organizations, maintaining a healthy diet, and making sure that we are eating enough protein each day. Alcohol, tobacco, and junk food consumption should be avoided. We need to engage in a variety of hobbies, such as yoga and painting, to maintain our mental health. Additionally, the optimistic outlook would work as a boon to support long-term human health.

References

Carapeto PV, Aguayo-Mazzucato C (2021) Effects of exercise on cellular and tissue aging. Aging 13(10):14522–14543

Cipriani G, Danti S, Nuti A, Carlesi C, Lucetti C, Di Fiorino M (2020) A complication of coronavirus disease 2019: delirium. Acta Neurol Belg 120(4):927–932

Contini C, Caselli E, Martini F, Maritati M, Torreggiani E, Seraceni S, Vesce F, Perri P, Rizzo L, Tognon M (2020) Covid-19 is a multifaceted challenging pandemic which needs urgent public health interventions. Microorganisms 8(8):1228

Galea S, Merchant RM, Lurie N (2020) The mental health consequences of covid-19 and physical distancing: the need for prevention and early intervention. JAMA Intern Med 180(6):817–818

https://www.niddk.nih.gov/health-information/weight-management/keep-active-eat-healthy-feel-great

https://youaremom.com/parenting/family-communication/

Kato TA, Sartorius N, Shinfuku N (2020) Forced social isolation due to covid-19 and consequent mental health problems: Lessons from hikikomori. Psychiatry Clin Neurosci 74(9):506–507

Ohrnberger J, Fichera E, Sutton M (2017) The relationship between physical and mental health: a mediation analysis. Soc Sci Med 195:42–49

Saxena R, Batra J (2020) Aging: oxidative stress and dietary antioxidants, 2nd ed. Arthritis as a disease of ageing and changes in antioxidant status, Chapter 8. Academic Press Elsevier publications, London, pp. 83–92

Saxena R, Ghosh S, Sharma D (2021) Hope to cope: issues and trends in the post covid era, 1st edn. Advance trends in Telehealth, Chapter 11. Inspira Publications, Jaipur, India, pp. 75–82

From Ideation to Innovation: Integrating Pharmaceutical Innovation and Sustainable Development

Heba M. Mansour and Aiman S. El-Khatib

Abstract In recent years, there has been a growing focus on sustainability in the pharmaceutical industry by policymakers, consumers, and organizations. This includes concerns about incorporating sustainability into the development of innovative pharmaceuticals that are environmentally friendly. This book chapter will provide a brief historical overview of the pharmaceutical industry. It will explain the definition, types, pros, cons, and limitations of pharmaceutical innovation. It will also explore the connection between each type of pharmaceutical innovation and sustainable development. The role of pharmaceutical innovation in achieving sustainable development goals is highlighted. Limitations and challenges hindering pharmaceutical innovation are extensively discussed, along with strategies to tackle these limitations. The chapter will also highlight potential areas for future research and discuss the challenges and future perspectives of achieving sustainable development in pharmaceutical innovation and usher in a new era of drug development.

Keywords Sustainable development · Pharmaceutical innovation · Regulatory science · Drug discovery · Sustainable development goal 3

List of Abbreviations

AI	Artificial intelligence
API	Active pharmaceutical ingredients
EVs	Extracellular vesicles

H. M. Mansour (✉)
General Administration of Innovative Products, Central Administration of Biological, Innovative Products, and Clinical Studies (Bio-INN), Egyptian Drug Authority (EDA), Giza, Egypt
e-mail: Heba.mo.mansour@std.pharma.cu.edu.eg

A. S. El-Khatib
Department of Pharmacology and Toxicology, Faculty of Pharmacy, Cairo University, Giza, Egypt

H. M. Mansour
Egyptian Drug Authority (EDA), Giza, Egypt

P. K. Prabhakar and W. Leal Filho (eds.), *Preserving Health, Preserving Earth*, World Sustainability Series, https://doi.org/10.1007/978-3-031-60545-1_15

FDC	Fixed dose combination
GDB	Gross domestic product
HTS	High-throughput screening
NIH	National Institutes of Health
NME	New molecular entity
PROTAC	Proteolysis Targeting Chimera
R&D	Research and development
SD	Sustainable development
SDG	Sustainable development goal
WHO	World Health Organization
FDA	Food and Drug Administration

1 Introduction

Medicines play a crucial role in the healthcare system, providing high-quality medical care and ensuring access to qualified healthcare. They are used for diagnostic, therapeutic, prophylactic, and rehabilitative purposes. The pharmaceutical industry plays a vital role in developing these medicines, with its foundation built upon legal, socioeconomic, and medical measures. These measures aim to enhance the protection and preservation of public health. The development of new pharmaceutical compounds is a fascinating innovation process that requires significant private investments in research and development (R&D) expenses. The pharmaceutical sector has high R&D/sales ratios, and global R&D expenditures reached $2.107 trillion. The United States is the leading investor in R&D, surpassing all other countries. In 2018, China funded more R&D than Japan, Germany, South Korea, and France combined (Tajaddini and Gholipour 2021). The connections between academic science and government research are important. Patent protection is especially significant in this field. New products must pass market acceptance and government regulatory scrutiny. In wealthier countries, insurance covers a large portion of consumer purchases, leading to issues like moral hazard and inequality. However, there is strong evidence that the introduction of new pharmaceutical products has prolonged human lives and alleviated disease burdens, achieving sustainable development goal 3 (SDG3) (Lichtenberg 2001, 2007, 2003). The approval of new active substances has significantly declined in Europe, Japan, and the US in recent years. This has raised concerns among regulatory authorities regarding the promotion of innovation (Karampli et al. 2014).

The pharmaceutical industry develops new drugs through smaller innovations rather than starting from scratch. A me-too drug is a pharmacologically active chemical that has a similar structure to a first-in-class compound. It is thought to be of the same medicinal class as the original chemical and is used for the same indications. However, it may differ in other ways, such as pharmacological activity specificity, adverse effects, or drug-drug interactions (Aronson and Green 2020a) Critics argue that "me-too" drugs are simply copies of existing molecules. However, the industry's R&D process is unique and risky, not a deliberate recycling of chemicals. Addressing the global disease burden has socioeconomic benefits. Pharmaceutical innovation can address changing patient needs, budget constraints, health problems, and technological changes. It improves patient outcomes, life expectancy, quality of life, treatment options, affordability, and system efficiency, achieving SDGs (Wakutsu et al. 2023).

In this chapter, we will give a brief historical overview of pharmaceutical industry. We will explain in detail definitions and added value of pharmaceutical innovation. Different types of pharmaceutical innovation, including new molecular entity, new fixed dose combination, new stereoisomer, repurposed drugs, and new drug delivery system. In addition, this chapter will discuss sustainable pharmaceutical innovation and explain role of each type of pharmaceutical innovation in achieving sustainable development. Also, this chapter will highlight the limitations that hinder development of sustainable pharmaceutical innovation and how to tackle these limitations. Collectively, this chapter will highlight potential areas for future research and discuss the challenges and future perspectives of achieving sustainable development in pharmaceutical innovation.

2 Emergence of Pharmaceutical Industry

In the 1800s, significant advancements in pharmacology occurred in the United Kingdom and Germany through the extraction and purification of compounds like atropine, morphine, and colchicine from plants. This period also witnessed the development of the first chemically synthesized drug. Progress in understanding human physiology and pathology enabled the discovery of new compounds with biological activity and the ability to test for toxicity (Sepkowitz 2011). The pharmaceutical industry originated in the late nineteenth century as a branch of the expanding chemical sector. German and Swiss companies, such as Bayer, Ciba, Hoechst, and Sandoz, played a pioneering role in this field. These companies leveraged their knowledge in organic chemicals and dyestuffs, which were cultivated in German educational institutions. Before World War I, American companies relied on European firms for technology and pursued two distinct paths. The first group focused on manufacturing and marketing their patented drugs, which were available over the counter. The second group focused on developing prescription drugs, known as "ethical" drugs, which were sold to pharmacists and doctors (Chandler 2005). Prior to World War II, pharmaceutical advances were developed by regular corporations using standardized techniques. The advent of large-scale production methods for penicillin was a significant milestone in the industry, considerably improving the treatment of infectious diseases. However, because penicillin was not patented, a slew of new entrants flooded the market, resulting in harsh competition. The cost of producing 100,000 units of conventional penicillin fell from US$20 in 1943 to 4.5 cents in 1950, making manufacture unprofitable (Moors et al. 2014). Pharmaceutical corporations invested heavily in research and patents to avoid prior difficulties. Publicly supported biomedical research also contributed to industrial growth. As a result, the pharmaceutical sector expanded, with many valuable medications now protected by patents and fewer hurdles to entry. Between 1951 and 1961, 4562 new prescription drugs were released, with 10% of them including novel chemical entities. According to analysts, 70 cents of every dollar spent on pharmaceuticals in 1961 went towards obtaining medications that were not accessible ten years before (Comanor 1964).

3 Pharmaceutical Innovation

Innovation is the advancement of technology that leads to the development of new products, reduction in production costs, or improvement in the therapeutic value of existing drugs. In the pharmaceutical industry, innovation refers to the entire process of discovering, developing, manufacturing, and distributing medical products, with the goal of making them more widely available and accessible. Pharmaceutical innovation is influenced by time, as the value of a pharmaceutical innovation changes over time due to competition and technological advancements (Morgan et al. 2008). The definition of pharmaceutical innovation varies among organizations and nations. Table 1 summarizes different definitions and characteristics of pharmaceutical innovation according to different regulatory authorities. Aligning these definitions can speed up the development of new treatments, reduce costs, and streamline drug research and approval processes (Wakutsu et al. 2023).

4 Types of Pharmaceutical Innovation

Pharmaceutical innovative products can be categorized into several groups: new chemical entities, novel formulations, novel drug delivery systems, novel manufacturing technologies, novel fixed dose combinations, and new indications for existing drugs (Karampli et al. 2014).

4.1 New Molecular Entity (NME)

A breakthrough innovation refers to a completely new approach in preventing, treating, or curing a specific disease (Morgan et al. 2008). Despite the pharmaceutical industry's significant investment in R&D, the number of new drugs approved by the Food and Drug Administration (FDA) remains relatively low. The FDA approved more new medicines in 2011 compared to the previous year. However, the pharmaceutical industry has not experienced significant growth in the past decade, despite using traditional methods for discovering and developing new products. In 2022, the FDA

Table 1 Summary of definition of innovative products and their characteristics according to different regulatory authorities

Regulatory authority	Definition of pharmaceutical innovative product	Characteristics of innovative product	The added therapeutic value	References
EMA	• Technological progress that results in the development of a completely new product, a decrease in production costs, or an increase in the therapeutic value of an existing product • Innovative products are biological or medicinal products that have curative or preventive properties for human diseases, restoring, correcting, or modifying physiological functions through pharmacological, immunological, or metabolic action	• New molecular entity • Incremental innovation • Me-too drugs • Reformulation • Decreases production cost • Process innovation • Drug repositioning of exiting drugs • New route of administration • Follow-on medications are defined as a new entry into a therapeutic class that was previously established by a distinct drug entity that was the first in the class to acquire regulatory approval for marketing • New drug delivery system	• A medical product is any substance or combination of substances that is used to treat or prevent diseases in humans • It can also be used to restore, correct, or modify physiological functions or make a medical diagnosis • Treatment of untreatable disease • Pronounced reduction in disability • Increase of life expectancy • Decrease of side effects • Increase of administration convenience • Novelty • Newness	(Charles River Associates 2004) • Directive 2004/27/EC of the European Parliament and of the Council of 31 March 2004 amending Directive 2001/83/EC on the Community code relating to medicinal products for human use • Regulation [EC] No 1394/2007 of the European Parliament and of the Council of 13 November 2007
FDA	Innovative drugs frequently translate to new treatment choices for patients and advancements in healthcare	• **Radical innovation**: A novel technical foundation with significant consumer advantage • **Technological breakthroughs**: Novel technological foundation with reduced customer benefit • **Market breakthroughs**: Minor adjustments in the technical basis that result in significant consumer benefit • **Incremental innovations**: Minor modifications in the technical foundation that provide little advantage to customers • New indication or drug repurposing • Fixed dose combination • Stereoisomer • Innovation in precision medicine or personalized medicine	• Superiority over existing treatments in case of radical innovation • Providing better improvement of clinical outcomes • Assessed based on novelty, newness • Availability of existing drugs • Reduced cost • Reduces side effects • Added life year • Safety	(Sternitzke 2010; Scott et al. 2022)

(continued)

Table 1 (continued)

Regulatory authority	Definition of pharmaceutical innovative product	Characteristics of innovative product	The added therapeutic value	References
UK	• Innovative medication is one that is novel, improves on existing medicines, and provides a step-change in terms of health outcomes • Step changes in clinical usefulness can occur in a variety of ways. The most successful pharmaceutical innovations are those that address previously incurable diseases • Drugs enhance safety, selective action, selective pharmacological changes, and convenience in treatment	The innovative pharmaceutical products can be classified according to degree of innovation as following: – **High innovation:** (New target, novel mechanism of action, novel application, and well-known pharmacogenetics profile) – **Moderate innovation:** (New type of compounds, fewer side effects, fewer interactions, and novel structure) – **Slight (health related) innovation**: Improved pharmacokinetics or improved formulation – **Slight (non-health related) innovation:** Improved production or novel structure	• Meets unmet therapeutic need • Improves access to effective medicine • Safe treatment • Less side effect • Convenience of treatment • Novelty • Administration	(Ferner et al. 2010; Wakutsu et al. 2023)

(continued)

Table 1 (continued)

Regulatory authority	Definition of pharmaceutical innovative product	Characteristics of innovative product	The added therapeutic value	References
France	• A drug innovation is a product that brings something new via radical change, possibly offering therapy or enhancing clinical outcomes while having wide medical benefits, an acceoted cost, and favorable benefit-risk ratio • The level of innovation brought to the market by a medicine is assessed by the drug's improvement in medical benefit or (amélioration du service médical rendu) ASMR over the current standard of care	• New molecular entity • Enhance route of administration • Improvement of formulation • New antiretroviral with improved method of administration • Therapeutic advances • Drugs with added value • Follow on drugs • Repurposed drugs • Drug may target disease, symptoms, or risk factor • Precision medicine • New drug delivery system • New mechanism of action • The innovative pharmaceutical products can be classified according to degree of innovation as following: – **Highly innovative:** modulation of a new Target, new mechanism of action, improved pharmacodynamic, the more precise pharmacogenetics profile, a new pro therapeutic or repurposed drugs) – **Moderately innovative:** A new therapeutic class, fewer side effects, less drug interactions, a new Structure, a new method of synthesis with added therapeutic value – **Slightly** innovative (from a medical perspective): improvement of pharmacokinetics, or improvement in formulation – **Slightly innovative (from a non-medical perspective):** new structure without added therapeutic value	The ASMR assigns ratings from I through V based on the extent of improvement; an ASMR level I, II, or III drug is regarded innovative • Newness • Novelty • Therapeutic benefit	(Andrade et al. 2016; Wakutsu et al. 2023)

(continued)

Table 1 (continued)

Regulatory authority	Definition of pharmaceutical innovative product	Characteristics of innovative product	The added therapeutic value	References
Germany	• Novel therapeutic entities that provide extra benefits over existing medicines and increase quality of life while also being cost effective • A leap innovation is a drug with a new active ingredient, while a step innovation is a further development or improvement of a known active ingredient, leading to better mode of action, fewer side effects, and expanded treatment range. Sham innovations are drugs with no additional benefit compared to equivalent therapy options, but comparable in effect but not superior	• New molecular entities with added therapeutic value • Products increase quality of life • Cost-effective products • New materials • New devices • New manufacturing process • New drug delivery system • New mechanism of action	Early benefit assessment under the law within the statutory health care system (Arzneimittelmarktneuordnungsgesetz, AMNOG, which means "Pharmaceuticals Market Reorganisation Act") by the German Federal Joint Committee determines whether innovative medicines have additional benefits • Novelty; newness • The German Federal Joint Committee determines whether innovative medications offer added therapeutic value	(Hörn et al. 2014; Gandjour 2018)

(continued)

Table 1 (continued)

Regulatory authority	Definition of pharmaceutical innovative product	Characteristics of innovative product	The added therapeutic value	References
Japan	Any pharmaceuticals developed by R&D and the use of modern technologies that address the demands of patients and clinical needs. The goal is to enhance drug discovery efficiency through enhanced R&D infrastructure and pharmaceutical regulatory reforms, thereby ensuring the long-term stability of the drug pricing system	• New molecular entities • New technologies (gene therapy, regenerative therapy, big data in drug discovery…etc)	• Adress unmet clinical need • Cost	(Wakutsu et al. 2023)
Australia (TGA)	Pharmaceutical products are considered innovative if they are new and their success is measured by sales. Higher sales are assumed to indicate the value of the innovation	• New molecular entity • New indication • Technological innovation • Novel formulation (although it may not result in health benefits) • New dosage form • New fixed dose of combination	• The value of new medicines lies in their therapeutic benefits for patients and society, including extended life expectancy, improved quality of life, and enhanced tolerance	(Morgan et al. 2008)

(continued)

Table 1 (continued)

Regulatory authority	Definition of pharmaceutical innovative product	Characteristics of innovative product	The added therapeutic value	References
Italy (AIFA)	The therapeutic value is the key criterion for defining the innovativeness status of a new medical product using a multidimensional approach	• New molecular entity • Pharmacological innovation (new mechanism of action) • Technological innovation	• Maximum therapeutic value: greater efficacy than marketed drugs • Important: improves efficacy relying on clinical data, reduces life-threatening condition, has better benefit/risk ratio, changes history of disease, or improves quality of life • Moderate: Alternative therapeutic option, or slightly better efficacy • Poor: better alternative therapeutic options are available	(Vitry et al. 2013; Fortinguerra et al. 2021)
WHO Commission on Intellectual Property Rights, Innovation, and Health (CIPIH)	In the developed world, innovation is defined as a cycle that begins with the discovery of new lead products, progresses to the creation and delivery of new tools, and then returns to the discovery stage via effective demand based on health requirements and postmarketing research	• New molecular entity • Incremental innovation confers clinical improvements, manufacturing improvements, or therapeutic advantages	Significant therapeutic value over existing ones	(Organization 2006)

approved 44 new entities (Kinch et al. 2023). The industry's lack of scientific productivity is evident in the rising attrition rates in clinical trials (Arrowsmith and Miller 2013a). Stricter regulations and challenging market conditions further contribute to the problem (Pammolli et al. 2011). These hurdles drive up the costs per approved molecule to unsustainable levels, necessitating a revamp of the current drug development system. Additionally, many new medicines offer little added value compared to existing treatments (Moors et al. 2014). In the past, pharmaceutical companies could focus on addressing unmet health issues, but now there are very few such categories remaining, resulting in increased competition with the existing drugs. Consequently, public policies have been implemented to incentivize the development of innovative medications (Grootendorst et al. 2011a).

4.2 New Fixed Dose Combination (FDC)

A fixed dose combination is a pharmaceutical product that has at least two active components in a single tablet, inhaler, or injection. This concept excludes natural plant or animal material mixes. FDC, independent of formulation or brand, is defined by the World Health Organization (WHO) as a specified combination of two or more active components in a precise dosage ratio. Future FDC products are predicted to have an increasingly important role in public health (Oo and Sy 2018). FDCs significantly decrease the possibility of medication non-adherence, especially in patients with chronic diseases (Bangalore et al. 2007a). In addition, advantages of FDCs compared to separate component prescriptions or concerns with monotherapy include: (i) improved response rates, when monotherapy is ineffective, due to different mechanisms of action in the combination; (ii) faster achievement of desired effects with FDCs; (iii) reduced toxicity as one medicine can counteract adverse reactions of another; and (iv) the ability to combine subtherapeutic doses that are safe as monotherapy, as seen with combination medicines for human immunodeficiency virus (HIV)-infected patients. These benefits often result in lower healthcare costs (Bell 2013).

Multiple techniques are frequently required for disease therapy to be effective. Because of the possibility of toxicity and medication resistance, relying on a single agent for long-term therapy is not optimal. Adopting innovative FDCs that include new and effective drugs is a better approach. Over loose combinations, FDCs have various benefits. They are more convenient, which leads to improved compliance and efficacy, particularly with repeated treatments. FDCs also provide lesser drug resistance, synergistic effects, and lower dosages for increased safety. Furthermore, by simplifying the production, distribution, dispensing, and administration procedures, FDCs have the potential to cut costs (Oo and Sy 2018).

Regulatory authorities recognize the relevance of FDCs and have a responsibility to promote them. They should prioritize disease prevention over excessive business activity. They can give incentives and speed approval processes in addition to enforcement actions. FDCs development is a complicated procedure that usually utilizes factorial clinical designs. Combination ratios, physical and chemical stabilities, drug-drug interactions, and other pharmacokinetics must all be evaluated (Dash et al. 2018). If there is adequate rationale, the FDA may not necessarily need full clinical trials for clearance; some of clinical research may be excluded (Kwon and Lee 2017).

4.3 Drug Repositioning

Drug repositioning entails discovering new applications for medications that have previously been authorized or are under investigation. This strategy provides patients with treatment alternatives that are safe, inexpensive, and timely (Mansour et al. 2021). Traditional drug makers, on the other hand, are frequently hesitant to invest in research on authorized treatments in the absence of financial incentives or patent protection. As a result, the majority of medication repositioning research is undertaken in independent clinical trials directed by academics, research institutes, or collaborative organizations (Albuquerque et al. 2022). The National Institutes of Health (NIH) offers funding to repurpose unsuccessful medications that have shown potential in development, such as viagra and thalidomide, to accelerate their advancement. The FDA and the Medical Research Council in United Kingdom both have medication repurposing efforts. Furthermore, governments, corporations, and foundations provide grant financing and access to resources (Low et al. 2020).

Clinical observation and fundamental research can lead to the discovery of new applications. The chemical AZD0530, for example, targets Fyn kinase, which is implicated in both Alzheimer's disease and solid tumor cell development. This broadens the possible applications of AZD0530 (Nygaard et al. 2014). Furthermore, a chemical may have effects on targets that were previously unrelated to certain disorders. This is why many academic drug development programs begin by investigating FDA-approved drugs that target various molecules. For example, an NIH-funded study discovered that antimicrobial cephalosporins can alter glutamate transporters (Rothstein et al. 2005). This discovery prompted a clinical trial of ceftriaxone for amyotrophic lateral sclerosis; however, the experiment was halted due to a lack of efficacy (Strittmatter 2014).

4.4 Novel Drug Delivery System

Drug delivery is critical in current pharmaceutical research and translation because it bridges the gap between fundamental drug discovery and therapeutic use. This field's success has been shown in recent decades by formulations such as Doxil®, Ambisome®, and liposomal anticancer medicines (van der Meel et al. 2019). However, the number of products on the market does not fully reflect the extensive academic research and number of published manuscripts in drug delivery labs worldwide of laboratory systems, and a lack of substantial funding to bring carriers closer to the bedside. The reasons behind this disparity between published research and clinically used formulations are most likely due to the elusive fundamental mechanisms, such as the enhanced permeation and retention effect, biological and pharmaceutical barriers to clinical translation (Fuhrmann 2023). Many drug delivery systems are not problem-solving oriented since they are not created with a clear therapeutic need in mind. There are, however, laboratories and organizations that do well-controlled, therapeutically-oriented drug delivery and bioengineering research. The recently produced mRNA COVID-19 vaccines based on lipid nanoparticles and viral carriers are an excellent example (Trollmann and Böckmann 2022). Personalized treatment options, such as cancer vaccine development, show promise. However, with the increasing complexity of nanotherapeutics, it's important to reconsider sustainable approaches to pharmaceutical research and how we train future generations (Fuhrmann 2023).

Examples of new drug delivery systems, including the osmotically controlled-release oral methylphenidate HCl (Concerta®) is designed to overcome drug tolerance by regulating its pharmacokinetics (Adler et al. 2009). Ritonavir (Norvir®) is a protease inhibitor used to treat HIV. It has been modified with thiazole to enhance its metabolic stability and solubility in water. Benazepril (Lotensin®) is an alkyl ester prodrug designed to conceal ionizable groups, thereby increasing its lipophilicity and enhancing its permeability (Jana et al. 2010). Naloxegol (Movantik®) is a modified form of naloxone that prevents off-target toxicity by not crossing the blood–brain barrier (Chey et al. 2014). Desmopressin (DDAVP) is an improved version of vasopressin, with non-natural amino acid substitution to enhance its half-life and stability (Garona et al. 2015). Semaglutide (Rybelsus®) is an oral GLP-1 agonist that contains SNAC for better stomach absorption, bypassing biological barriers (Suzuki et al. 2020). Belatacept (Nulojix®) is a fusion protein that has been modified to enhance its selectivity for CD86 and CD80. This modification improves its ability to target these specific proteins (Larsen et al. 2005). Trastuzumab and hyaluronidase-oysk is a subcutaneous depot that uses hyaluronidases to release Herceptin over time. This non-invasive administration method provides sustained release of the medication (Duco et al. 2020). To overcome biological barriers that restrict the delivery of complex therapeutic molecules, it is crucial to develop drug delivery systems that can efficiently release biomolecules in specific tissues, at specific times and concentrations, in response to environmental cues (Vargason et al. 2021).

Drug delivery systems have improved patient health by improving therapeutic delivery, decreasing off-target accumulation, and increasing patient compliance. These methods have been modified to allow for the use of nucleic acids, peptides, proteins, and antibodies, so broadening therapeutic choices. Delivery tactics and technology modified to address evolving demands as the therapeutic landscape expanded. Small-molecule medications were once the primary treatment, but additional roles faced obstacles such as protein and peptide stability, nucleic acid intracellular delivery needs, and live cell viability (Vargason et al. 2021).

4.5 Novel Technology in Manufacturing

The FDA is dedicated to modernizing pharmaceutical manufacturing to protect and promote public health. In 2002, the FDA launched the "Pharmaceutical cGMPs for the 21st Century: A Risk-Based Approach" initiative to promote the implementation of a modern, science- and risk-based quality assessment system for pharmaceuticals. The initiative aims to encourage early adoption of new technologies by the industry and ensure that regulatory policies are based on the latest pharmaceutical science. The launch of CDER's Office of Pharmaceutical Quality in 2015 is a significant step towards ensuring the availability of quality medicines for the American public. To promote the use of innovative technologies, they have established an emerging technology team as the main contact for pharmaceutical companies interested in utilizing emerging technologies for product development and market approval. This team helps in implementation of new technologies by improving communication between the FDA and sponsors. This allows for more discussion about design and development issues before applications are submitted. They also lead or co-lead the quality assessment team for submissions involving emerging technology. They also develop and document resolutions to scientific and policy issues, which can inform future FDA approaches and recommendations for similar technologies (O'Connor et al. 2016). Successful modernization of the pharmaceutical manufacturing sector requires collaboration among industry, academia, government, funding agencies, and regulatory bodies to address quality challenges.

4.6 Novel Stereoisomers

If a carbon atom has four different substituents, it can have varying conformations with surrounding atoms or molecules. Other atoms, like phosphorus, can also be asymmetric. This results in different forms with optical activities that rotate polarized light either left or right. A substance that rotates polarized light to the right is called dextrorotatory (d), while a substance that rotates polarized light to the left is called levorotatory (l). The configuration around an asymmetric atom is designated as R or S, or D or L for right- and left-handedness when the spatial arrangement of the molecules is known. Stereoisomers are examples of me-too drugs, such as escitalopram, esketamine, and esomeprazole. These drugs are S-stereoisomers of racemic mixtures.. However, there may be some differences between the two, such as the specificity of pharmacological action, adverse reactions profile, or drug-drug interactions. Approximately 40% of drugs used in clinical practice are chiral, with 90% of them being sold as racemic mixtures. Some drugs, such as naproxen and levodopa, are marketed as single stereoisomers. Stereoisomers can have significant pharmacological differences, as seen with S-warfarin being more potent than R-warfarin and L-sotalol having different effects than D-sotalol. However, it is challenging to produce individual isomers, and using a single isomer often leads to conversion back to the racemic form in the body. Some drug companies have extended patent lives by marketing isomers separately, but these formulations usually offer no advantages over racemic mixtures (Aronson and Green 2020b). Table 2 summarizes the classifications, pros, cons, and limitations of innovative pharmaceutical products.

5 Sustainable Pharmaceutical Innovation

The United Nations' Brundtland Commission coined the term "sustainable development" as development that meets present needs without compromising future generations' ability to meet their own needs. This widely accepted definition encompasses various concepts, including the SDGs and the One-Health approach (Wynendaele et al. 2021). Sustainable innovation involves developing new products, processes, services, and technologies that meet human needs and promote societal well-being, while also respecting and preserving natural resources and their regeneration. Innovations that enhance products can have positive effects on the economy, environment, and society in the short- and long-term. These innovations can create beneficial impacts on society and the environment (Library et al. 2010).

The pharmaceutical industry is a complicated and dynamic sector that includes the research, development, manufacturing, and marketing of novel chemical entities and bioproducts (Sharma et al. 2023). The pharmaceutical industry is crucial to the healthcare system as it manufactures and markets medicines, biological products, and medical devices for diagnosing and treating diseases. It also conducts research to develop new products for improving human health. Therefore, it is essential to uphold strict standards in pharmaceutical production to avoid any potential health risks, especially since many of these products are life-saving (Bose 2009). Establishing a pharmaceutical manufacturing unit is crucial for achieving the sustainable development goals of the international organization. Improving population well-being, fostering abilities, and reducing poverty are crucial for developing new, effective, and affordable medications. This objective is in line with the primary SDGs (Lakner et al. 2019). New pharmaceutical technologies are essential for conserving water and energy (Argiyantari et al. 2020). Efforts to make pharmaceuticals more affordable aid in the reduction of inequality. The pharmaceutical industry's growing emphasis on "green chemistry" is critical for decreasing carbon emissions, encouraging ethical manufacture and consumption, and addressing climate change. Green chemistry is increasingly moving beyond the concept of pure curiosity and into large-scale therapeutic applications in industries. However, industries, as one of the most dynamic areas, are constantly at the forefront of any significant developments. These developments in terms of novel concepts, traditional feed stocks, safer raw materials, and alternative methods in laboratories on a pilot scale appear to be exciting to implement in industries (Mishra et al. 2021). Finally, increasing the pharmaceutical industry is critical to meeting the United Nation's SDG (Lakner et al. 2019).

The increasing global population and aging population will result in a greater demand for medicines. However, this will have long-term effects on the environment

Table 2 Summary of innovative pharmaceutical products: classifications, pros, cons, and limitations of sustainable development

Type of pharmaceutical innovation	Definition	Advantages	Drawbacks/Limitations	References
Breakthrough/radical pharmaceutical innovation				
New molecular entity	– Radical innovations are usually the initial derivatives of a drug class – Breakthrough drugs generate significant societal value by uniquely improving health outcomes that would otherwise be impossible – For a drug to be considered a breakthrough, it must provide substantial advancements compared to current treatments, even in cases of urgent unmet needs	– Increases competitiveness – Increases profit – Increases health outcomes – Getting patent and market exclusivity – Opens new markets and increases opportunities for importation	– Increased cost and R&D expenditure – Restricted regulation – Time-consuming – Increased cost of clinical trials – A lack of scientific productivity in the industry – Market saturation of drugs foe most diseases – Market conditions are becoming more challenging – Lack of added value compared with existing drugs	(Morgan et al. 2008; Munos 2009; Pammolli et al. 2011; Arrowsmith and Miller 2013b; Oo and Sy 2018; Kinch et al. 2023)
Incremental pharmaceutical innovation				
Fixed dose combination (FDC)	A drug product that combines two active ingredients in a single oral pill, inhaler, or co-packaged injectable. This definition excludes mixtures of natural plant or animal materials	– Suitable to target multi-etiological diseases – Improves patient compliance – Decreases single drug toxicity – Decreases drug resistance – Better effectiveness – Synergistic effect – Reduces cost of manufacturing – Achieves SDG3 by improving patient compliance, and SDG9 by innovation – Not all clinical trials phases are required if evidence provided	– Lack of dose flexibility – Potential change in individual drug bioavailability – Different dose and dose regimen – Ineffective in case of drugs with narrow therapeutic index – Unnecessary broader coverage – Increased toxicity and side effects – Increased resistance in case of combined antibiotics	(Kwon and Lee 2017)

(continued)

Table 2 (continued)

Type of pharmaceutical innovation	Definition	Advantages	Drawbacks/Limitations	References
Drug repositioning (new indication)	Developing new therapies is a challenging, time-consuming, and expensive process. Additionally, safety concerns often arise during the development of many compounds. Drug repurposing involves testing and developing drugs that have already been proven safe in humans for their effectiveness in treating a different disease than the one they were originally intended for	– Known safety and tolerability – Time saving – Overcoming complex drug development process – Financial saving – Increasing success in regulatory approval	– Limited intellectual property protection – Lack of clinical trials – Variable success rate – Limited market exclusivity	(Mansour et al. 2021, 2023)
Novel drug delivery technology	Drug delivery technologies have improved patient health by enhancing the delivery of therapeutics through drug modification, or microenvironment modification, or creating interface	– Improves the solubility – Controls drug release – Enhanced drug activity – Promotes drug stability – Improves pharmacokinetics – Improves permeability – Suppresses off-target side effects – Non-invasive administration – Reduces immunogenicity in case of proteins, peptides, anti-bodies, and nucleic acid – Enhances target selection – Improves patient compliance	– Increased manufacturing cost – Potential toxicity of used materials – Nano formulations can cause unintended toxicity when they penetrate and translocate to the blood–brain barrier and lungs – Nanocarriers can change shape and size, leading to diverse physicochemical interactions and activity	(Vargason et al. 2021; Adepu and Ramakrishna 2021)
Novel technology in manufacturing	Modernization pf pharmaceutical manufacturing to improve product quality	– Improves product quality	– Technical, regulatory, and business risk – Capital investment in buying new equipment – Upgrading facilities – Extra effort and time	(O'Connor et al. 2016)

(continued)

Table 2 (continued)

Type of pharmaceutical innovation	Definition	Advantages	Drawbacks/Limitations	References
New stereoisomers	If a carbon atom has four different substituents, it can exhibit various conformations with surrounding atoms or molecules. Similarly, other atoms such as phosphorus can also possess asymmetry. Consequently, these variations result in different forms that exhibit optical activities, causing the rotation of polarized light either to the left or right. A substance that rotates polarized light to the right is referred to as dextrorotatory (d), while a substance that rotates polarized light to the left is known as levorotatory (l). The configuration around an asymmetric atom is denoted as R or S, or D or L to indicate right- or left-handedness, respectively, when the spatial arrangement of the molecules is known	• Specific pharmacological action • Different metabolism and toxicity profile	• Challenges to separate stereoisomers	(Aronson and Green 2020b)

and our health. Currently, the concept of sustainability in the pharmaceutical industry is fragmented. To address this, a consortium of researchers from five European universities is advocating for a global, systematic approach that prioritizes sustainability from the early stages of drug development, such as drug discovery. The researchers believe that competent authorities, universities, research institutions, and industrial organizations all need to consider sustainability more seriously (Wynendaele et al. 2021).

5.1 Sustainability from the Cradle: Drug Discovery

5.1.1 Drug Discovery and Ecological-Environmental Impact

Human and veterinary pharmaceuticals are entering and accumulating in the environment worldwide. There is growing awareness of the impact of active pharmaceutical ingredients (APIs) and their byproducts on the environment. It is crucial to understand and predict the environmental effects of these compounds. Each year, thousands of tons of APIs are produced, and considering the vast ocean volume, it is evident that these bioactive APIs continuously enter the global sea at parts per billion levels. Diclofenac residues in cattle carcasses have been proven to be lethal to vultures, and exposure of naproxen to light breaks down the drug into more toxic compounds for aquatic life. Wastewater treatment facilities, if available, do not effectively remove pharmaceuticals, allowing drugs and their metabolites to pass through freely (Belew et al. 2021; Smale et al. 2021).

When considering the environmental impact of a medicine throughout its lifecycle, it is important to note that current medicine costs do not account for the environmental burden associated with its discovery, development, packaging, and clinical use. If sustainability concepts are not incorporated into the early drug discovery phase, future generations will bear this burden (Wynendaele et al. 2021). The European Commission (EC) has outlined actions in its communication titled "The European Union's Strategic Approach to Pharmaceuticals in the Environment". These actions include incorporating environmental aspects into medical training and professional development programs (Kohl 2019). Also, nature is a significant source of new lead structures in modern drug development, accounting for over half of approved new medicines. Therefore, it is crucial to preserve global biodiversity, especially considering the widespread use of plant and animal materials as medicines. Innovative tissue culture technologies can help prevent the extinction of original plants. Additionally, the impact of pharmaceuticals on biodiversity in the environment, including bioremediation, is an area of ongoing research (Newman and Cragg 2020).

5.1.2 Drug Discovery and Unmet Medical Needs

The medical field should prioritize important health challenges, such as life-threatening, incurable, therapy-resistant, and neglected diseases. Currently, there are approximately 6,500 diseases without approved therapies (Roessler et al. 2021). Repurposing opportunities, utilizing real-world data and artificial intelligence (AI) exploration, can help address these unmet clinical needs and promote equitable healthcare distribution worldwide. However, determining what to prioritize is not a neutral decision. Various factors, including developmental status, types of diseases, profitability of pharmaceutical companies, and more, must be considered. It is crucial to critically evaluate the agenda, decision-makers, objectives, and outcomes of these prioritization efforts (Wynendaele et al. 2021).

5.1.3 Drug Discovery and Green Chemistry

Green chemistry is a valuable approach to reducing waste and mitigating climate change. The 12 principles of green chemistry provide a comprehensive toolkit for achieving these goals (Sheldon 2007). This toolkit includes concepts such as the E-factor, bioconversions, flow chemistry, and green analytics, including chips and biosensors (Erythropel et al. 2018). Currently, synthesis processes generate excessive amounts of organic waste, primarily in the form of organic solvents. Additionally, we are depleting our limited natural resources, particularly endangered metals like platinoids, which are extensively used in catalysis. Furthermore, the pharmaceutical industry has a significant ecological footprint, contributing to substantial greenhouse gas emissions, surpassing even the automobile industry in CO_2 production (Clarke et al. 2018). This issue is not only present in large-scale syntheses, but also in the labor-intensive and time-consuming search for new small- APIs in the drug discovery phase.

Currently, sustainability aspects are rarely considered during the discovery phase, unlike in the later development phases, and even then, only to a limited extent. This overall effect can be described as insufficient and delayed. The concept of green chemistry can also be applied to pre-formulation work and dosage form design, where the synthesis and use of environmentally friendly excipients should be prioritized. As awareness of the environmental impact grows, it is important to explore strategies for making synthetic chemistry more sustainable. Two such strategies are flow chemistry and synthetic chemistry in water. These methods aim to reduce the use of solvents. To promote greener synthetic chemistry, parameters have been introduced to assess the efficiency and impact of a reaction, including chemical yield and waste production, such as atom economy (Trost 1991). Recombinant DNA technology has made enzymes and combinations of enzymes more popular among synthetic organic chemists and biotechnologists. This is due to the knowledge of biosynthetic pathways and the ability to use enzymes to tackle limitations such as narrow substrate scope, poor selectivity, and instability (Li et al. 2018).

5.1.4 Drug Discovery and Artificial Intelligence

The use of automation, artificial intelligence (AI), and big data in drug discovery can improve speed and efficiency. This includes synthesis planning and predicting properties and outcomes. By doing so, it can reduce the need to test animals and lessen the burden on humans and the environment (Wynendaele et al. 2021). Automating and parallelizing chemical synthesis enhances speed, throughput, reproducibility, and reduces material consumption. This, combined with innovative scaffold design, allows for the exploration of larger areas of chemical space in a shorter time compared to manual, sequential compound synthesis. Advancements in AI have aided in planning chemical syntheses. Growing availability of compound activity data allows for deep learning strategies to predict activity and bioavailability. Deep learning has been successful in image analysis and natural language processing and is now being used to predict aqueous solubility of pharmaceutical compounds (Cui et al. 2020). Big data approaches are crucial for drug discovery and personalized drug prescriptions. Pharmacogenetics and metabolomics provide guidance for finding the best outcomes for patients (Klont et al. 2021).

5.1.5 Drug Discovery and Targeting the Root Cause of Diseases

The focus should be on drugs that address the root causes of diseases, rather than just treating symptoms. Currently, many drugs only alleviate symptoms, such as mucolytic agents for chronic bronchitis. The healthcare system is primarily reactive, waiting until a patient is sick before addressing the issue, and is mainly focused on diagnosis and symptom management. To improve patient outcomes, future therapies should prioritize targeting the underlying cause of diseases, rather than just managing symptoms. While the use of new drugs and technologies in the antiviral field is well established, their application in other fields is less established. However, new technologies such as Proteolysis Targeting Chimera (PROTACs), messenger RNA (mRNA), and CRISPR/Cas9 are expected to revolutionize therapy in the future. PROTACs, which induce targeted protein degradation, have advantages over traditional inhibitors in terms of dosing, side effects, drug resistance, and targeting "undruggable" proteins (Xiao et al. 2021). mRNA-based drugs are a promising class of biologics that can encode any protein of interest for vaccination or therapeutic purposes (Alabi and Crews 2021). The affordability of mRNA-based drugs is also improving. CRISPR technology has the potential to edit genetic mutations and cure diseases caused by these mutations, such as tumors and cystic fibrosis (Mansour and El-Khatib 2023a).

5.1.6 Drug Discovery and Decision-Taking Models

To make optimal sustainable decisions, it is important to consider the complete life cycle of the drug. In drug discovery, the focus is usually on finding selective ligands for specific targets. However, there has been progress in understanding how these ligands affect cellular networks, which can improve the translation of preclinical discoveries into clinical benefits. New computational approaches, like sustainability metrics, are expected to help guide important decisions. It is also important to consider social concerns and apply the precautionary principle when making environmental and health decisions (Garcia et al. 2021).

5.1.7 Drug Discovery and Bioinformatics

The use of biomarkers and bioinformatics can support precision medicine, reducing the burden on patients, society, and the environment by minimizing ineffective medicine use. A precise and rapid diagnosis, preferably confirmed at the molecular level, can improve the effectiveness and cost-efficiency of therapies (Jaber et al. 2016). This includes identifying disease-causing gene mutations. During the drug discovery phase, it is important to incorporate genetic and metabolomic biomarkers that can help stratify patients based on disease risk and determine the most effective drug therapies. Additionally, drug non-adherence should be considered, as it leads to inefficient use of medications. Non-adherence to drugs and vaccines can compromise efforts to combat deadly diseases like corona, TB, and malaria. Therapeutic drug monitoring and advancements in adherence-enhancing devices, such as smart drug delivery systems, can help optimize drug usage (Chimeh et al. 2020).

5.1.8 Drug Discovery and Cost-Effectiveness

Medicines are an essential part of the healthcare system, and social justice means ensuring equal access to them. This includes transparent discussions about costs, profits, and effectiveness, which should lead to prices that are acceptable to society. To achieve this, we need adjusted reimbursement policies, intellectual property regulations, and data protection measures. The profitability and fairness of a medicine throughout its life cycle, including any variations or extensions, should be considered. The price of a medicine is influenced by organizational models and the societal-political environment. While the cost of discovering a medicine is relatively small, it ultimately determines the price (Chaudhuri 2021).

5.1.9 Drug Discovery and Sustainable Innovation

The productivity of drug discovery has decreased due to the difficulty and cost of finding and validating new therapeutic targets. Research institutions and universities play a crucial role in identifying these targets, which can lead to the development and marketing of new medicines. It is important for academia to have a sustainable position and to translate their findings into industrial development and commercialization through spin-offs or external partners. The government is expected to support and stimulate innovative and sustainable research, as well as ensure fair access to compensate for its investment (Jaehn et al. 2017).

5.2 *New Fixed Dose Combination and Sustainability*

The SDGs aim to a one-third decrease in morbidity and mortality linked to non-communicable diseases like cardiovascular and respiratory diseases (Mahipala et al. 2019). The World Health Organization (WHO) supports the use of FDCs for infectious diseases like HIV, malaria, Hepatitis B, and TB to enhance treatment effectiveness and prevent resistance (Sawicki-Wrzask et al. 2015). This is crucial in sub-Saharan Africa, where there is a high prevalence of both non-communicable diseases like hypertension and diabetes, and infectious diseases like HIV, TB, and malaria. Additionally, there is a higher prevalence of patients with multiple co-existing conditions compared to other continents (Haacker et al. 2019). A previous Cochrane review found no difference in outcomes between FDCs and single-drug formulations for managing newly diagnosed pulmonary TB patients (Gallardo et al. 2016). A meta-analysis has shown that using FDCs can reduce the risk of medication non-compliance in patients with chronic conditions like hypertension. This can lead to improved medication compliance and better clinical outcomes (Bangalore et al. 2007b). Ensuring sufficient therapeutic compliance is also economically sustainable in terms of Pharmacoeconomics (Golay 2011).

5.3 *Drug Repositioning and Sustainable Development*

The global population using regular medicines has doubled in recent decades. Alongside this, life expectancy has significantly increased, and there is a rise in lesser understood diseases. However, the lengthy process of designing and approving drugs (10–12 years), higher rates of approval failure, and recalls after launch, increased resources for new drug discovery and development, and a shift towards green chemistry have made the traditional drug discovery process inadequate (Bhar and Bhar 2023).

Drug repositioning, also known as repurposing, is the preferred strategy for introducing sustainability in drug discovery and development. Repurposing offers several advantages over developing new treatments, including reduced risk of failure, lower development costs, and a shorter time frame for registration (Bhar and Bhar 2023). The main reason for the failure or withdrawal of an approved drug is adequately addressed by considering the potential repurposing of a drug that has already undergone preclinical testing and early-stage human clinical trials (phase I & II). This ensures its safety and makes subsequent efficacy trials more predictable. Furthermore, by avoiding the repetition of preclinical testing, safety assessment, and formulation development, drug development timelines can be significantly shortened. This approach is also economically advantageous, as it requires minimal investment. The regulatory and phase III costs can result in substantial savings compared to preclinical, phase I, and phase II expenses (Mansour et al. 2021; Mansour and El-Khatib 2023b). Also, repurposing off-patent medicines for cancer treatment could address unmet needs in a cost-effective way, especially for rare cancers, which account for around 22% of new cases in Europe. This non-commercial approach is particularly beneficial in areas that are not attractive for industry (To and Cho 2022).

Previously, repurposing relied on ad hoc discovery, but it now relies on organized, methodical, data-driven processes to uncover viable candidates. The Drug Repurposing Hub is a public repository containing almost 7000 chemicals, many of which have been authorized by the FDA. The Hub's researchers vetted and confirmed these chemicals, which are now being tested against disease cell lines. They also use this database to acquire fresh insights into disease features, which may lead to the development of novel drug discovery programs. The Broad Institute in the United States created the Hub. There are many sustainable drug repurposing strategies, including signature mapping, in silico screening and molecular docking, pathway matching, genetic association, drug repurposing, high-throughput screening, retrospective analysis of clinical data, and so on (Pushpakom et al. 2019).

5.3.1 Signature Mapping

Signature mapping process involves comparing distinct features, known as the 'signature', of one drug to another. This can be done through comparing disease pathways, clinical phenotypes, or using a drug-disease similarity approach. The signature pattern of two drugs can be determined using three databanks: transcriptomic (RNA), metabolomic, or proteomic data, chemical structures, and adverse event profiles. For instance, prednisolone and topiramate were identified as two drugs using the drug-disease similarity approach by correlating their gene expression signature with that of the disease (Wu et al. 2017).

5.3.2 High-Throughput Screening

High-throughput screening (HTS) is a widely used method for lead discovery in Pharma and Biotech companies. HTS involves screening large chemical libraries to identify compounds that show activity against specific biological targets. This is achieved through the use of automation, miniaturized assays, and extensive data analysis. HTS aims to find prospective hits or lead compounds with desirable features or activities that might be turned into novel medicines, probes, or research tools (Mayr and Bojanic 2009). The discovery of disulfiram, originally used to treat alcohol abuse, has been found to also have selective antineoplastic properties. This has been confirmed through genome-wide gene expression studies (Beaudry et al. 2023).

5.3.3 Pathway/Network Matching

In certain disease pathways, the genes involved cannot be targeted with drugs. In these cases, a comparison of genes that are either downstream or upstream in the pathway, or a genome-wide association study (GWAS)-related target, can be used for repurposing purposes. Network analysis utilizes available information on gene expression patterns, disease pathology, protein interactions, or GWAS data to construct a drug or disease network, which can help identify potential candidates for repurposing. For instance, pranlukast, an asthma drug that acts as a leukotriene receptor 1 antagonist, and amrinone, a phosphodiesterase inhibitor, are both used for congestive heart failure (Greene and Voight 2016).

5.3.4 In Silico Screening and Molecular Docking

To find possible drug-target interactions, in silico techniques that depend primarily on computational methods integrate knowledge mining with molecular modelling. Algorithms are used in these procedures to screen compounds and assess how they interact with target proteins. This computational approach uses structure similarity to predict the compatibility of a with a therapeutic target. For instance, molecular docking studies confirmed that mebendazole, an anti-parasitic drug and inhibitor of vascular endothelial growth factor receptor 2 (VEGFR2), is also involved in angiogenesis (Bao et al. 2014).

5.3.5 Retrospective Analysis of Clinical Data

We have access to, gather, integrate, and analyze enormous volumes of historical clinical data, such as hospital records, clinical trial data, and post-marketing surveillance data. These databases contain both organized and unstructured information about patient responses and outcomes (Bhar and Bhar 2023).

5.3.6 Genetic Association Studies

By evaluating the relationship between disease status and genetic variation, genetic association studies aim to find genes or genomic areas that may contribute to a given disease. A greater frequency of a single nucleotide polymorphism (SNP) allele or genotype in a group of patients may indicate that the variation raises the risk of the disease. Genes that are assumed to be correlated with a disease pathogenesis should be considered as possible drug targets (Cordell and Clayton 2005).

5.4 *Sustainable Innovative Drug Delivery System*

New and innovative methods for manufacturing improved nano pharmaceuticals that can address previous issues rely on green technologies. Green chemistry focuses on the synthesis of chemical compounds and engineering processes using natural materials in industrial applications. It also applies to laboratory experiments to create a safer environment. By minimizing the use of hazardous materials during reaction and synthesis, green chemistry aims to produce renewable substances and reduce waste. The main goals of green chemistry include designing efficient reactions, using renewable materials and energy sources, employing safe solvents and reactants, and preventing waste production. Green nanotechnology applies the principles of green chemistry to design new nanomaterials for economic, social, health, and environmental benefits (Jahangirian et al. 2017).

5.4.1 Sustainable Biopolymeric Nanoparticles

Cancer therapy is a global concern, and recent advances in nanomedicine are crucial. Nanotechnology offers targeted delivery systems for treating diseases using natural bioactive compounds, gaining widespread attention in cancer treatment. Natural bioactive compounds have potential as chemotherapeutic agents, in addition to their nutritional benefits. Biopolymeric nanoparticles have been effectively used in delivering therapeutics to specific sites. The main aim of biopolymer-based nanomedicine is to deliver drug specifically to cancer cells, increasing effectiveness and reducing toxicity. Biopolymer composites combine the qualities of multiple materials to enhance the physiological and mechanical needs of biomedical areas. The chemical substances used in standard processes to produce polymeric nanomaterials can have long-term stability and environmental impact. In contrast, biopolymers are often made from non-toxic monomers and are carbon–neutral, making it easier to create nanomaterials from them at the nanoscale (Kanaoujiya et al. 2022; Pathak et al. 2022).

Biopolymers can be found in plants, algae, fungi, bacteria, and animals, including macromolecules such as starch, dextran alginate, chitin, and chitosan. Other compounds like pectin, lignin, cellulose and hyaluronic acid are also used to create nanomaterial composites. In the medical field, nano biocomposites have shown promising results, such as using biodegradable substances and chitosan nanoparticles to deliver anticancer drugs (Kanaoujiya et al. 2022). Biopolymers have diverse components and different physiological properties and can be used to create molecular capsules by utilizing intramolecular hydrogen bonding. For example, starch can be used to create polymeric nanocomposites by combining metals or metal oxides, and chitosan is widely used in nanotechnology applications due to its extensive compatibility. Nanomaterials can be incorporated into biopolymers through impregnation or absorption, and their physical features can differ from macro-sized bulk materials due to their greater surface area and reactivity (Raveendran et al.

2003). These biopolymer-based systems offer a cost-effective and environmentally friendly solution to improve treatment efficacy, providing effective drug delivery with increased selectivity and lower toxicity (Pathak et al. 2022).

5.4.2 Sustainable Green Nanotechnology

Synthetic nanomaterials have numerous uses and advantages today, but historically, their production was costly and associated with the production of environmentally harsh products. The use of synthetic nanostructures in medicine is limited due to complications and risks. Nanoparticles, which are smaller than 100 nm, can potentially pose health risks by attracting to body systems. For example, nanomaterials like carbon nanotubes and metal oxides can have harmful impacts on the respiratory system similar to asbestos particles or quartz dust. Studies show that the body's defense mechanisms treat nanoparticles like microorganisms, but they can bond together to form fibers that are too large to be processed by macrophages. Nanoparticles with a high ratio of transition metals may have reactive areas that can change the toxicity of particles and damage human cells. To address these concerns, scientists are now focusing on using green methods to synthesize nanomaterials to prevent such complications. Principles and methods of green chemistry have been developed to reduce the effects and dangers of chemicals (Morose 2010).

The purpose of greener nanotechnology, also known as sustainable nanotechnology, is to design processes and products at the molecular level to improve product performance and minimize environmental and health impacts (Hutchison 2008). The goal of the green nanotechnology center is to reduce product risks by designing materials with inherent safety, minimizing exposure, and minimizing the impact of nanomaterial production. Green chemistry principles are applied to nanotechnology throughout the product life cycle, from production to end-of-life (Morose 2010). This integration of green chemistry principles emphasizes the role of design and incorporates principles that work together to achieve the same design goal. Modeling greener nanoscale products involves combining several key principles to: (1) reduce risk while maintaining performance, (2) design nanomaterials for end-of-life scenarios (degradation, reuse, or repurposing), and (3) incorporate intrinsic safety (Gilbertson et al. 2015).

Green chemistry aims to reduce the risks associated with materials and chemicals, and one way to achieve this is by designing at the molecular level to minimize exposure. Nanomaterials are synthesized using wet chemical methods, so creating greener combinations of nanomaterials can be guided by strategies developed for molecular species. Key strategies include using safer reagents and solvents, reducing or eliminating solvents, improving atom efficiency to decrease waste, and eliminating the need for purifications. The ultimate goal of greener nanotechnology is to maximize the benefits to society while minimizing negative effects on the environment and health (Nabipour and Hu 2020).

5.4.3 Sustainable Extracellular Vesicles: Biogenic Carriers

Extracellular vehicles (EVs) are biogenic nanocarriers. EVs are small particles (50–200 nm) naturally produced by eukaryotic and prokaryotic cells and play a role in intercellular communication (Fuhrmann et al. 2015). The interaction of membrane proteins and phospholipids in vesicles has sparked interest in employing vesicles for drug delivery. Cell-derived vesicles have various benefits over manufactured lipid nanocarriers, including natural composition, non-toxicity, possible targeting capabilities, and increased stability under physiological conditions owing to immune evasion. EVs provide a viable approach for drug carrier development due to their natural origin and the ability to cultivate producer cells under large-scale settings utilizing proven biotechnological technologies. However, there are several obstacles to using both human EVs and bacterial Outer Membrane Vesicles (OMVs) for biomedical applications, including EV heterogeneity and complexity, reproducible upscale production appropriate preclinical models that match EVs with appropriate comparator carriers, and optimized formulation and stability (Fuhrmann 2023).

6 Toward Sustainable Pharmaceutical Innovation

The pharmaceutical industry thrived from the 1960s to the 1990s due to advances in synthetic chemical technology, which permitted the synthesis of new compounds with comparable mechanisms of action. Patent protection enabled high prices, which resulted in higher turnover and profitability. The rise of the generic medication sector, on the other hand, shortened sales cycles and increased pressure on innovation. Growing societal focus on medication safety has resulted in a rise in regulatory demands, which has added to the challenges of developing innovative medications. The top three obstacles to pharmaceutical innovation were linked to drug development costs, time; and regulatory limitations (Eichler et al. 2009). Universities began patenting their innovations, making them competitors instead of partners. The current regulatory system in the developed world, for patent regulation and market approval, has led to higher medicine costs. This has created access problems, especially in the developing countries, further damaging the industry's reputation. The responsibility for resolving issues with expensive or unavailable medicines lies with many actors in the healthcare system, including the pharmaceutical industry (Hogerzeil 2013). Pharmaceutical companies have grown in size through mergers and acquisitions, gaining control of the market. They have lobbied for patent extensions and high drug prices, leading to an unpopular industry focused on marketing and eroding innovative products. The question is How can these problems be resolved to achieve future sustainability in pharmaceutical innovation? Table 3 summarizes factors that cause declining in sustainable pharmaceutical innovation and our recommendations.

Table 3 Factors affecting declining innovative products: insights, key findings, and recommendations

No	Challenge	Key findings	Recommendations
1	Market saturation	• There is a lack of added value in many new medicines compared to existing treatments (Moors et al. 2014)	• Target unmet clinical needs • Exploring new fixed dose combinations • Redirecting toward precision medicine • Invest more in incremental innovation instead of breakthrough innovation
2	Patent system and intellectual property challenges	• The role of patents in drug innovation is to provide temporary protection for companies to recover their costs by charging higher prices • However, the current drug patent system has drawbacks. High drug prices limit access for those without comprehensive insurance or enough money to pay. Certain aspects of the patent system also increase the cost of finding new therapies, decrease sales revenue, and reduce the financial motivation to innovate (Gurgula 2020)	• Improving Intellectual Property Protection and Encouraging Innovation • Intellectual property protection is crucial for fostering innovation and creativity. By safeguarding the rights of inventors, creators, and innovators, we can incentivize them to continue pushing boundaries and developing new ideas • To enhance intellectual property protection, it is essential to establish robust legal frameworks that effectively address copyright, patents, and trademarks. These frameworks should provide clear guidelines and procedures for registering and enforcing intellectual property rights • Additionally, it is important to strengthen international cooperation and collaboration in the fight against intellectual property infringement. By working together, countries can share best practices, exchange information, and coordinate efforts to combat piracy and counterfeiting • Furthermore, promoting public awareness and education about intellectual property rights is crucial. By educating individuals and businesses about the value of intellectual property and the consequences of infringement, we can foster a culture of respect for innovation and creativity • Incentivizing innovation requires a balanced approach. While strong intellectual property protection is necessary, it is also important to ensure that patent and copyright laws do not stifle competition or hinder access to knowledge and technology. Striking the right balance between protection and access is key to fostering innovation and benefiting society as a whole • Strong competition in the market for innovative products is expected to have two long-term effects. Firstly, it will encourage companies to invest more in research and development (R&D) for new products that can quickly gain acceptance, in order to maintain a competitive product portfolio. Secondly, it will also incentivize companies to focus on incremental innovations, which will extend their period of market exclusivity • Public funding of fundamental research and clinical studies (Grootendorst et al. 2011b) • Royalty- and reward-based schemes (Grootendorst et al. 2011b)

(continued)

Table 3 (continued)

No	Challenge	Key findings	Recommendations
3	Complex and time-consuming regulatory processes hinder the pace of innovation	• Long time before market authorization	• Streamlining regulatory processes to expedite approval • The accelerated procedure allows for faster market access for products that offer significant added therapeutic benefits • Breakthrough treatments may be eligible for conditional approval • Simplifying the regulatory process and shifting the regulatory authorities' focus to offering scientific advice and support to industry can enhance certainty and streamline procedures for companies seeking guidance on development issues in specific therapeutic and technological fields • Increasing marketing exclusivity period for innovative products • Communication between drug authorities and pharmaceutical companies needs improvement • Establishment of rewarding reference pricing system • Development of systematic methodology to evaluate the degree of innovation • Accessibility of drugs before marketing authorization
4	Cost of developing new innovative products	• The complexity of the product affects research and development costs, with more complex products being more expensive to develop (Charles River Associates 2004) • The number of trials required to support a new product has increased. This is due to the need for comparative studies to support marketing, formulary negotiations, and reimbursement decisions (Charles River Associates 2004) • Regulatory requirements have not been a major factor in the long-term cost increase, but they may have led to increased costs in the late 1990s due to product withdrawals (Charles River Associates 2004) • New technology has temporarily increased research and development costs. It is uncertain how quickly these costs will be recouped and whether they will result in a significant reduction in development costs (Charles River Associates 2004) • Cost-effectiveness studies may lead to higher R&D expenses due to the need for additional data on the pharmacoeconomic value of products. This increase in costs can potentially decrease the motivation to invest in innovative products (Charles River Associates 2004)	• By offering incentives to pharmaceutical companies for investing in cost-effective products with social benefits, we can improve the efficiency of R&D allocation • Novel payment and pricing models, when used appropriately and tailored to the situation, can accelerate patient access, allowing payers to manage clinical uncertainty, budget impact and sustainability of the healthcare system, whilst providing sufficient incentives for innovation

(continued)

Table 3 (continued)

No	Challenge	Key findings	Recommendations
5	Technological limitations blocking breakthrough innovations	• Acquiring new technologies needs new equipment and upgrading facilities • This process is risky • Acquiring new technology requires extra expenditure, effort, and time	• Advancing regulatory science for innovative pharmaceutical manufacturing • Establishing a public–private institute for supporting pharmaceutical innovation • Making incentives for investment in innovative technology and upgrade of facilities and manufacturing processes • Acquiring external expertise in drug development • Expediting international regulation of innovative pharmaceutical products
6	Lack of collaboration between pharmaceutical companies and knowledge sharing	–	• Increasing collaboration through merging and acquisition
7	Lack of collaboration between industry and academia	–	• Improved cooperation is needed between private and public research organizations, such as universities, research institutions, and the pharmaceutical industry, to address fragmented research systems, bringing finances, scientific knowledge, and centers of expertise
8	Lack of pharmaceutical innovation in the developing countries	–	• Enhancing developing countries' appeal as an Innovative Hub • Identifying industry capacity and key bottlenecks in development resources is crucial • Increasing research staff, awareness of patients to participate in clinical trials, and management of these trials • Offering pricing advantages to the pharmaceutical innovative products to encourage pharmaceutical companies to compensate R&D expenses • Coordinating tax credits at the developing country level is necessary to fully consider spillover effects • A conceptual framework for Equity-Based Tiered Pricing aims to consider the ability to pay across countries when determining the prices of innovative medicines. This framework is based on the principle of solidarity between countries, with the goal of reducing the unavailability of new medicines and access delays

6.1 Organizational and Technological Changes

Drug development rationalization strives to reduce attrition rates, uncover failures early, and enable innovation. However, this strategy does not reduce the cost of medications or enhance access to treatments in developing countries (Kaitin 2010). Repurposing existing drugs for new indications is a more cost-effective approach than developing new drugs (Mansour et al. 2022). However, it can substantially increase costs. Additionally, medicalizing pharmaceutical research could enhance innovation but also comes with higher development risks (EA 2009). Pharmaceutical companies have made major organizational changes by merging and acquiring other companies in order to increase their size and market share. Unfortunately, these changes are often focused only on cutting costs, which leads to less investment in research and innovation, as well as less competition. This could potentially result in higher drug prices (Lamattina 2011).

Organizational reforms like public–private cooperation, precompetitive collaboration, and open-access innovation networks increase the potential for innovation but do not lessen the risks associated with medication development. Concentrating drug discovery in nonprofit, publicly supported organizations such as universities might greatly alleviate access issues since researchers will be motivated by medical need rather than market potential (Schuhmacher et al. 2013).

6.2 Addressing Patent Challenge: Balancing Pharmaceutical Innovation and Access

Proposed changes to patent and data protection systems are varied and not always consistent. Some advocate for both shorter and longer durations of intellectual property protection. However, allowing companies to set their own prices means these changes may not lead to reduced costs. If the period of product protection is shortened, companies may raise prices to maintain income. Conversely, there is no incentive to lower prices when exclusivity is extended (Higgins and Graham 2009).

A more relaxed patent system in developing countries would increase access to drugs in underdeveloped countries by allowing for the production of local generic drugs. However, this change would not impact other aspects of sustainable drug development. Limiting pharmaceutical companies' ability to pay generic competitors to delay the release of their products would result in earlier availability of generic drugs and lower costs (Musselwhite and Andrews 2010).

In the pharmaceutical sector, patents constrain pricing competitiveness. Only patenting significant innovation would restrict the development and marketing of follow-on medications, hence encouraging innovation. However, because of their increased failure probability, developing more novel medications is dangerous. This, along with a lack of competition from follow-on treatments, results in exorbitant drug prices that are out of reach for many patients. Participating in patent pools and abstaining from patenting by publicly financed research organizations will encourage innovation and access to medications for a broader range of diseases. These alternatives, however, would not reduce the hazards associated with drug development (Yancey and Stewart 2010).

6.3 Alteration of Reimbursement

In both developed and developing countries, creative solutions have been proposed to improve access to medications. Price controls, whether implemented directly or indirectly through limitations on reimbursement per quality-adjusted life year, face challenges due to ideological opposition or difficulties in implementation due to variations in national health systems. While price controls could lower costs, they may also negatively impact other aspects of a sustainable drug development system (Miller and Henderson 2007). A tiered pricing system could impact access in the developing countries but would not affect the risks of drug development or innovation. Awarding prizes for innovation instead of a patent could make new drugs more accessible and provide an incentive for innovation. However, the uncertainty of receiving an award may still make drug development a risky endeavor (DiMasi and Grabowski 2007). Reimbursement schemes that limit payment for effective drugs in individual cases or reward manufacturers based on reducing disease burden in a population do not decrease the risks and costs of drug development. However, they can positively impact innovation and access. An intriguing reimbursement system has been implemented by the Melissa and Bill Gates Foundation, which ensures markets for new vaccines and drugs, especially in developing countries. This system offers access to specific drugs and vaccines for designated populations but does not establish a sustainable drug development system (Shaffer et al. 2023).

6.4 Changes in Regulatory Systems

The regulatory system has contradictory advice. Both restricting and raising regulatory burdens have been proposed. To stimulate innovation, both business and policymakers are advocating early medication approval before the entire therapeutic benefit has been established (Boon et al. 2010). Well-designed regulatory and pharmacovigilance strategies are essential for optimizing risk management on a case-by-case basis. These strategies should determine the best toolkit for establishing the benefit: risk profile of innovative drugs (Ebbers et al. 2011). In addition, using real-life data from electronic health records is considered an innovative way to compare and evaluate the performance of innovative drugs after they are approved for market. These methods could reduce the time it takes to bring a drug to market, while also ensuring safety through ongoing post-marketing surveillance (Moors et al. 2014). Earlier approval would not impact the risks of drug development but would extend the exclusive marketing period and increase the costs of these drugs (Kesselheim 2011).

The Affordable Care Act in the United States has the potential to drastically alter the pharmaceutical industry by transitioning from unit pricing to value-based buying. This means that innovative pharmaceuticals will be priced depending on the value assigned to them by healthcare payers. Many supporters believe that the value of innovative treatments in comparison to established therapies should be considered throughout the marketing authorization process (Russo and Balekdjian 2008).

6.5 Increase Collaboration Between Academia and Industry

Biomedical research is crucial for economic growth, according to national and international organizations. Funding for this research often relies on public–private partnerships, with the pharmaceutical industry playing a significant role. In the EU's Innovative Medicines Initiative (IMI) program, pharmaceutical companies not only determine the research topics but also lead the public–private consortia of research groups (https://www.imi.europa.eu/). The IMI is a public–private partnership between the European Commission and the European Federation of Pharmaceutical Industries and Associations (EFPIA). Launched in 2008, the IMI has a budget of €2 billion and aims to enhance the development of new medicines in Europe by fostering collaboration between large pharmaceutical companies, academic institutions, small and medium enterprises, patients, and regulatory authorities. IMI consortia projects have already yielded significant results, demonstrating the effectiveness of this collaborative model (Goldman 2012).

7 Future Directions

Considering the current status of pharmaceutical innovation, there are many aspects to promote sustainable development in pharmaceutical sector, including:

1. Promoting essential medicine for universal accessibility is a key aspect of healthcare system sustainability. The establishment of a national essential medicine system aims to enhance social justice and harmony by ensuring the availability, quality, and appropriate use of essential medicines. The essential medicine experts should consist of clinicians, pharmacists, payer representatives, and health economists. Their responsibility is to select and regularly update essential medicines. They should establish principles, standards, procedures, and methods for selecting essential medicines. The selection process should consider factors such as drug efficacy, safety, quality, stability, cost-effectiveness, and production and supply security. The committee should determine the number of chemicals, biological medicines, and traditional medicines

on the essential medicine list based on scientific criteria for safety, effectiveness, and cost. The selection process should be open and transparent, with input from medical staff. The essential medicines chosen should meet basic medical needs.

2. Essential innovative drugs should be added to medical insurance's prescription reimbursement system. Levels of reimbursement will be increased, and efforts will be made to boost publicity, education, and training. This is to ensure universal access and the priority usage of important innovative medications.
3. Encourage drug industrialization by instituting laws such as a drug development fund, enabling businesses to contribute funding as donations rather than taxes, and promoting the transfer of findings from research and intellectual property to businesses. Relevant departments will give substantial assistance in novel medicine evaluation and approval, price, reimbursement, tax breaks, and financial services.
4. Reform industrial policy to assure long-term growth of the pharmaceutical business, which is critical to national economic development, social stability, public health, and national security. Scientific development principles should lead pharmaceutical industry strategy, with a defined direction and concentration on pharmaceutical innovation that matches with related taxation, finance, resource, and environmental protection laws. This would enable the sector to fully contribute to the national economic and social development, while also fostering long-term growth and increasing international competitiveness. The objective is for developing countries to transform from a significant pharmaceutical manufacturer to a pharmaceutical powerhouse.
5. Prioritization of the development of new drugs and encourage social investment in research and development should be conducted.
6. Establishment of innovation fund through tax incentives to encourage enterprise participation.
7. Improvement of the drug discovery platform and fully mobilize researchers to form a world-class drug innovation system.
8. Developing innovative medicines with independent intellectual property rights to treat major diseases threatening public health.
9. Implement a pricing system for drugs based on value.
10. Achieving a more equitable sharing of risks and rewards between governments and industry.
11. Encouraging the repurposing of drugs.
12. Invest in research and development programs for neglected diseases and sustainable methods for producing innovative drugs.
13. Natural products are a potential source of new APIs, offering the opportunity for more sustainable and affordable innovative drugs. This underscores the importance of preserving biodiversity and utilizing it for economic development.
14. Increase collaboration between academia and industry.

8 Conclusion

No single reform can substantially improve the societal costs of drugs or the risks of drug development. A balanced combination of suggested reforms is necessary for a sustainable drug development system. The failure of the current system can be attributed to the start of drug patenting by pharmaceutical companies in the mid-twentieth century, the introduction of strict regulatory systems after the thalidomide tragedy in 1961, and the implementation of data protection by the Hatch–Waxman Act of 1984. These mechanisms have allowed the pharmaceutical industry to maintain an exclusive and successful position in drug development for decades. However, with the end of the pharmaceutical industry's golden age, this monopoly should be reconsidered. Transitioning to a sustainable drug development system requires reforming the regulatory and patent systems and shifting the responsibility for discovering and clinically evaluating new drugs to the public sector (Moors et al. 2014).

Towards sustainable development in pharmaceutical innovation, we propose a structural reform where the medical and scientific community in universities and public research institutions would have a greater responsibility for discovering, developing, and evaluating the effectiveness and safety of new drugs. Public funding should support scientific research, and all findings should be accessible to the public. The biomedical field has the necessary resources to take on this responsibility, as demonstrated in other areas of medical intervention. The pharmaceutical industry must consider these aspects early in the development of a potential medicine. Global sustainability evaluation should be a mandatory and significant part of the "efficacy-risk" balance when authorizing medicines. Currently, environmental risk assessment is not even considered in the approval process for marketing authorization. Academic institutions also have a responsibility to include sustainability aspects in drug discovery education. Finding effective treatments for unmet diseases, accessible to all at a fair price, while also protecting the environment, is a significant challenge that requires better preparation.

References

Adepu S, Ramakrishna S (2021) Controlled drug delivery systems: current status and future directions. Molecules 26. https://doi.org/10.3390/MOLECULES26195905

Adler LA, Zimmerman B, Starr HL et al (2009) Efficacy and safety of OROS methylphenidate in adults with attention-deficit/hyperactivity disorder: a randomized, placebo-controlled, double-blind, parallel group, dose-escalation study. J Clin Psychopharmacol 29:239–247. https://doi.org/10.1097/JCP.0B013E3181A390CE

Alabi SB, Crews CM (2021) Major advances in targeted protein degradation: PROTACs, LYTACs, and MADTACs. J Biol Chem 296:100647. https://doi.org/10.1016/J.JBC.2021.100647

Albuquerque PC, Zicker F, Fonseca BP (2022) Advancing drug repurposing research: trends, collaborative networks, innovation and knowledge leaders. Drug Discov Today 27:103396. https://doi.org/10.1016/J.DRUDIS.2022.103396

Chandler Jr AD (2005) Shaping the industrial century: the remarkable story of the evolution of the modern chemical and pharmaceutical industries. Harvard University Press

Andrade LF, Sermet C, Pichetti S (2016) Entry time effects and follow-on drug competition. Eur J Health Econ 17:45–60. https://doi.org/10.1007/S10198-014-0654-9

Argiyantari B, Simatupang TM, Basri MH (2020) Pharmaceutical supply chain transformation through application of the Lean principle: a literature review. J Ind Eng Manage 13:475–494. https://doi.org/10.3926/jiem.3100

Aronson JK, Green AR (2020a) Me-too pharmaceutical products: history, definitions, examples, and relevance to drug shortages and essential medicines lists. Br J Clin Pharmacol 86:2114. https://doi.org/10.1111/BCP.14327

Aronson JK, Green AR (2020b) Me-too pharmaceutical products: history, definitions, examples, and relevance to drug shortages and essential medicines lists. Br J Clin Pharmacol 86:2114. https://doi.org/10.1111/BCP.14327

Arrowsmith J, Miller P (2013a) Phase II and phase III attrition rates 2011–2012. Nat Rev Drug Discov12(8):569. https://doi.org/10.1038/nrd4090

Arrowsmith J, Miller P (2013b) Phase II and phase III attrition rates 2011–2012. Nat Rev Drug Discov 12(8):569. https://doi.org/10.1038/nrd4090

Bangalore S, Kamalakkannan G, Parkar S, Messerli FH (2007a) Fixed-dose combinations improve medication compliance: a meta-analysis. Am J Med 120:713–719. https://doi.org/10.1016/J.AMJMED.2006.08.033

Bangalore S, Kamalakkannan G, Parkar S, Messerli FH (2007b) Fixed-dose combinations improve medication compliance: a meta-analysis. Am J Med 120:713–719. https://doi.org/10.1016/J.AMJMED.2006.08.033

Bao J, Zhou N, Luo K et al (2014) In silico discovery of potential VEGFR-2 inhibitors from natural derivatives for anti-angiogenesis therapy. Int J Mol Sci 15:15994. https://doi.org/10.3390/IJMS150915994

Beaudry A, Jacques-Ricard S, Darracq A et al (2023) Repurposing disulfiram, an alcohol-abuse drug, in neuroblastoma causes KAT2A downregulation and in vivo activity with a water/oil emulsion. Sci Rep 13(1):1–13. https://doi.org/10.1038/s41598-023-43219-2

Belew S, Suleman S, Wynendaele E et al (2021) Environmental risk assessment of the anthelmintic albendazole in Eastern Africa, based on a systematic review. Environ Pollut 269:116106. https://doi.org/10.1016/J.ENVPOL.2020.116106

Bell DSH (2013) Combine and conquer: advantages and disadvantages of fixed-dose combination therapy. Diabetes Obes Metab 15:291–300. https://doi.org/10.1111/DOM.12015

Bhar S, Bhar S (2023) Role of drug repurposing in sustainable drug discovery. Drug Repurposing Adv Scopes Oppor Drug Discov. https://doi.org/10.5772/INTECHOPEN.110621

Boon WPC, Moors EHM, Meijer A, Schellekens H (2010) Conditional approval and approval under exceptional circumstances as regulatory instruments for stimulating responsible drug innovation in Europe. Clin Pharmacol Ther 88:848–853. https://doi.org/10.1038/CLPT.2010.207

Bose R (2009) Advanced analytics: opportunities and challenges. Ind Manag Data Syst 109:155–172. https://doi.org/10.1108/02635570910930073/FULL/PDF

Charles River Associates (2004) Innovation in the pharmaceutical sector

Chaudhuri S (2021) Patent protection and access to covid-19 medical products in developing countries. SSRN Electron J. https://doi.org/10.2139/SSRN.3902314

Chey WD, Webster L, Sostek M et al (2014) Naloxegol for opioid-induced constipation in patients with noncancer pain. N Engl J Med 370:2387–2396. https://doi.org/10.1056/NEJMOA1310246

Chimeh RA, Gafar F, Pradipta IS et al (2020) Clinical and economic impact of medication non-adherence in drug-susceptible tuberculosis: a systematic review. Int J Tuberc Lung Dis 24:811–819. https://doi.org/10.5588/IJTLD.19.0754

Clarke CJ, Tu WC, Levers O et al (2018) Green and sustainable solvents in chemical processes. Chem Rev 118:747–800. https://doi.org/10.1021/ACS.CHEMREV.7B00571/ASSET/IMAGES/MEDIUM/CR-2017-005715_0055.GIF

Comanor WS (1964) Research and competitive product differentiation in the pharmaceutical industry in the United States. Economica 31:372. https://doi.org/10.2307/2550516

Cordell HJ, Clayton DG (2005) Genetic association studies. The Lancet 366:1121–1131. https://doi.org/10.1016/S0140-6736(05)67424-7

Cui Q, Lu S, Ni B et al (2020) Improved prediction of aqueous solubility of novel compounds by going deeper with deep learning. Front Oncol 10:519625. https://doi.org/10.3389/FONC.2020.00121/BIBTEX

Dash RP, Rais R, Srinivas NR (2018) Key Pharmacokinetic essentials of fixed-dosed combination products: case studies and perspectives. Clin Pharmacokinet 57:419–426. https://doi.org/10.1007/S40262-017-0589-2/FIGURES/1

DiMasi JA, Grabowski HG (2007) Should the patent system for new medicines be abolished? Clin Pharmacol Ther 82:488–490. https://doi.org/10.1038/SJ.CLPT.6100393

Duco MR, Murdock JL, Reeves DJ (2020) Trastuzumab/hyaluronidase-oysk: a new option for patients with HER2-positive breast cancer. Ann Pharmacother 54:254–261. https://doi.org/10.1177/1060028019877936

Ebbers HC, Mantel-Teeuwisse AK, Moors EHM et al (2011) Today's challenges in pharmacovigilance: what can we learn from epoetins? Drug Saf 34:273–287. https://doi.org/10.2165/11586350-000000000-00000

Eichler HG, Abadie E, Raine JM, Salmonson T (2009) Safe drugs and the cost of good intentions 360:1378–1380. https://doi.org/10.1056/NEJMP0900092

Erythropel HC, Zimmerman JB, De Winter TM et al (2018) The Green ChemisTREE: 20 years after taking root with the 12 principles. Green Chem 20:1929–1961. https://doi.org/10.1039/C8GC00482J

Ferner R, Hughes DA, Aronson J (2010) NICE and new: appraising innovation. BMJ 340:245–247. https://doi.org/10.1136/BMJ.B5493

Fortinguerra F, Perna S, Marini R et al (2021) The assessment of the innovativeness of a new medicine in Italy. Front Med (lausanne) 8:793640. https://doi.org/10.3389/FMED.2021.793640/BIBTEX

Fuhrmann G (2023) Drug delivery as a sustainable avenue to future therapies. J Control Release 354:746–754. https://doi.org/10.1016/J.JCONREL.2023.01.045

Fuhrmann G, Herrmann IK, Stevens MM (2015) Cell-derived vesicles for drug therapy and diagnostics: opportunities and challenges. Nano Today 10:397–409. https://doi.org/10.1016/J.NANTOD.2015.04.004

Gallardo CR, Rigau Comas D, Valderrama Rodríguez A et al (2016) Fixed-dose combinations of drugs versus single-drug formulations for treating pulmonary tuberculosis. Cochrane Database Syst Rev 2016. https://doi.org/10.1002/14651858.CD009913.PUB2

Gandjour A (2018) Underuse of innovative medicines in Germany: A justification for government intervention? Health Policy (new York) 122:1283–1286. https://doi.org/10.1016/J.HEALTHPOL.2018.08.009

Garcia E, Ly N, Diep JK, Rao GG (2021) Moving from point-based analysis to systems-based modeling: integration of knowledge to address antimicrobial resistance against MDR bacteria. Clin Pharmacol Ther 110:1196–1206. https://doi.org/10.1002/CPT.2219

Garona J, Pifano M, Orlando UD et al (2015) The novel desmopressin analogue [V4Q5]dDAVP inhibits angiogenesis, tumour growth and metastases in vasopressin type 2 receptor-expressing breast cancer models. Int J Oncol 46:2335. https://doi.org/10.3892/IJO.2015.2952

Gilbertson LM, Zimmerman JB, Plata DL et al (2015) Designing nanomaterials to maximize performance and minimize undesirable implications guided by the principles of green chemistry. Chem Soc Rev 44:5758–5777. https://doi.org/10.1039/C4CS00445K

Golay A (2011) Pharmacoeconomic aspects of poor adherence: can better adherence reduce healthcare costs? J Med Econ 14:594–608. https://doi.org/10.3111/13696998.2011.597808

Goldman M (2012) The innovative medicines initiative: a European response to the innovation challenge. Clin Pharmacol Ther 91:418–425. https://doi.org/10.1038/CLPT.2011.321

Greene CS, Voight BF (2016) Pathway and network-based strategies to translate genetic discoveries into effective therapies. Hum Mol Genet 25:R94–R98. https://doi.org/10.1093/HMG/DDW160
Grootendorst P, Hollis A, Levine DK et al (2011a) New approaches to rewarding pharmaceutical innovation. CMAJ 183:681–685. https://doi.org/10.1503/CMAJ.100375
Grootendorst P, Hollis A, Levine DK et al (2011b) New approaches to rewarding pharmaceutical innovation. CMAJ 183:681–685. https://doi.org/10.1503/CMAJ.100375
Gurgula O (2020) Strategic patenting by pharmaceutical companies–should competition law intervene? IIC Int Rev Ind Prop Copyr Law 51:1062. https://doi.org/10.1007/S40319-020-00985-0
Haacker M, Bärnighausen T, Atun R (2019) HIV and the growing health burden from noncommunicable diseases in Botswana: modelling study. J Glob Health 9. https://doi.org/10.7189/JOGH.09.010428
Hogerzeil HV (2013) Big pharma and social responsibility–the access to medicine index. N Engl J Med 369:896–899. https://doi.org/10.1056/NEJMP1303723
Hörn H, Nink K, McGauran N, Wieseler B (2014) Early benefit assessment of new drugs in Germany-results from 2011 to 2012. Health Policy 116:147–153. https://doi.org/10.1016/J.HEALTHPOL.2013.12.008
Higgins MJ, Graham SJH (2009) Intellectual property: balancing innovation and access: patent challenges tip the scales. Science 326. https://doi.org/10.1126/SCIENCE.1176116
Hutchison JE (2008) Greener nanoscience: a proactive approach to advancing applications and reducing implications of nanotechnology. ACS Nano 2:395–402. https://doi.org/10.1021/NN800131J/ASSET/IMAGES/LARGE/NN-2008-00131J_0004.JPEG
Jaber Z, Aouad P, Al Medawar M et al (2016) Role of systems biology in brain injury biomarker discovery: neuroproteomics application. Methods Mol Biol 1462:157–174. https://doi.org/10.1007/978-1-4939-3816-2_10
Jaehn P, Inhoffen J, Scheer J et al (2017) Possibilities for universities to influence global pricing of medicines. Gesundheitswesen 79:530–534. https://doi.org/10.1055/S-0043-113249
Jahangirian H, Lemraski EG, Webster TJ et al (2017) A review of drug delivery systems based on nanotechnology and green chemistry: green nanomedicine. Int J Nanomedicine 12:2957–2978. https://doi.org/10.2147/IJN.S127683
Jana S, Mandlekar S, Marathe P (2010) Prodrug design to improve pharmacokinetic and drug delivery properties: challenges to the discovery scientists. Curr Med Chem 17:3874–3908. https://doi.org/10.2174/092986710793205426
Kaitin KI (2010) Deconstructing the drug development process: the new face of innovation. Clin Pharmacol Ther 87:356. https://doi.org/10.1038/CLPT.2009.293
Kanaoujiya R, Saroj SK, Srivastava S, Chaudhary MK (2022) Renewable polysaccharide and biomedical application of nanomaterials. J Nanomater 2022. https://doi.org/10.1155/2022/1050211
Karampli E, Souliotis K, Polyzos N et al (2014) Pharmaceutical innovation: impact on expenditure and outcomes and subsequent challenges for pharmaceutical policy, with a special reference to Greece. Hippokratia 18:100
Kesselheim AS (2011) An empirical review of major legislation affecting drug development: past experiences, effects, and unintended consequences. Milbank Q 89:450. https://doi.org/10.1111/J.1468-0009.2011.00636.X
Kinch MS, Kraft Z, Schwartz T (2023) 2022 in review: FDA approvals of new medicines. Drug Discov Today 28:103622. https://doi.org/10.1016/J.DRUDIS.2023.103622
Kitsis EA (2009) Rx for the pharmaceutical industry: call your doctors. Hastings Cent Rep 39:18–21. https://doi.org/10.1353/HCR.0.0160
Klont F, Kremer D, Gomes Neto AW et al (2021) Metabolomics data complemented drug use information in epidemiological databases: pilot study of potential kidney donors. J Clin Epidemiol 135:10–16. https://doi.org/10.1016/J.JCLINEPI.2021.02.008
Kohl S (2019) EAHP opinion on pharmaceuticals in the environment. Eur J Hosp Pharm 26:237–238. https://doi.org/10.1136/EJHPHARM-2019-002004

Kwon KC, Lee C (2017) Analysis of fixed-dose combination products approved by the US food and drug administration, 2010–2015: implications for designing a regulatory shortcut to new drug application. Ther Innov Regul Sci 51:111–117. https://doi.org/10.1177/2168479016663263/METRICS

Lakner Z, Kiss A, Popp J et al (2019) From basic research to competitiveness: an econometric analysis of the global pharmaceutical sector. Sustainability 11:3125. https://doi.org/10.3390/SU11113125

Lamattina JL (2011) The impact of mergers on pharmaceutical R&D. Nat Rev Drug Discov 10:559–560. https://doi.org/10.1038/NRD3514

Larsen CP, Pearson TC, Adams AB et al (2005) Rational development of LEA29Y (belatacept), a high-affinity variant of CTLA4-Ig with potent immunosuppressive properties. Am J Transplant 5:443–453. https://doi.org/10.1111/J.1600-6143.2005.00749.X

Li R, Wijma HJ, Song L et al (2018) Computational redesign of enzymes for regio- and enantioselective hydroamination. Nat Chem Biol 14(7):664–670. https://doi.org/10.1038/s41589-018-0053-0

Library WO, Elke H, Bos-Brouwers J (2010) Corporate sustainability and innovation in SMEs: evidence of themes and activities in practice. Bus Strategy Environ 19:417–435. https://doi.org/10.1002/BSE.652

Lichtenberg FR (2007) The impact of new drugs on us longevity and medical expenditure, 1990–2003: evidence from longitudinal, disease-level data. Am Econ Rev 97:438–443. https://doi.org/10.1257/AER.97.2.438

Lichtenberg FR (2001) The benefits and costs of newer drugs: evidence from the 1996 medical expenditure panel survey

Low ZY, Farouk IA, Lal SK (2020) Drug repositioning: new approaches and future prospects for life-debilitating diseases and the covid-19 pandemic outbreak. Viruses:12. https://doi.org/10.3390/V12091058

Mahipala P, Dorji G, Tisocki K, Rani M (2019) A critical review of addressing cardiovascular and other non-communicable diseases through a primary health care approach in the South-East Asia Region. Cardiovasc Diagn Ther 9:150. https://doi.org/10.21037/CDT.2018.09.03

Mansour HM, El-Khatib AS (2023a) Exploring parkinson-associated kinases for CRISPR/cas9-based gene editing: beyond alpha-synuclein. Ageing Res Rev:102114. https://doi.org/10.1016/J.ARR.2023.102114

Mansour HM, El-Khatib AS (2023b) Repositioning of receptor tyrosine kinase inhibitors: from cancer to neurodegenerative and psychiatric disorders. Recept Tyrosine Kinases Neurodegener Psychiatr Disorders:353–401. https://doi.org/10.1016/B978-0-443-18677-6.00010-5

Mansour HM, Fathi AM, El-Khatib AS, MahmoudM K (2023) Kinases control of regulated cell death revealing druggable targets for Parkinson's disease. Ageing Res Rev 85:101841. https://doi.org/10.1016/J.ARR.2022.101841

Mansour HM, Fawzy HM, El-Khatib AS, Khattab MM (2021) Potential repositioning of anti-cancer EGFR inhibitors in alzheimer's disease: current perspectives and challenging prospects. Neuroscience 469:191–196. https://doi.org/10.1016/J.NEUROSCIENCE.2021.06.013

Mansour HM, Fawzy HM, El-Khatib AS, Khattab MM (2022) Repurposed anti-cancer epidermal growth factor receptor inhibitors: mechanisms of neuroprotective effects in Alzheimer's disease. Neural Regen Res 17:1913. https://doi.org/10.4103/1673-5374.332132

Mayr LM, Bojanic D (2009) Novel trends in high-throughput screening. Curr Opin Pharmacol 9:580–588. https://doi.org/10.1016/J.COPH.2009.08.004

Miller HI, Henderson DR (2007) Governmental influences on drug development: striking a better balance. Nat Rev Drug Discov 6:532–539. https://doi.org/10.1038/NRD2323

Mishra M, Sharma M, Dubey R et al (2021) Green synthesis interventions of pharmaceutical industries for sustainable development. Curr Res Green Sustain Chem 4:100174. https://doi.org/10.1016/J.CRGSC.2021.100174

Moors EHM, Cohen AF, Schellekens H (2014) Towards a sustainable system of drug development. Drug Discov Today 19:1711–1720. https://doi.org/10.1016/J.DRUDIS.2014.03.004

Morgan S, Lopert R, Greyson D (2008) Toward a definition of pharmaceutical innovation. Open Medicine 2:e4
Morose G (2010) The 5 principles of "design for safer nanotechnology." J Clean Prod 18:285–289. https://doi.org/10.1016/J.JCLEPRO.2009.10.001
Munos B (2009) Lessons from 60 years of pharmaceutical innovation. Nat Rev Drug Discov 8(12):959–968. https://doi.org/10.1038/nrd2961
Musselwhite LW, Andrews J (2010) Protect pharmaceutical innovation. Science 328:1354. https://doi.org/10.1126/SCIENCE.328.5984.1354-A
Nabipour H, Hu Y (2020) Sustainable drug delivery systems through green nanotechnology. Nanoeng Biomater Adv Drug Deliv:61–89. https://doi.org/10.1016/B978-0-08-102985-5.00004-8
Newman DJ, Cragg GM (2020) Natural products as sources of new drugs over the nearly four decades from 01/1981 to 09/2019. J Nat Prod 83:770–803. https://doi.org/10.1021/ACS.JNATPROD.9B01285/SUPPL_FILE/NP9B01285_SI_009.PDF
Nygaard HB, Van Dyck CH, Strittmatter SM (2014) Fyn kinase inhibition as a novel therapy for Alzheimer's disease. Alzheimers Res Ther 6. https://doi.org/10.1186/ALZRT238
O'Connor TF, Yu LX, Lee SL (2016) Emerging technology: a key enabler for modernizing pharmaceutical manufacturing and advancing product quality. Int J Pharm 509:492–498. https://doi.org/10.1016/J.IJPHARM.2016.05.058
Oo C, Sy SKB (2018) Fixed-dose combinations: a potential means to boost drug development for selected drugs. Drug Discov Today 23:457–459. https://doi.org/10.1016/J.DRUDIS.2017.11.001
Organization WHealth (2006) Public health-innovation and intellectual property rights: report of the commission on intellectual property rights, innovation and public health, vol 228
Pammolli F, Magazzini L, Riccaboni M (2011) The productivity crisis in pharmaceutical R&D. Nat Rev Drug Discov 10(6):428–438. https://doi.org/10.1038/nrd3405
Pathak N, Singh P, Singh PK et al (2022) Biopolymeric nanoparticles based effective delivery of bioactive compounds toward the sustainable development of anticancerous therapeutics. Front Nutr 9:963413. https://doi.org/10.3389/FNUT.2022.963413/BIBTEX
Public Policy and the Economics of Entrepreneurship (2003) Public Policy and Innovation in the U.S. Pharmaceutical Industry. Public Policy and the Economics of Entrepreneurship
Pushpakom S, Iorio F, Eyers PA et al (2019) Drug repurposing: progress, challenges and recommendations. Nat Rev Drug Discov 18:41–58. https://doi.org/10.1038/NRD.2018.168
Raveendran P, Fu J, Wallen SL (2003) Completely "green" synthesis and stabilization of metal nanoparticles. J Am Chem Soc 125:13940–13941. https://doi.org/10.1021/JA029267J/SUPPL_FILE/JA029267JSI20030324_031335.PDF
Roessler HI, Knoers NVAM, van Haelst MM, van Haaften G (2021) Drug repurposing for rare diseases. Trends Pharmacol Sci 42:255–267. https://doi.org/10.1016/J.TIPS.2021.01.003
Rothstein JD, Patel S, Regan MR et al (2005) Beta-lactam antibiotics offer neuroprotection by increasing glutamate transporter expression. Nature 433:73–77. https://doi.org/10.1038/NATURE03180
Russo MJ, Balekdjian D (2008) Weighing the outcomes. Nat Biotechnol 26:173–182. https://doi.org/10.1038/NBT0208-173
Sawicki-Wrzask D, Thomsen M, Bjerrum OJ (2015) An analysis of the fixed-dose combinations authorized by the european union, 2009–2014: a focus on benefit-risk and clinical development conditions. Ther Innov Regul Sci 49:553–559. https://doi.org/10.1177/2168479014567322/METRICS
Schuhmacher A, Germann PG, Trill H, Gassmann O (2013) Models for open innovation in the pharmaceutical industry. Drug Discov Today 18:1133–1137. https://doi.org/10.1016/J.DRUDIS.2013.07.013
Scott KA, Ropek N, Melillo B et al (2022) Stereochemical diversity as a source of discovery in chemical biology. Curr Res Chem Biol 2:100028. https://doi.org/10.1016/J.CRCHBI.2022.100028

Sepkowitz KA (2011) One hundred years of salvarsan 365:291–293. https://doi.org/10.1056/NEJMP1105345

Shaffer J, Alenichev A, Faure MC (2023) The Gates Foundation's new AI initiative: attempting to leapfrog global health inequalities? BMJ Global Health 8:e013874

Sharma D, Patel P, Shah M (2023) A comprehensive study on Industry 4.0 in the pharmaceutical industry for sustainable development. Environ Sci Pollut Res 30:90088–90098. https://doi.org/10.1007/S11356-023-26856-Y/FIGURES/4

Sheldon RA (2007) The E factor: fifteen years on. Green Chem 9:1273–1283. https://doi.org/10.1039/B713736M

Smale EM, Egberts TCG, Heerdink ER et al (2021) Waste-minimising measures to achieve sustainable supply and use of medication. Sustain Chem Pharm 20:100400. https://doi.org/10.1016/J.SCP.2021.100400

Sternitzke C (2010) Knowledge sources, patent protection, and commercialization of pharmaceutical innovations. Res Policy 39:810–821

Strittmatter SM (2014) Overcoming drug development bottlenecks with repurposing: old drugs learn new tricks. Nat Med 20(6):590–591. https://doi.org/10.1038/nm.3595

Suzuki R, Brown GA, Christopher JA et al (2020) Recent developments in therapeutic peptides for the glucagon-like peptide 1 and 2 receptors. J Med Chem 63:905–927. https://doi.org/10.1021/ACS.JMEDCHEM.9B00835/ASSET/IMAGES/MEDIUM/JM9B00835_0009.GIF

Tajaddini R, Gholipour HF (2021) Economic policy uncertainty, R&D expenditures and innovation outputs. J Econ Stud 48:413–427. https://doi.org/10.1108/JES-12-2019-0573/FULL/XML

To KKW, Cho WCS (2022) Drug repurposing for cancer therapy in the era of precision medicine. Curr Mol Pharmacol 15:895–903. https://doi.org/10.2174/1874467215666220214104530

Trollmann MFW, Böckmann RA (2022) MRNA lipid nanoparticle phase transition. Biophys J 121:3927–3939. https://doi.org/10.1016/J.BPJ.2022.08.037

Trost BM (1991) The atom economy—a search for synthetic efficiency. Science 254:1471–1477. https://doi.org/10.1126/SCIENCE.1962206

van der Meel R, Sulheim E, Shi Y et al (2019) Smart cancer nanomedicine. Nat Nanotechnol 14(11):1007–1017. https://doi.org/10.1038/s41565-019-0567-y

Vargason AM, Anselmo AC, Mitragotri S (2021) The evolution of commercial drug delivery technologies. Nat Biomed Eng 5(9):951–967. https://doi.org/10.1038/s41551-021-00698-w

Vitry AI, Shin NH, Vitre P (2013) Assessment of the therapeutic value of new medicines marketed in Australia. J Pharm Policy Pract 6:2. https://doi.org/10.1186/2052-3211-6-2

Wakutsu N, Hirose E, Yonemoto N, Demiya S (2023) Assessing definitions and incentives adopted for innovation for pharmaceutical products in five high-income countries: a systematic literature review. Pharmaceut Med 37:53–70. https://doi.org/10.1007/S40290-022-00457-5/TABLES/8

Wu H, Huang J, Zhong Y, Huang Q (2017) DrugSig: a resource for computational drug repositioning utilizing gene expression signatures. PLoS ONE 12:e0177743. https://doi.org/10.1371/JOURNAL.PONE.0177743

Wynendaele E, Furman C, Wielgomas B et al (2021) Sustainability in drug discovery. Med Drug Discov 12:100107. https://doi.org/10.1016/J.MEDIDD.2021.100107

Xiao Z, Song S, Chen D et al (2021) Proteolysis targeting chimera (PROTAC) for macrophage migration inhibitory factor (MIF) has anti-proliferative activity in lung cancer cells. Angew Chem Int Ed 60:17514–17521. https://doi.org/10.1002/ANIE.202101864

Yancey A, Stewart CN (2010) Patent reform in the US: what's at stake for pharmaceutical innovation? Expert Opin Ther Pat 20:603–608. https://doi.org/10.1517/13543771003709775

Artificial Intelligence for Advanced Sustainable Development Goals: A 360-Degree Approach

Rahul Joshi, Krishna Pandey, and Suman Kumari

Abstract The phenomenon referred to by the phrase "artificial intelligence" has the potential to have both positive and negative impacts on individuals and the environment, and there is now a substantial amount of attention put on the research and assessment of these effects. The United Nations 2015 developed a set of seventeen Sustainable Development Goals (SDGs), which included social, economic, and environmental objectives. This chapter introduces two theoretical models that may be used to determine how Artificial intelligence fits in with overall objectives. It argues that the current concentration on AI applications risks leading to disappointment, even as AI, in every sense, is likely to play a crucial role in efforts to accomplish the SDGs. First, it frames the issue within the larger one of aligning international Research & Development activities with the SDGs. This chapter focuses on presenting and discussing up-to-date information on the degrees of alignments and disarray and the potential for alternate configurations to create more precise routes. These are starting to appear from the food business to the energy industry, but AI and the digital world have some catching up to do. Second, it shows how different forms of AI may play different helpful and required roles in achieving Sustainable Development Goals in different contexts, such as cities, regions, and nations.

Keywords AI · SDG · UNDP · Healthcare · Development · Research · Economic · Social · Environment

1 Introduction

Artificial intelligence has immense promise, and there is a growing consensus, supported by evidence, that it can enhance the world (Vinuesa et al. 2020). The AI4Good campaign, spearheaded by the United Nations, exemplifies the perception of AI as a positive force (Sætra 2021). Despite widespread excitement, there are still

R. Joshi (✉) · K. Pandey · S. Kumari
School of Media Studies and Humanities, Manav Rachna International Institute of Research and Studies, Faridabad, Haryana, India
e-mail: rahuljoshi.785@gmail.com

P. K. Prabhakar and W. Leal Filho (eds.), *Preserving Health, Preserving Earth*, World Sustainability Series, https://doi.org/10.1007/978-3-031-60545-1_16

many people who need to be more cautious about the research that claims to show the effectiveness of AI, as well as the many possible negative impacts of AI. The study of AI ethics has emerged as a dynamic area. However, it is still in its early stages, needing more consensus on its scope and the appropriate approach to explore. "The United Nations (UN) established 17 Sustainable Development Goals (SDGs) in 2015 to accomplish them by 2030" (Nations 2015). These include environmental, social, and economic objectives. With the growing presence of AI in today's cultures, the SDGs provide a valuable framework for examining and classifying the possible advantages and disadvantages it generates. Here, the SDGs are used as a framework to explore the broad impacts of AI on sustainability matters, and a notable aspect of the article is its demonstration of the SDGs' usefulness in assessing the ethical implications of AI.

In this perspective, AI is seen as a technology that is not neutral and apart from its surroundings. This involves seeing AI as an integral component of a socio-technical system, including diverse structures and economic and political systems, rather than examining this technology in isolation (Barley 2020). Contemporary artificial intelligence is closely connected to the limited availability of ever-larger volumes of data and the necessary computational capacity to use them. The concentration of power in a few governments and firms in the field of AI raises significant concerns about the possible adverse effects of AI on the Sustainable Development Goals. Artificial Intelligence can address several challenges and execute various jobs, making it a versatile technology. AI has several applications that contribute to the betterment of society. They cover all industries, and new ones are constantly emerging (Floridi et al. 2021a). Nevertheless, it is well-recognized that AI might be excessively used or employed for immoral objectives (King et al. 2020). Various stakeholders, from commercial enterprises to national governments, advocate incorporating ethical standards into AI applications as part of their national agendas. The use of AI for social benefit must align with the ethical principles that govern the overall design and assessment of AI systems. Specifically, the implementation of the concept of goodwill is crucial. The principle asserts that AI should benefit individuals and the environment. AI for social good should strive to provide ecologically and socially sustainable results.

2 How AI Works in SDGs

The United Nations introduced the SDGs in the 2015 publication "Transforming our World: The 2030 Agenda for Sustainable Development" (Nations 2015). The 17 objectives were a progression of the Millennium Development Objectives (MDG), a framework of 8 goals set in 2000 to achieve them by 2015 (Sachs 2012). The SDG framework is intricately linked to a diverse array of human rights. However, they remain separate since the SDGs prioritize individuals, the environment, economic well-being, peacefulness, and collaboration. The Sustainable Development Goals are characterized by their ambitious nature and extensive scope (Pekmezovic 2019). These are referred to as stretch objectives, which may seem very difficult to achieve but are nevertheless pursued to encourage and promote innovative and pioneering techniques and attempts to make progress (Gabriel and Gauri 2019). The SDGs' lack of vagueness and practicality are advantageous rather than drawbacks.

An inherent critique of the SDGs is the possible divergence between addressing climate change and achieving the other goals. Nerini et al. (2019), demonstrate the potential impact of climate change on achieving 16 Sustainable Development Goals while acknowledging that efforts to address climate change might hinder the progress of 12 other SDGs. The interconnections among the different objectives are crucial, and this forms the essence of the classification and examination of AI in its context, focusing on the direct and indirect impacts of AI on these objectives. The top-level objectives of the SDGs are often classified into more comprehensive categories. The United Nations proposes a classification of sustainability into three dimensions: economics, society, and environment. One drawback of this category is that politics is primarily divisive. The categorization of the Sustainable Development Goals has been based on the framework of ESG (environment, social, governance), which is often used in finance and investment (Berenberg 2018). Highlighting the political aspect of sustainability and the SDGs is crucial for making substantial advancements in fulfilling the SDGs, as shown in this essay. Furthermore, it is essential to understand and mitigate the possible adverse consequences of AI on the Sustainable Development Goals.

In recent years, the UNDP and other organizations have devised a strategy to use various types of intelligence to support the SDGs. This involves using information collected by sensors, satellites, and mobile phones and employing open innovation approaches to access novel ideas. The job of inventing SDGs in each area may be broken further into four essential components:

1. ***Problem comprehension: comprehension***: To effectively conclude, it is crucial to use a diverse array of sources, including both empirical evidence and factual material. Artificial intelligence is vital in identifying patterns like pests, diseases, and migration. "Some innovative models integrate collective intelligence, artificial intelligence, and data. These examples encompass various initiatives such as Action Insight Data in Uganda, the LICCI project, which aims to harness implicit knowledge from farmers to enhance climate change models, and other endeavors addressing the broader challenge of natural language processing (NLP) in dialects, minority languages, and indigenous languages in non-literate societies" (Reyes-García et al. 2020). This may involve adapting BERT and BART models to accommodate different languages. A further instance of conflict between people and wildlife is the use of artificial intelligence to analyze remote sensing data and the geographical and chronological attributes of crop raiding. This is done to forecast and chart the likelihood of such conflicts occurring.
2. ***Request for a solution***: The central concept is to seek ideas and solutions from various sources, including operations, new businesses, societies impacted by issues, innovators, and other sectors. This can be achieved by utilizing tools such as open technology platforms, challenge prizes, pursuit and suggestion functions, and novel forms of citizen science to discover a more diverse range

of solutions. "We Farm is a commendable illustration of connecting more than a million farmers in East Africa." It enables people to submit their issues via SMS, use artificial intelligence to identify possible individuals within the community who might provide solutions, and then disseminate this information (Omulo and Kumeh 2020). Collective thinking and computational intelligence mutually reinforce one another in this manner.

3. ***Judgment making***: Subsequently, there will be an increased reliance on collective intelligence to inform choices and facilitate the implementation of policies. An instance of this is the use of Polis AI technologies in the context of democracies to steer discussions towards achieving agreement.
4. ***Acquiring knowledge or skills***: Continuing learning is necessary to comprehend trends and use evidence sources, such as Microsoft Graph, 'What Works centres,' and platforms for review and learning among peers.

The following section illustrates the function of Artificial Intelligence in each of the 17 Sustainable Development Goals.

1. ***No Poverty (SDG-1)***: Artificial intelligence systems may detect those living below the poverty threshold and provide specific aid, such as nourishment, medical care, water, and electricity provisions.
2. ***Zero Hunger (SDG-2)***: Artificial intelligence can improve food distribution, minimize food waste, and raise agricultural output using precision farming methods. It can forecast agricultural production, assess soil conditions, and suggest the most suitable periods for planting.
3. ***Ensure healthy lives and Promote well-being (SDG-3)***: Artificial intelligence facilitates medical diagnostics, reduces medication development, and enables personalized treatment programs. It can analyze extensive healthcare information to detect disease patterns and enhance healthcare service delivery. Furthermore, using diverse self-technologies and self-tracking methods might facilitate individuals in adopting healthier lifestyles. AI is used in therapy to enhance the mental well-being of those lacking access to human therapists. Nevertheless, self-tracking and quantifying one's own experiences may be less beneficial. Moreover, social media and the other domains where AI is prominent promote mental unwellness instead of enhancing it, while the data remains inconclusive.
4. ***Quality Education (SDG-4)***: AI-driven educational solutions can customize learning experiences, adjust to individual requirements, and close educational disparities. Education is enhanced by virtual instructors, language translation, and interactive material, which improve accessibility. Delivering education in even the most isolated regions of the globe has become much more cost-effective and highly efficient. However, this necessitates the availability of cost-effective Intelligent Tutoring Systems (ITSs) designed and developed to cater to diverse languages and disciplines. Suppose the corporations prioritize their domestic markets, which are very lucrative, and concentrate on building these systems. In that case, the consequences will result in a rise in educational disparities rather than the desired outcome of reducing them.

5. ***Gender Equality (SDG-5)***: Artificial intelligence may mitigate gender prejudice by advocating for equitable recruitment procedures, ensuring equal remuneration, and fostering impartial decision-making. The system can analyze data linked to gender, detect any discrepancies, and provide recommendations for necessary measures to address them.
6. ***Clean Water and Sanitation (SDG-6)***: Artificial intelligence can forecast water availability, supervise water quality, and enhance the efficiency of water distribution networks. It facilitates the early identification of waterborne illnesses and enables effective water resource management.
7. ***Affordable and Clean Energy (SDG-7)***: Artificial intelligence enhances the efficiency of energy networks, forecasts energy use, and incorporates sustainable energy sources. Smart grids, demand response, and energy-efficient technologies all play a role in promoting sustainable energy use.
8. **Encourage accessible, sustainable economic growth and decent work for everyone:** *(SDG-8)*: The use of AI-driven automation has the potential to increase productivity and provide novel employment prospects and better labor conditions. It facilitates economic expansion by optimizing procedures and promoting creativity.
9. ***Industry, Innovation, and Infrastructure (SDG-9)***: Artificial intelligence expedites research, development, and innovation across many industries. It facilitates the process of strategizing, upkeeping, and enhancing infrastructure.
10. ***Reduced Inequalities (SDG-10)***: Artificial intelligence can detect discrepancies and provide strategies to mitigate differences. It fosters inclusive decision-making and fair distribution of resources. Implementing AI can potentially worsen disparities between the affluent and underprivileged in several domains, and the risk of AI impeding the achievement of Sustainable Development Goal 10 is significant. This phenomenon is seen at national and international levels since disparities within countries (such as the United States, where major technology companies are located) and worldwide disparities are unlikely to be resolved by artificial intelligence.
11. ***Sustainable Cities and Communities (SDG-11)***: Artificial intelligence facilitates the implementation of intelligent urban solutions, such as optimizing traffic flow, minimizing waste generation, and enhancing energy conservation. It enhances the durability and quality of urban settings. AI may play a significant role in ensuring safety via monitoring and predictive policing. Nevertheless, many factions within the AI ethics community express considerable apprehension over the "prison-to-tech pipeline." They consistently advocate for the elimination of technologies like face recognition, which they claim sustains racism and other negative prejudices. The glossy facade of advanced AI technology now conceals these biases. Furthermore, there is no evidence to substantiate the claim that AI promotes inclusive communities or directly enhances resilience and sustainability broadly. In contrast, the more sinister cultures of the present-day world actively use surveillance and artificial intelligence to exert control over their inhabitants akin to absolute dominance. This development is incompatible with SDG11 in every aspect. Although governments can use AI to control

their populace, it is also linked to heightened polarization, which may hinder the establishment of secure and inclusive communities.

12. ***Responsible Consumption and Production (SDG-12)***: Artificial intelligence improves supply chains, minimizes waste, and promotes the efficiency of resources. It facilitates the implementation of circular economy principles and promotes sustainable manufacturing.
13. ***Climate Action (SDG-13)***: AI models predict climate change impacts, monitor deforestation, and optimize renewable energy deployment. It aids in achieving emission reduction targets and climate resilience. However, Big Tech has yet to show a solid commitment to encouraging a shift towards environmental sustainability. However, efforts should be made to create eco-friendly information technologies, as individuals progressively recognize that widespread use and manufacturing of technology incur significant ecological expenses. AI and Big Tech are integral components of an economic framework where traditional expansion is the primary motivator. AI must be disentangled from this system to fully recognize AI's substantial beneficial impact on addressing climate change.
14. ***Life Below Water (SDG-14)***: Artificial intelligence protects aquatic life by analyzing ocean data, identifying illegal fishing, and preserving marine ecosystems. AI may facilitate the dissemination of information and promote cooperation across different entities, including governments, academics, civil society, and the corporate sector, to accomplish SDG 14. Nevertheless, artificial intelligence presents significant hazards and difficulties, including ethical, legal, and societal consequences, concerns around data accuracy and accessibility and the ability of humans to adapt and be included. Hence, it is crucial to guarantee the responsible, transparent, and human-rights-based development and use of AI by the principles and values of the United Nations.
15. ***Life on Land (SDG-15)***: Artificial intelligence helps manage animal habitats, prevent deforestation, and monitor biodiversity. The initiatives to conserve land and sustainably utilize land are supported.
16. ***Promote Peace, Justice, and Strong Institutions (SDG-16)***: Artificial intelligence may strengthen law enforcement, identify fraud, and enhance legal procedures. It helps ensure that institutions are accountable and contribute to transparent government. One potential approach to promoting democracy via AI would be exploring how an AI technocracy may ironically contribute to a rejuvenated democratic system. Implementing these remedies would require significant changes to our political systems. However, there needs to be more evidence to support their effectiveness or to suggest that they would not result in critical adverse outcomes. Regarding the negative consequences, it is widely argued that AI exacerbates polarization via processes like filter bubbles and echo chambers. Artificial intelligence-driven social media platforms are also associated with disseminating false information. Moreover, AI facilitates the more effective tampering or influencing of persons.
17. ***Partnerships for the Goals (SDG-17)***: Using data analysis, linking stakeholders, and promoting information exchange, artificial intelligence helps to

enhance cooperation. International collaborations for sustainable development are strengthened as a result.

3 Related Work

Following the establishment of the Sustainable Development Goals and initiatives like AI4Good, people have examined the precise ramifications of artificial intelligence. Chui et al. (2018), advocated for a comprehensive and impartial evidence-based examination of the possible advantages of AI, which is also connected to the Sustainable Development Goals. Other recent examples of research related to the aim of this study include the study conducted by Di Vaio et al. (2020), which primarily examines SDG12 and sustainable business models, and the research carried out by Khakurel et al. (2018), which extensively investigates the technical aspects of AI and the SDG framework. A paradigm was presented to assess the capacity of AI to provide value across the three dimensions of sustainability: economic, environmental, and social. In addition, they deliberated on the difficulties and prospects associated with AI-powered sustainable business models, including concerns around data accuracy, privacy, and ethical considerations. Previous studies have investigated the sustainability of artificial intelligence without explicitly connecting it to Sustainable Development objectives or by just focusing on specific goals, as shown by Yigitcanlar and Cugurullo (2020). Nevertheless, different people have tackled comparable inquiries without directly engaging in the SDG framework. Floridi et al. (2021b) proposes an ethical framework known as "AI4People," which is closely linked to considerable research on responsible AI (Dignum 2019).

In their work, Gill and Germann (2022) put up a theoretical and prescriptive framework for the regulation of artificial intelligence in a worldwide digital environment that promotes the Sustainable Development Goals. The researchers highlighted four crucial aspects of AI governance: ethical, legal, social, and technological. In addition, they proposed certain principles and processes for governing AI, including protecting human rights, involving multiple stakeholders, and establishing accountability. Visvizi (2022) investigated the potential of AI in assisting companies in addressing common obstacles to achieving SDG-related benchmarks. These problems include disruptions in production and supply chains, inventory management, budget planning, and personnel management. He contended that artificial intelligence has the potential to bolster enterprises' ability to withstand and recover from challenges, improve operational effectiveness, foster creativity, and make valuable contributions toward achieving the Sustainable Development Goals. While Thaler and Sunstein (2009), advocate encouraging it to be used for positive purposes, it is highly improbable that individuals with alternative motives will refrain from employing these techniques to advance their self-interest at the expense of other citizens, consumers, and individuals. The potential consequences of adverse AI influence on SDG16, such as heightened polarization and favoritism toward specific groups, can have detrimental effects on multiple SDGs, including SDG1, SDG2, and, notably, the significant SDG17.

4 Research Deficiencies on the Role of AI in SGDs

Investing in AI safety research becomes increasingly crucial as we continue implementing AI applications, such as "autonomous vehicles, AI-powered healthcare solutions, and smart electrical grids." (Bonnefon et al. 2016; De Fauw et al. 2018). This research is necessary to ensure these systems remain solid and beneficial and prevent malfunctions or hacking incidents. An essential area of study for ensuring the safe integration of AI is comprehending catastrophic events that may arise from systemic failures in AI technology. Specifically, the integration of AI in the financial industry has raised concerns, as highlighted in a recent study by the World Economic Forum (WEF). Hence, it is crucial to enhance consciousness about the dangers linked to potential malfunctions of AI systems in an increasingly reliant society on this technology (Uykur 2018). In addition, while we discovered a multitude of studies indicating that AI has the potential to facilitate various SDG targets and indicators, a substantial portion of these studies were carried out in controlled laboratory settings, relying on restricted datasets, or utilizing prototypes (Gandhi et al. 2017). Therefore, using this knowledge to assess the consequences often proves difficult. This is especially true when evaluating the influence of AI on larger scales, including both temporal and geographical dimensions. We recognize that doing controlled experimental trials to determine the real-world effects of AI may only provide a limited view since AI tools are customized for the context of the trial. Nevertheless, owing to the dynamic nature of society and the influence of other elements, such as non-AI technical advancements, the criteria for AI are also subject to change.

This creates a reciprocal relationship where society and AI continuously interact and influence each other. Another overlooked element in current literature is the society's capacity to adapt and withstand changes brought forth by AI technology. Hence, it is essential to develop innovative approaches to evaluate the ramifications of emerging technologies in terms of effectiveness, morality, and environmental responsibility before implementing extensive AI initiatives. Research focused on understanding the causes of AI system failures and implementing methods that mix human and machine analysis is crucial for developing responsible AI technology, considering the significant risks of such shortcomings (Nushi et al. 2018). While our research uncovered more published data supporting AI as a facilitator rather than a hindrance to the SDGs, it is crucial to consider two significant factors. Initially, it is anticipated that self-interest would influence the AI research community and businesses to favor the publication of favorable outcomes. Furthermore, identifying the adverse effects of AI may need extended research, and as previously said, there is a scarcity of recognized assessment tools to do this. The inclination to prioritize the publication of favorable outcomes is especially evident in the Sustainable

Development Goals related to the Environment category. In this context, machine-learning algorithms can provide optimal solutions by considering many characteristics to determine the most suitable locations to include in conservation networks. Nevertheless, while the solutions may be mathematically optimal within a specific range of chosen parameters, further investigation is required to evaluate the enduring effects of these algorithms on equity and fairness. This is necessary due to unseen factors that could potentially influence the outcomes. Regarding the second argument mentioned earlier, the AI initiatives with the most significant potential for generating profit will probably get funding. Without regulation, research on artificial intelligence is anticipated to prioritize AI applications that align with funding and commercial objectives. This might lead to a rise in inequality. If the predicted economic impact of technology based on AI is not considerable, they may not be prioritized, even if they can contribute to Sustainable Development Goals.

Nevertheless, a lack of study exists in the practical implementation of these systems in real-life scenarios, particularly in the context of governmental use. Institutions face many obstacles when incorporating AI systems into their decision-making processes, such as the need to establish robust cybersecurity measures and safeguard the privacy of individuals and their data. Both features have ramifications for human rights, including monitoring, tracking, communication, data storage and operations automation without stringent ethical standards. Addressing these deficiencies is crucial to guarantee the usefulness and feasibility of AI technology for governments. Understanding the long-term implications of AI and regulating its usage to mitigate any bias are contingent upon meeting this criterion (Courtland 2018).

Many AI systems used in agriculture, such as those that automate harvesting or optimize scheduling, are primarily found in affluent countries. Our literature search found a few instances where AI technologies address SDG-related problems in countries with little AI research. Furthermore, if AI technologies are specifically created and used in highly sophisticated settings, they may worsen issues in less affluent countries, particularly regarding food production. This discovery raises a significant apprehension that advancements in AI technology might exacerbate disparities across and within nations in ways that undermine the overarching objectives of the SDGs. We urge academics and funding to prioritize the creation and advancement of AI solutions that address specific issues in economically disadvantaged countries and areas. Projects engaging in such work should guarantee that solutions are not just replicated from technologically advanced countries. Instead, it is necessary to construct them with a profound understanding of the specific place or culture to enhance the probability of acceptance and achievement.

5 AI Framework for SDGs and Transitioning Research and Development Environment

Historically, AI has been driven mainly by military research, surveillance efforts such as the NSA, academic research, and little commercial funding. During the 2010s, there was a significant increase in the magnitude of commercial research. In 1960, the United States Department of Defence supported one-third of all research and development activities worldwide. This advancement facilitated the use of subsequent technology such as microprocessors, GPS, touch displays, space launches, and satellites. In 2016, the same number was 3.6%. The Research and development investment of the top 5 technology companies in the United States is now tenfold more than that of the top 5 Defence companies. In 2019, the top five technology businesses in the USA allocated $106 billion to research and development (R&D), surpassing the total R&D expenditures of all European Union governments. These companies, including Alibaba, Tencent, and Huawei, have effectively gained significant influence in the worldwide regulation of several technological domains.

In the mid-2015s, there was a noticeable increase in interest in using artificial intelligence for positive purposes and specifically for achieving the Sustainable Development Goals. This was accompanied by the organization of several conferences, programmes, and funds dedicated to this cause. These studies examined the use of AI in various domains such as pest management, refugee job matching, personalized education, and healthcare. Microsoft, for instance, allocated several hundred million dollars annually to programmes focused on 'AI for Good.' According to a recent study, AI was shown to have potential advantages in 42 out of the 60 Sustainable Development Goal objectives, which accounts for 70% of the targets. However, adverse effects were seen in 20 targets, representing 33% of the total. Nevertheless, this research focused on the possible consequences rather than the existing ones. Furthermore, ongoing 'AI for Good' initiatives are promptly scrutinized for their limited influence, lack of systemic impact, and their peripheral role in the constant pursuit of developing AI for commercial or military/security objectives. In addition, they often need to pay more attention to the broader strategic considerations related to data.

6 AI, SDG, Research and Development: An Indian Scenario

According to the Stanford AI Index 2021, India has attained the top position in "AI Adoption by Organizations". It ranks eighth in the "Number of newly funded AI companies" from 2013 to 2021. The Indian government has implemented concrete measures to promote the responsible use of artificial intelligence and instill public trust in applying this technology. The government has undertaken these projects with the central focus of "AI for All" as a critical component of its National Strategy for AI. The user's text is a single character, a hyphen followed by an exclamation mark. Artificial intelligence can potentially affect many Sustainable Development Goals, such as poverty reduction, health, education, gender equality, climate action, etc. The following instances of artificial intelligence applications for the Sustainable Development Goals in India are:

1. MyGov Helpdesk is an AI-powered chatbot on WhatsApp that provides information and facilitates Covid-related vaccines.
2. Bhashini, a natural language translation system, enhances the accessibility of the internet and digital governance.
3. The government app Umang has just launched a voice-activated chatbot that may be used for a range of government tasks.
4. The national AI site of India, known as "INDIA" ai, has published Forecast 2023, which is a collection of 23 articles discussing the applications of artificial intelligence in different sectors such as agriculture, healthcare, education, manufacturing, and more.

To enhance India's AI ecosystem, it is vital to reinforce its six foundational components:

1. ***Academics***: Indian universities serve as centers for innovation and are prominent providers of skilled individuals for technology companies. Universities should integrate artificial intelligence into their academic programs, enhance the teaching and learning of AI, and actively support and nurture AI startups, all while ensuring that these initiatives align with national objectives.
2. ***New Startups***: Startups often need help with financing and technical and human resources. As a result, they prefer using pre-existing AI models rather than developing new ones. Hence, it is imperative to provide incentives to Indian companies for the development of indigenous AI applications.
3. ***Policy makers and governmental institutions***: It is necessary to determine the most appropriate policymakers and regulatory organizations to establish rules about AI governance and guarantee they possess a strong understanding and knowledge of AI. The efficacy of established precedents for selecting regulators may be compromised due to the industry's dynamic and frequently changing nature. Instead, there is a need for versatile teams with a flexible attitude to develop and oversee rules.
4. ***Multinational corporations***: India's leading investors in artificial intelligence are technology multinationals and private entities. Hence, they wield significant influence in shaping the regulatory framework for AI in the nation, and their opinions and viewpoints should be included in formulating policies.
5. ***Public–private partnerships (PPPs)***: Further examination of PPPs might be undertaken to enhance their role in facilitating AI governance. Partnerships between governments, the corporate sector, and civil society organizations (CSOs) via public–private partnerships enable the implementation of industry self-regulation measures, the establishment of AI ethical advisory boards, and the joint development of AI policies. Additionally, these collaborations could

enhance the adoption of AI by the public sector. India can benefit from training data partnerships since AI algorithms, which typically serve as decision-makers, exhibit biases influenced by the training datasets used.

6. ***AI regulations***: Another crucial element in establishing a robust AI ecosystem is bolstering the safeguards provided to emerging developers via new rules, especially those with little financial resources. However, extensive teams have historically supported significant AI models and platforms, and groundbreaking advancements have mainly originated from smaller organizations. Hence, AI legislation should prioritize major applications and primarily address giant corporations with the capacity and magnitude to inflict extensive damage.

India should carefully deliberate and establish suitable regulations for artificial intelligence, leveraging its current domestic technology policies and integrating pertinent elements from global efforts like the OECD AI Principles, the EU AI Act, and the G7 Guiding Principles (Galindo et al. 2021; Veale and Zuiderveen Borgesius 2021; Snoswell 2023). This will further guarantee the continuous organic evolution of India's AI ecosystem, avoiding the possible stagnation that may arise from stringent rules. Conversely, several experts argue that regulations about the swiftly advancing field of AI would need ongoing adjustments or substitutions, therefore necessitating an emphasis on adaptability in these legislations. This situation is like the phone issue, which resulted in extensive disputes among operators, politicians, and regulators. These problems were only settled after the technology became fully developed.

7 The Disparity Between Research and Development Efforts and the Sustainable Development Goals

A significant disparity between worldwide research efforts and the Sustainable Development Goals is an essential factor. Jeff Hammerbacher, the former leader of statistics at Facebook, famously said, "The most brilliant individuals of my era are focused on devising strategies to encourage individuals to interact with advertisements." However, this is indicative of a somewhat more extensive trend. The recent findings from the STRINGS project reveal the discrepancy and lack of synchronization within individual countries. "There needs to be more distribution of research and development activities, with most of such work being carried out in middle and high-income countries. Specifically, 90% of the science, technology, and innovation work related to sustainable development goals is published or patented in high- and upper-middle-income countries" (Ciarli et al. 2022).

Furthermore, there are disparities within each field, such as the well-documented bias in the pharmaceutical industry toward drugs that require repeat prescriptions and are more prevalent in wealthy countries. However, this bias has been partially addressed in recent decades. Additionally, there is a continued preference for R&D in large corporations and universities, while alternative and cost-effective models, like those observed in Shenzhen, India, and East Africa, need to be more noticed. By using STRINGS data, it is now possible to examine these misalignments on a country-specific basis. Furthermore, investing exhibits a significant degree of concentration. The top five firms with the most influential intellectual property (IP) portfolios in artificial intelligence have 14% of the AI-related IP portfolio, mainly patents.

These same corporations are also leading research and development investors. Some industries have tried to address these disparities by forming partnerships and alliances and consolidating budgets. CGIAR, an organization focused on agriculture, is a remarkable illustration of longevity, having existed for over 50 years. GAVI and its derivatives, such as COVAX, endeavored to bring about a comparable transformation in the pharmaceutical industry. They acknowledged that the current research and development methods and economic models could have helped create feasible solutions for developing nations. The digital sectors must be more active in developing something equivalent instead of opting for superficial endeavours. AI has also shown this characteristic, prompting a preference for localized solutions rather than comprehensive movements towards research and development.

Nevertheless, in the realm of AI, like other domains, there needs to be more comprehensive data about the correlation between robust investment practices and achieving Sustainable Development Goals. Moreover, the transparency of commercial operations is progressively diminishing instead of improving. There has been an increasing interest in alternate routes and orientations in several disciplines, such as altering the whole direction of science, technology, and innovation (STI) away from dependence on fossil fuels or large-scale farming. More substantial research is needed on alternative approaches to the Fourth Industrial Revolution (4IR) and Artificial Intelligence. Instead, the focus has primarily been on regulations and limitations regarding cross-border data traffic. "This has diverted attention from exploring different paths, such as government-controlled systems utilizing social credit systems, US commercial models centered around data harvesting or models that promote citizen ownership and control of data, and increased algorithms' transparency."

In order to tackle these issues, a prerequisite is:

1. ***Enhanced information and evaluation***: It refers to accurately documenting the present investment patterns, the specific areas allocated, and identifying any significant deficiencies or gaps.
2. ***Clusters of financiers***: The objective is to intentionally synchronize financing and assistance in domains such as AI for public health, AI for education, or AI for agriculture. Here, valuable models facilitate enhanced cooperation between funders and practitioners, including exchanging shared data.
3. ***Shared expenditures***: It has been shown to significantly enhance the effectiveness of both food and vaccination initiatives. This is especially true when public, charitable, and commercial financing can be combined and coordinated. Once again, AI needs to catch up to the most effective methods.
4. ***Shared governance***: It will increasingly need the establishment of norms and governance structures at regional or global scales. Although there have been

significant advancements via efforts like the GPAI, there has been minimal actual implementation in this regard, despite the occasional mention of international charters or rights.

8 Transitioning from a Supply Driven Approach to Demand Driven One

These strategies seek to reverse the conventional approach to technology completely. Frequently, emerging technologies such as machine learning (ML) or blockchain actively pursue potential applications. Their designers often grow obsessed with the methodology. However, an alternative strategy involves determining the specific intelligence required by various actors, such as organizations, communities, or organizations. This approach then proceeds by assembling the necessary components, likely to encompass data, comprehension, estimation, and verification of effective strategies, peer knowledge, and other relevant factors. This results in the emergence of intelligence assemblies or collective intelligence methodologies. The problem is that only a limited number of organizations consistently curate intelligence of this kind. However, efforts are already being made to use this approach to oceans. This involves integrating digital replicas, predictions, and public participation to develop and execute effective remedies. Crucial assets like Copernicus are also starting to shift in this trajectory. When faced with these situations, it is essential to go from pondering 'How can AI contribute to the SDGs?' to contemplate 'How can we effectively harness many forms of intelligence to contribute to the SDGs, and how can AI fit into this larger endeavour?'.

9 Intersection of Global Warming, Statistics, and AI

Data and modeling have enabled us to accurately quantify the extent of climate change. Over many years, the meticulous gathering of meteorological data and sea temperatures has provided valuable input for models that assess, forecast, and elucidate the impact of human activities on our climate. However, the specific contribution of data and various models in resolving the problem still needs to be determined. They can play a significant role, if we successfully implement substantial changes in data management, moving away from the prevailing commercial proprietary paradigms dominating our economies. Utilizing digital technologies may have a positive impact on the environment. For instance, using Zoom for work instead of traveling can significantly reduce emissions. However, it accounts for just fifty percent of the narrative. Digital and Internet activity contributes around 3.7% of emissions, equivalent to the emissions generated by air travel. Data centers in the United States contribute around 2% to the overall power consumption. The statistics about AI could be more favourable. Based on an estimate, training a machine learning algorithm consumes a remarkable amount of 626 pounds of carbon dioxide (CO2), five times a vehicle's total fuel consumption over its lifespan and 60 times more than a transatlantic journey. Several projections anticipate a significant increase in these emission levels (Stein 2020). Blockchain, the underlying technology of Bitcoin, is notorious for its excessive energy consumption and detrimental environmental effects: Bitcoin's carbon footprint is almost like that of New Zealand.

However, artificial intelligence may be used to reduce carbon emissions, having the most significant potential in buildings, power, transportation, and agriculture. Electricity is the most sophisticated. It constitutes around 25% of greenhouse gas emissions, and many large-scale corporations regulate it. They have discovered that AI is highly suitable for optimizing intricate electrical networks, including complicated inputs such as the intermittent impact of renewable energy sources like wind power and intricate use patterns (Victor 2019). Google DeepMind undertook a study to enhance wind pattern prediction to optimize power generation in the United States. AI can improve traffic management and optimize agricultural practices by monitoring and forecasting weather conditions and insect behavior. Nevertheless, the digital sectors have been reluctant to participate actively. Apple has a well-known lack of concern for environmental issues for a long time. Similarly to other technology companies, they contributed to large amounts of electronic garbage due to the need to constantly upgrade to the newest iPhone, iPad, or Galaxy device. Facebook has refrained chiefly from addressing environmental concerns, except for delayed initiatives such as its recent endorsement of a Climate Science Data Center.

Jeff Bezos, the CEO of Amazon, has pledged $10 billion in funding for environmental and mental health organizations in 2020. This is a significant commitment, considering his prior lack of public engagement on these matters. Bill Gates has shown a higher level of involvement, but with the conventional perspective of the technology industry, innovation alone can resolve the issues at hand. Microsoft has a superior track record, which includes taking significant measures to address past carbon emissions. The analogy to the realm of investing is remarkable. Under the influence of influential personalities such as Larry Fink from BlackRock and former Bank of England Governor Mark Carney, significant investors have reluctantly begun to assess environmental implications and carbon emissions and alter their investment strategies. Despite the ongoing influx of substantial investments into carbon-intensive sectors, the focus of the discussion has changed. What are the necessary conditions for data and AI to be more prominent in achieving net zero? The primary resource that powers AI is data, which is the first challenge. Most data that provide information on the operations of electricity networks, buildings, or transit systems are owned by businesses and kept confidential. To fully capitalize on its potential and educate future generations of AI, it is necessary to open, standardize, and share it. Significant efforts are now being made in this area, including the development of Carbon Tracker, a dashboard initiative that utilizes satellite data to represent coal emissions visually, and the Icebreaker One project, which seeks to empower investors by providing them with the ability to monitor the overall carbon footprint resulting from their choices. However, these initiatives need to be expanded in scope and more cohesion. Ultimately, political will is required to make this data accessible to the public. If the platform were structured as a common, it might be used to commission artificial intelligence systems to assist whole cities or nations reduce their emissions. Abundant ideas are available, such as those presented in a recent comprehensive analysis of machine learning applications for addressing climate change, which provides a glimpse into potential possibilities (Rolnick et al. 2022). Additional sites, such as the climate change and AI wiki, also provide valuable insights. However, this brings us to the next obstacle: data or algorithm ownership or governance. Here, there remains a conspicuous deficiency. In the next decade, we will likely need novel and diverse forms of data trust to manage and distribute data. This may include establishing public–private partnerships in areas such as transportation and energy, where data like intelligent meter readings are collected.

Additionally, there may be a need for entirely public entities dedicated to research purposes. The absence of such establishments is a contributing cause to the failure of several innovative city initiatives, such as Google's Sidewalk Labs in Toronto and Replica in Portland, since they need help convincing the public of their reliability. Additional regulations will also be necessary. As previously mentioned, there is a rising interest in establishing worldwide declarations of rights concerning artificial intelligence. The European Union is now developing an extensive framework to regulate artificial intelligence by conducting risk assessments. This framework includes prohibitions on some applications of AI, such as face recognition and credit scoring. In parallel, China is also implementing its regulations for AI. An alternative approach would be to mandate data sharing and grant customers the authority to share their data with a third party as the standard practice. As a prerequisite for obtaining a public license, any private entity, such as a provider of a 5G network, Uber, electricity supply, or a supermarket with local planning permission, would be obliged to furnish pertinent data in a standardized, anonymized, and machine -readable format. These are a limited number of necessary modifications to develop the digital aspect of strategies aimed at reaching net zero emissions and attaining other environmentally Sustainable Development Goals.

10 Engagement of Energy Efficiency and AI

AI plays a crucial role in energy systems, encompassing various energy practices such as electricity, hydrogen-based combustibility, wind, nuclear power, solar energy, and other renewable sources, as well as the perspective of end-use in electronic devices, heating, and transportation production, business, and other sectors. Artificial intelligence can support the establishment of various energy sources and localizing energy production for infrastructure planning. It can also assist in predicting energy consumption and implementing intelligent control systems. AI approaches are used to enhance the creation, production, and optimization of energy-related elements, as well as to identify the most suitable materials and improve the safety of energy utilization. Artificial intelligence can potentially enhance the implementation of smart grids, hence improving the efficiency of both local and global power systems. Two aspects facilitate the implementation of an intelligent grid. Firstly, deploying contemporary IoT devices enables the increase of the amount and quality of data acquired from the network.

Furthermore, artificial intelligence can already efficiently analyze the vast amount of recorded data, enabling rapid decision-making that surpasses the capabilities of human operators (Ali and Choi 2020). "The electrical grid is a complex and dynamic system that operates under decentralized control, with several uncertainties. Incorporating renewable energy sources, such as solar and wind farms and electric and plug-in hybrid cars, introduces additional intricacy and difficulties to the many tiers of the power grid. Considerable efforts have been dedicated to advancing intelligent grid technology to harmonize the concerns of electric users, utilities, and environmentalists" (Venayagamoorthy 2009). The integration of real-time data from buildings and weather predictions, together with the implementation of intelligent systems, has the potential to accurately anticipate the optimal timing for heating and cooling operations. This significantly enhances the system's efficiency (Yan et al. 2021). Di Santo et al. (2018) argued, "Artificial intelligence can do active demand-side management for families in intelligent grids, including distributed solar photovoltaic power and energy storage." Thoughtful demand response has the potential to provide 185 gigawatts (GW) of flexibility to the electrical system, which is about equivalent to the current combined energy supply capacity of Australia and Italy (I. E. Agency 2017). "Experts worldwide agree that our future energy source should be cost-effective, environmentally friendly, and secure. Consequently, if employed with careful consideration, this will contribute to sustainable development via the provision of inexpensive and accessible energy, the reduction of greenhouse gas emissions, the optimization of grid operations, and the preservation of dependable power infrastructure" (Gielen et al. 2019).

11 The Relevant Connection of AI to Low Carbon Energy System

The shift towards a low-carbon future is a worldwide need. In our pursuit of achieving carbon neutrality, it is necessary for conventional power generation technologies, such as coal-fired plants and gas turbine combined cycles, to undergo adaptations to fulfill new functions within the developing power grid. Renewable energies such as solar and wind power are becoming more critical, requiring power energy systems that can effectively incorporate these fluctuating sources (Change 2014). This is the point at which artificial intelligence becomes involved. AI algorithms optimize the functioning of power energy systems, effectively dealing with the uncertainties associated with renewable energy sources.

The following are the fundamental applications of low energy carbon systems:

1. ***Renewable Energy Forecasting***: Artificial intelligence plays a crucial role in forecasting the availability of renewable energy. Methods such as neural

networks, time series analysis, and ensemble models improve the precision of results. These predictions enhance the power grid integration, energy storage planning, and energy consumption control from the demand side.

2. ***Smart Grid Management***: AI algorithms dynamically optimize the balance between supply and demand, considering weather conditions, load levels, and grid reliability. This reduces dependence on non-renewable energy sources. Artificial intelligence identifies deviations from standard patterns, averting grid malfunctions and guaranteeing continuous provision of electricity. An AI-powered load control system optimizes energy use during periods of high demand, alleviating power grid stress (Omitaomu and Nilu, 2021).
3. ***Energy Storage Optimization***: Artificial intelligence enhances the efficiency of determining the optimal timing for charging and discharging energy storage systems (ESS). It prolongs the duration of battery life and improves overall efficiency. Artificial intelligence calculates the most efficient combination of renewable energy sources and energy storage system components to provide a dependable power supply (Abdalla et al., 2021).
4. ***Carbon Emission Reduction***: Artificial intelligence detects and identifies the origins of emissions in industrial operations. It proposes more environmentally friendly options and oversees adherence. Artificial intelligence aids in the surveillance and enhancement of carbon capture systems.
5. ***Building Energy Management***: Artificial intelligence optimizes heating, ventilation, and air conditioning (HVAC) systems by considering occupancy, weather conditions, and energy costs. Artificial intelligence optimizes the use of lighting in both commercial and residential structures (Cheng and Lee 2019).
6. ***Economic Dispatch***: AI algorithms effectively address the economic dispatch issue by considering various restrictions, such as transmission losses and network balancing. The convergence speed and global optimization are enhanced by techniques such as the improved bat algorithm (IBA) (Chawla and Duhan 2015).
7. ***Expert Systems***: Artificial intelligence expert systems are used to identify and analyze malfunctions in power equipment, reducing the time that the equipment is out of service and lowering the expenses associated with maintenance. Artificial intelligence uses predictive algorithms to forecast equipment malfunctions, facilitating prompt maintenance interventions.
8. ***Plug-in Electric Vehicles (PEVs)***: Artificial intelligence enhances the performance of plug-in electric vehicle charging stations, guaranteeing optimal use, and maintaining the stability of the power grid. Artificial intelligence devises the most efficient paths for plug-in electric vehicles, considering the availability of charging stations and the prevailing traffic circumstances.
9. ***Microgrids and Decentralized Systems***: Artificial intelligence oversees the control and coordination of distributed energy resources (DERs) inside microgrids. Microgrids increase resilience in the face of power grid disruptions or natural calamities.
10. ***Energy Trading and Pricing***: Artificial intelligence models enhance the efficiency of energy trading in decentralized marketplaces. Artificial intelligence

uses predictive algorithms to forecast energy costs, therefore assisting producers and consumers in making informed decisions.

12 Open Challenges in Research and Development in AI and the SDGs

The potential of AI to advance the SDGs is significant. However, it is crucial to solve these obstacles to create an AI society that aligns with sustainable development objectives.

The following are some of the important challenges:

1. Utilizing artificial intelligence to enhance societal welfare by enhancing healthcare, education, and overall well-being, mitigating disparities, and advocating for justice and human rights.
2. Utilizing artificial intelligence to promote environmental sustainability by addressing climate change, preserving, and rehabilitating biodiversity, and effectively managing natural resources and waste.
3. Utilizing AI to promote economic growth by improving productivity, boosting innovation and competitiveness, generating employment opportunities and skill development, and encouraging entrepreneurship and social innovation.
4. Creating and using artificial intelligence to promote peace and security by avoiding and resolving conflicts, improving humanitarian aid and disaster relief efforts, and countering terrorism and cybercrime.
5. Creating and implementing artificial intelligence to govern and establish policies, focusing on improving the delivery of public services, engaging citizens, ensuring accountability, and facilitating decision-making and policy development based on solid evidence.
6. Creating and using artificial intelligence to support cultural and creative endeavors, such as safeguarding and advancing cultural heritage, diversity, and expression and fostering artistic and scientific innovation and exploration.
7. Creating and implementing artificial intelligence systems prioritizing ethics and values, including guaranteeing justice, transparency, explainability, and accountability, while upholding human dignity, autonomy, and rights.
8. Creating and implementing artificial intelligence to facilitate interaction and collaboration between humans and AI. This includes creating AI systems that prioritize human needs and are easy to use, promoting trust, cooperation, and the mutually beneficial relationship between humans and AI.
9. Creating and using artificial intelligence technology to improve learning and education by boosting educational results, increasing accessibility and quality, and promoting continuous learning and digital literacy.
10. Developing and applying AI for research and innovation, this includes pushing the boundaries of AI knowledge and technology and promoting multidisciplinary collaboration and participatory approaches in research and innovation.

To promote and expedite research and development in artificial intelligence and the sustainable development goals, several aspects and suggestions that may aid in this process are:

1. Enhancing the standardization and incorporation of artificial intelligence and the Sustainable Development Goals via the development and implementation of AI technologies that adhere to the ideals and principles of the SDGs. Additionally, AI is used to aid in achieving and tracking progress towards the SDGs.
2. The aim is to enhance the alignment and cooperation between artificial intelligence and the Sustainable Development Goals. This will be achieved by establishing and facilitating platforms, networks, and partnerships that unite many fields of study, industries, and individuals involved. The primary objectives of these collaborations are to exchange information, distribute resources, and collectively develop innovative solutions.
3. Enhance and facilitate platforms, networks, and partnerships that unite many fields of study, industries, and individuals. The primary objectives of these collaborations are to exchange information, distribute resources, and collectively develop innovative solutions.
4. To enhance the governance and regulation of AI and the SDGs, we will create and enforce policies, standards, and guidelines that guarantee the responsible and ethical development and deployment of AI and the SDGs. Additionally, we will establish mechanisms for oversight, accountability, and redress.
5. To enhance the influence and assessment of artificial intelligence and the Sustainable Development Goals, we aim to create and use techniques and resources for quantifying and evaluating the results and effects of AI and the SDGs. Additionally, we will provide feedback and opportunities for learning to enhance progress and innovation.

13 Conclusion

Artificial intelligence has immense promise, although this potential is often exaggerated. This chapter highlights the need to consider AI within a broader framework. It points out that prior studies on AI and the SDGs have consistently underestimated the possible adverse impacts of AI on the SDGs. Firstly, research has shown that while AI may have specific good impacts on different objectives, it is also an integral component of a system that concurrently undermines many of these outcomes. Furthermore, the chapter highlights the need to acknowledge that several proofs-of-concept about the promising potential of AI also inadvertently reveal how AI might be used in ways that result in adverse impacts on the SDGs. We must consider these elements to avoid overlooking the actual good possibilities of AI and instead fall into the domain of AI hype. “AI hype” often refers to exaggerating AI’s capabilities, a critical issue examined here. It is important to note that the limited accomplishments achieved in controlled experimental environments may only sometimes apply to real-life open systems. One method of generating excitement about AI is to examine the

SDGs as distinct and independent objectives without considering their significant interconnections. To accurately assess the significance of AI, it is crucial to include the indirect impacts of achieving specific key SDGs and keep the evaluation of the same factors distinct.

Furthermore, a comprehensive interpretation of the SDGs necessitates considering aspects of inclusion, sustainability, universality, inequality, and fairness. Adopting this approach significantly increases the difficulty of categorizing AI as a facilitator of several objectives. Furthermore, the excitement around AI may be exacerbated by disregarding the specific circumstances in which AI has been created, continues to be developed, and is now used. The control over AI is mainly concentrated within a limited group of entities, including corporations and nations, and its close association with the prevailing capitalist system is significant and cannot be disregarded. The contemporary period has seen substantial advancements and expansion, although it has also facilitated unprecedented disparity and deterioration of the environment. To fully understand the implications of AI, it is necessary to examine it within a broader framework. Such a study reveals that AI is closely linked to significant risks that threaten most SDGs. To dismiss this and contend that AI is a detached and impartial technology capable of rescuing the world is, at best, uninformed and, at worst, profoundly negligent and dangerous. These concerns emphasize the need for an impartial and thorough investigation into AI's ethical consequences and long-term viability. It is prudent to exercise caution when considering the study conducted inside the same system being discussed as the ultimate authority on the sustainability of AI.

References

Abdalla AN, Nazir MS, Tao H, Cao S, Ji R, Jiang M, Yao L (2021) Integration of energy storage system and renewable energy sources based on artificial intelligence: an overview. J Energy Storage 40:102811

Ali SS, Choi BJ (2020) State-of-the-art artificial intelligence techniques for distributed smart grids: a review. Electronics 9(6):1030

Barley SR (2020) Work and technological change. Oxford University Press, USA

Berenberg J (2018) Understanding the SDGs in sustainable investing

Bonnefon JF, Shariff A, Rahwan I (2016) The social dilemma of autonomous vehicles. Science 352(6293):1573–1576

Brown M, Woodhouse S, Sioshansi F (2019) Digitalization of energy. In: Consumer, prosumer, prosumager: how service innovations will disrupt the utility business model. pp 3–25

Change IC (2014) Mitigation of climate change. In: Contribution of working group III to the fifth assessment report of the intergovernmental panel on climate change, vol. 1454, p 147

Chawla M, Duhan M (2015) Bat algorithm: a survey of the state-of-the-art. Appl Artif Intell 29(6):617–634

Cheng CC, Lee D (2019) Artificial intelligence-assisted heating ventilation and air conditioning control and the unmet demand for sensors: Part 1. Problem formulation and the hypothesis. Sensors 19(5):1131

Chui M, Harryson M, Valley S, Manyika J, Roberts R (2018) Notes from the AI frontier applying AI for social good

Ciarli T, AlDoh A, Arora S, Arza V, Asinsten J, Assa J, Yegros A (2022) Changing directions: steering science, technology and innovation towards the sustainable development goals
Courtland R (2018) The bias detectives as machine learning infiltrates society, scientists grapple with how to make algorithms fair. Nature 558(7710):357–360
De Fauw J, Ledsam JR, Romera-Paredes B, Nikolov S, Tomasev N, Blackwell S, Ronneberger O (2018) Clinically applicable deep learning for diagnosis and referral in retinal disease. Nat Med 24(9):1342–1350
Di Santo KG, Di Santo SG, Monaro RM, Saidel MA (2018) Active demand side management for households in smart grids using optimization and artificial intelligence. Measurement 115:152–161
Di Vaio A, Palladino R, Hassan R, Escobar O (2020) Artificial intelligence and business models in the sustainable development goals perspective: a systematic literature review. J Bus Res 121:283–314
Dignum V (2019) Responsible artificial intelligence: how to develop and use AI in a responsible way, vol 2156. Springer, Cham
Floridi L, Cowls J, King TC, Taddeo M (2021) How to design AI for social good: seven essential factors. Ethics Govern Policies Artif Intell 125–151
Floridi L, Cowls J, Beltrametti M, Chatila R, Chazerand P, Dignum V, Vayena E (2021) An ethical framework for a good AI society: opportunities, risks, principles, and recommendations. In: Ethics, governance, and policies in artificial intelligence. pp 19–39
Fuso Nerini F, Sovacool B, Hughes N, Cozzi L, Cosgrave E, Howells M, Milligan B (2019) Connecting climate action with other sustainable development goals. Nat Sustain 2(8):674–680
Gabriel I, Gauri V (2019) Towards a new global narrative for the sustainable development goals. In: Sustainable development goals: harnessing business to achieve the SDGs through finance, technology, and law reform. pp 53–70
Galindo L, Perset K, Sheeka F (2021) An overview of national AI strategies and policies
Gandhi N, Armstrong LJ, Nandawadekar M (2017) Application of data mining techniques for predicting rice crop yield in semi-arid climatic zone of India. In: 2017 IEEE technological innovations in ICT for agriculture and rural development (TIAR). IEEE, pp 116–120
Gielen D, Boshell F, Saygin D, Bazilian MD, Wagner N, Gorini R (2019) The role of renewable energy in the global energy transformation. Energ Strat Rev 24:38–50
Gill AS, Germann S (2022) Conceptual and normative approaches to AI governance for a global digital ecosystem supportive of the UN sustainable development goals (SDGs). AI and Ethics 2(2):293–301
Khakurel J, Penzenstadler B, Porras J, Knutas A, Zhang W (2018) The rise of artificial intelligence under the lens of sustainability. Technologies 6(4):100
King TC, Aggarwal N, Taddeo M, Floridi L (2020) Artificial intelligence crime: an interdisciplinary analysis of foreseeable threats and solutions. Sci Eng Ethics 26:89–120
Nations U (2015) Transforming our world: the 2030 agenda for sustainable development. United Nations, Department of Economic and Social Affairs, New York
Nushi B, Kamar E, Horvitz E (2018) Towards accountable AI: hybrid human-machine analyses for characterizing system failure. In: Proceedings of the AAAI conference on human computation and crowdsourcing, vol. 6. pp 126–135
Omitaomu OA, Niu H (2021) Artificial intelligence techniques in smart grid: a survey. Smart Cities 4(2):548–568
Omulo G, Kumeh EM (2020) Farmer-to-farmer digital network as a strategy to strengthen agricultural performance in Kenya: a research note on 'Wefarm'platform. Technol Forecast Soc Chang 158:120120
Pekmezovic A (2019) The UN and goal setting: from the MDGs to the SDGs. In: Sustainable development goals: harnessing business to achieve the SDGs through finance, technology, and law reform. pp 17–35
Reyes-García V, Fernández-Llamazares Á, García-del-Amo D, Cabeza M (2020) Operationalizing local ecological knowledge in climate change research: Challenges and opportunities of citizen

science. In: Changing climate, changing worlds: local knowledge and the challenges of social and ecological change. pp 183–197
Rolnick D, Donti PL, Kaack LH, Kochanski K, Lacoste A, Sankaran K, Bengio Y (2022) Tackling climate change with machine learning. ACM Comp Surv (CSUR) 55(2):1–96
Sachs JD (2012) From millennium development goals to sustainable development goals. The Lancet 379(9832):2206–2211
Sætra HS (2021) AI in context and the sustainable development goals: factoring in the unsustainability of the sociotechnical system. Sustainability 13(4):1738
Snoswell A (2023) Draft G7 Guiding principles for organizations developing advanced AI systems
Stein AL (2020) Artificial intelligence and climate change. Yale J Reg 37:890
Thaler RH, Sunstein CR (2009) Nudge: improving decisions about health, wealth, and happiness. Penguin
Uykur D (2018) The new physics of financial services-understanding how artificial intelligence is transforming the financial ecosystem-part of the future of financial services series| Prepared in collaboration with Deloitte. WEF: World Economic Forum
Veale M, Zuiderveen Borgesius F (2021) Demystifying the draft EU artificial intelligence act—analysing the good, the bad, and the unclear elements of the proposed approach. Comp Law Rev Int 22(4):97–112
Venayagamoorthy GK (2009) Potentials and promises of computational intelligence for smart grids. In: 2009 IEEE power & energy society general meeting. IEEE, pp 1–6
Victor DG (2019) How artificial intelligence will affect the future of energy and climate
Vinuesa R, Azizpour H, Leite I, Balaam M, Dignum V, Domisch S, Fuso Nerini F (2020) The role of artificial intelligence in achieving the sustainable development goals. Nat Commun 11(1):1–10
Visvizi A (2022) Artificial intelligence (AI) and sustainable development goals (SDGs): exploring the impact of AI on politics and society. Sustainability 14(3):1730
Yan B, Hao F, Meng X (2021) When artificial intelligence meets building energy efficiency, a review focusing on zero energy building. Artif Intell Rev 54:2193–2220
Yigitcanlar T, Cugurullo F (2020) The sustainability of artificial intelligence: an urbanistic viewpoint from the lens of smart and sustainable cities. Sustainability 12(20):8548

www.ingramcontent.com/pod-product-compliance
Ingram Content Group UK Ltd.
Pitfield, Milton Keynes, MK11 3LW, UK
UKHW022213051026
15624UKWH00001B/148

* 9 7 8 3 0 3 1 6 0 5 4 4 4 *